Supercapacitor Technology
Materials, Processes and Architectures

Edited by

Inamuddin[1,2,3], Rajender Boddula[4], Mohd Imran Ahamed[5] and Abdullah M. Asiri[1,2]

[1]Chemistry Department, Faculty of Science, King Abdulaziz University, Jeddah 21589, Saudi Arabia

[2]Centre of Excellence for Advanced Materials Research, King Abdulaziz University, Jeddah 21589, Saudi Arabia

[3]Department of Applied Chemistry, Faculty of Engineering and Technology, Aligarh Muslim University, Aligarh-202 002, India

[4]CAS Key Laboratory of Nanosystem and Hierarchical Fabrication, National Center for Nanoscience and Technology, Beijing 100190, PR China

[5]Department of Chemistry, Faculty of Science, Aligarh Muslim University, Aligarh-202 002, India

Published by **Materials Research Forum LLC**
Millersville, PA 17551, USA

Published as part of the book series
Materials Research Foundations
Volume 61 (2019)
ISSN 2471-8890 (Print)
ISSN 2471-8904 (Online)

Print ISBN 978-1-64490-048-2
eBook ISBN 978-1-64490-049-9

Distributed worldwide by

Materials Research Forum LLC
105 Springdale Lane
Millersville, PA 17551
USA
http://www.mrforum.com

Manufactured in the United States of America
10 9 8 7 6 5 4 3 2 1

Table of Contents

Preface

Nowadays, the development of clean, sustainable energy conversion and storage technology to cope with the fossil fuel crisis and environmental pollution issues has pulled much consideration in the energy research community. The advancement of human progress has straightforwardly identified the progress in energy science and technology. Today, somebody who could envision the life without electrochemical energy storage devices would see a world without electronic gadgets, for example, digital communications, laptops, iPods, mobile phones and furthermore without electric vehicles and planes which makes our life simpler and progressively agreeable. Therefore, in this energy-dependent world, electrochemical gadgets for energy storage play an essential role in defeating petroleum derivative exhaustion and thereby, environmental issues. Among different electrochemical storage devices, supercapacitors, have pulled remarkable enthusiasm from both the scholarly community and industry during last few years due to their unbeatable power density, quick charge/discharge rate and prolonged lifetime compared to batteries. Vast efforts have thus been put to make supercapacitors more competitive with existing options of rechargeable battery based energy storage. Material science has played an important role to develop exciting supercapacitor devices, which are conveying huge financial advantages over a wide scope of business sectors.

The book is dedicated to grasping the knowledge of supercapacitor innovation with an indepth literature on material syntheses, mechanistic processes, and device architectures. It is an important resource of comprehensive knowledge pertinent to supercapacitor innovation and related fields of research. This book includes a few most important topics of supercapacitor technology, including the use of inorganic, organic and gel electrolytes, electrodes and separators used in different types of supercapacitors. This invaluable guide will fascinate researchers, electrochemists, and postgraduate students working in the area of material science, nanotechnology, and solid-state electrochemistry. Futheremore, it provides knowledge of supercapacitor technology, including essential aspects of material synthesis, characterization, fundamental electrochemical properties, and most promising applications.

Inamuddin[1,2,3], Rajender Boddula[4], Mohd Imran Ahamed[5] and Abdullah M. Asiri[1,2]

[1]Chemistry Department, Faculty of Science, King Abdulaziz University, Jeddah 21589, Saudi Arabia

[2]Centre of Excellence for Advanced Materials Research, King Abdulaziz University, Jeddah 21589, Saudi Arabia

[3]Department of Applied Chemistry, Faculty of Engineering and Technology, Aligarh Muslim University, Aligarh-202 002, India

[4]CAS Key Laboratory of Nanosystem and Hierarchical Fabrication, National Center for Nanoscience and Technology, Beijing 100190, PR China

[5]Department of Chemistry, Faculty of Science, Aligarh Muslim University, Aligarh-202 002, India

Chapter 1

Organic Electrolytes for Supercapacitors

Ivy Heng, Chin Wei Lai[*], Joon Ching Juan

Nanotechnology & Catalysis Research Centre (NANOCAT), Level 3, Block A, Institute for Advanced Studies (IAS), University of Malaya (UM), 50603, Kuala Lumpur

*cwlai@um.edu.my

Abstract

Organic electrolytes can provide comprehensive opportunities for assembling high-energy supercapacitors because of their wide potential windows. The most conventional organic electrolytes are acetonitrile and propylene carbonate. The detailed review of different types of solvents and salts present in organic electrolytes are discussed through extensive analysis of the literature. This chapter intends to provide an overall assessment of organic electrolytes in use in present supercapacitor applications. Several possible oversights for the rational selection of future organic electrolytes are proposed for supercapacitor devices.

Keywords

Organic Electrolytes, Solvents, Solutes, Acetonitrile, Propylene Carbonate, Tetrafluoroborate

Contents

Supercapacitor Technology: Materials, Processes and Architectures   Materials Research Forum LLC
Materials Research Foundations **61** (2019) 1-10   https://doi.org/10.21741/9781644900499-1

1. Introduction

The properties of electrolytes play an important role in determining good electrochemical performance for a supercapacitor device. An electrolyte occurs inside both active material layers and separator. Different kinds of used electrolytes in supercapacitor applications are shown in Fig. 1. It is well known that a high potential window brings the significant effect of increasing capacitance and energy density. Although aqueous electrolytes are good in ion transfer and allowing higher conductivity, they are impeded by their narrow operation window (1V). Researchers have attempted to use organic electrolytes for improving energy storage capacity, because of their broad working voltage windows (2.5-2.7V). Most literature report showed the dependency of electrode performance on the size of ion, conductivity and viscosity of organic electrolytes. The following section is devoted to the common organic solvents, solutes and the properties of the organic electrolytes-based supercapacitors.

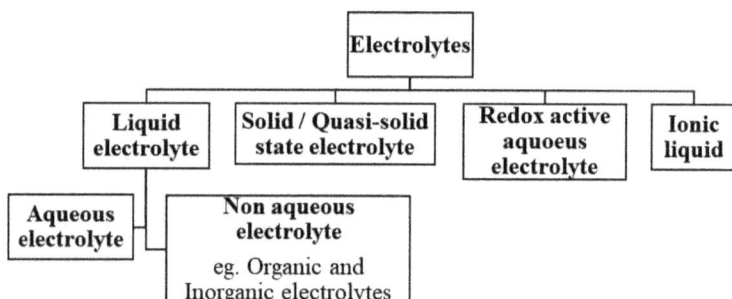

Figure 1. Four types of electrolytes in supercapacitor application.

2. Solvents

The notable organic solvents include propylene carbonate (PC) and acetonitrile (ACN). ACN shows better conductivity and electrochemical performance at a lower temperature compared to the electrolytes containing PC [1]. However, ACN is more flammable than PC, thus the performance is limited to the operating temperature of 60°C and above [2]. Besides that, PC has resistance against hydrolysis and is non-toxic while ACN can be hazardous to human and environment. Taking into safety consideration, PC-based electrolyte is a more desirable choice for supercapacitor devices.

It has been noticed that the long-running cycle life has significance among the requirements of global supercapacitor technology. Most studies have established that organic electrolytes have been able to run a broad range of voltage, higher than 2.8V, but the lifelong cyclic performance cannot be achieved due to their high viscosities. Alternative solvents like electrolytes containing higher electrochemical stability than that of PC and ACN have also been proposed.

Cyclic carbonate[3], sulfones [4, 5] and nitrile [6] have been demonstrated as alternative solvents nowadays. Viscosities are the critical issue to influence the performance and cycling stability. Table 1 presents physical and chemical data of different organic solvents. From Table 1, the alternate solvents offered higher viscosities than those of conventional systems. For example, an alkyl nitrile functional group, adiponitrile (ADN) can be used to run the voltage up to 3.5V due to higher viscosity than ACN and PC-based electrolytes [7]. In term of power density and cycling stability, alternative solvents are inferior to ACN and PC-based electrolytes.

High flammability, volatility and toxicity are the critical issues for organic solvents. Interestingly, the proposed alternative electrolytes exhibit higher flash and boiling points compared to ACN. In addition, the boiling and flash points of ADN transcend those of cyclic carbonate solvents [8].

Table 1. Various physical-chemical data of organic solvents in supercapacitors.

Electrolytes	Structural Formula	Viscosity (mPa s)	Conductivity (mS cm^{-1})	MP and BP(^{o}C)	Operating Voltage (V)	Ref.
Conventional						
ACN	CH_3CN	0.6	56	-30 to 60	2.6	[1]
PC	$C_4H_6O_3$	2.5	15	-30 to 243	2.7	[1]
Alternative						
ADN	$CN(CH_2)_4CN$	6.6	4.3	-2 to 290	3.5	[9]
EMS	$C_3H_8O_2S$	5.5	2.58	34 to 239.2	3.3	[3, 10]
EC/DMC	$C_3H_4O_3$ /$C_3H_6O_3$	0.118	12	EC: 39-243 DMC:2-90	3.0	[11, 12]

3.   Solutes

Solutes are capable of increasing solubility due to their dissymmetric structures that exhibit lower crystal lattice energy [13]. Ammonium-based and phosphine-based cations are presently used as commercial solutes in organic electrolytes. For example, the most widely use of conducting salts include tetraethylammonium tetrafluoroborate (TEABF$_4$), tetramethylammonium tetrafluoroborate (TMABF$_4$) tetraethylphosphonium tetrafluoroborate (TEPBF$_4$) and triethylmethylammonium tetrafluoroborate (TEMABF$_4$) [14].

Table 2. Conductivity and Viscosity of various solutes dissolved in AC or PC based electrolyte.

Solutes /Solvents with concentration (M)	Viscosity (mPa s)	Conductivity (mS /cm)	Research Findings	Ref
Single salt				
1M TEABF$_4$/PC	4.14	13.1	EMBP$_4$BF$_4$ has superior rate capability due to its highest conductivity, followed by TEMABF$_4$ and TEABF$_4$.	[18]
1M TEMABF$_4$/PC	4.06	13.6		
1M EMPF$_4$BF$_4$/PC	4.09	14.7		
1.5M TEMABF$_4$/PC	4.0	14.6	Higher mobility and diffusion rate of SBP cations into PC solvent because of the smaller SBP cation, therefore SBPBF$_4$ has better conductivity compared to TEMABF$_4$.	[19]
1.5M SBPBF$_4$/PC	4.51	17.0		
1M TEABF$_4$/AC	0.62	55.7	The mixing solutes bring increment in conductivity and decline in viscosity among those of single salt solutions.	[17]
1M TMEABF$_4$/AC	0.58	50.3		
Binary salts				
1M TEABF$_4$ + TMABF$_4$/AC	0.60	56.4		
1M TMEABF$_4$ + TMABF$_4$/AC	0.57	51.9		

The solutes performed at potential voltage were lower than 2.7 V and moderate salt solubility [15]. Some of the other solutes have been investigated in the interest of increasing solubility in organic electrolytes. Lately, pyrrolidiniunm cations such as spiro-(1,1)-bipyrrolidinium tetrafluoroborate (SBPBF$_4$) and 1,1-ethylmethyl-pyrrolidinium

(EMP) have been used as replacement of conventional $TEABF_4$ solutes. Cheng and co-authors [16] have reported $SBPBF_4$ as alternate to achieve more capacitance, lower down the equivalent series resistance (ESR) and expand voltage window compared to supercapacitor containing $TEABF_4$ solutes in PC electrolytes. Recently, the PC and ACN based alternative electrolytes made up of binary salts based on these cations and anions have been developed. For example, Sieun Park and co-authors [17] have reported combinations of $TMA\ BF_4$ and various solutes such as $TEABF_4$, $TEMABF_4$ and $SBPBF_4$. Table 2 lists the conductivity and viscosity of single and binary solutes dissolved in ACN or PC based electrolyte. In addition, Li^+, Na^+ and Mg^{2+} salts also have been widely used for pseudocapacitors and hybrid supercapacitors. Wise selection of suitable cations or anions is important for the realization of alternative salts. Besides that, more investigations are required to observe the interactions between salt and solvent on the operative voltage of electroactive materials.

4. Different electrode materials conjunction with their organic electrolytes

The important factors to be taken into account for organic electrolytes are the size of the pore and the size of the ion. Carbon-based electrodes have larger solvated ion than the organic ion, resulting in lower capacitance. Although the pores in carbon electrodes may provide more accessible surface area, it also can suppress some of the organic electrolyte ions access into the small pores, especially the larger organic ions. The optimized pore is 0.7-0.8 nm for the common organic ion. To boost the electrochemical supercapacitor performance, it is worth noting to synchronize the pore size of carbon electrodes and the size of ions. Therefore, it is important to study and optimize complementing between carbon with different pore size distributions and organic ion size.

Pseudocapacitors such as metal oxide have been also used in organic electrolyte. The lithium-containing organic electrolytes have been particularly employed for pseudocapacitors because of smaller size of Li-ion for fast ion intercalation-deintercalation purpose. As mentioned earlier, the common Li salts such as $LiPF_6$ and $LiClO_2$ have been used in organic electrolytes. The mixture electrolytes made up of EC and DMC have also been widely used for pseudocapacitor.

In view of energy and power capacity, asymmetric hybrid supercapacitors in organic electrolytes have been proposed in recent years. Lithium-ion capacitors (LICs) have drawn considerable attention among the hybrid devices. LICs consist of carbon material (cathode) and metal oxide (anode) in lithium-containing organic electrolyte. Because of LICs is able to operate approximately at voltage of 3.8-4.0 V in organic electrolyte, asymmetric supercapacitor delivers higher energy (>20 $Whkg^{-1}$) compared to reported asymmetric supercapacitors using aqueous electrolytes [20]. Table 3 summarizes the different electrode materials in conjunction with organic electrolytes used in EDLCs, pseudocapacitors and hybrid supercapacitors.

Table 3. Different electrode materials conjunction with their organic electrolytes.

No.	Materials used	Electrolyte	Measurement protocol	Electrode configurations	Maximum SC	Ref
Electric Double Layer Capacitor						
1.	Hierarchical porous carbon	1M TEABF$_4$/ACN	GCD	Two-electrode system	130 Fg^{-1} at 1 Ag^{-1}	[21]
2.	Activated carbon	1M TEABF$_4$/ACN	CV	Two-electrode system	87 Fg^{-1} at 10 mVs^{-1}	[22]
3.	N-rGO	1M TEABF$_4$/PC	CV	Three-electrode system	234 Fg^{-1} at 5 mVs^{-1}	[23]
Pseudocapacitor						
4.	MoS$_2$	0.5M TEABF$_4$/ ACN	GCD	Two-electrode system	14.75 Fg^{-1} at 0.75 Ag^{-1}	[24]
Asymmetric Supercapacitor						
5.	rGO// rGO/V$_2$O$_5$	1M LiClO$_4$/PC	GCD	Two-electrode system	384 Fg^{-1} at 0.1 Ag^{-1}	[25]
6.	rGO// MWCNT/Ti$_3$C$_2$T$_x$	1M TEABF$_4$/ACN	CV	Two-electrode system	83 Fg^{-1} at 10 mVs^{-1}	[26]

Note: rGO-reduced graphene oxide; SC-specific capacitance; MWCNT- multiwalled carbon nanotube; Ti$_3$C$_2$T$_x$- titanium carbide; V$_2$O$_5$- vanadium pentoxide; LiClO$_4$- Lithium perchlorate; ACN- Acetonitrile; PC- Propylene carbonate; TEABF$_4$-tetraethylammonium tetrafluoraborate.

Conclusion

Organic electrolytes achieve higher energy density than aqueous electrolytes due to organic electrolytes being able to provide a broad potential window. An ideal electrolyte should have a broad voltage window, low viscosity, high thermal stability, being safe to use and high electrochemical stability. The key challenges for organic electrolytes are the performance is limited by higher resistivity, low salt solubility in organic solvents and large size of organic molecules. Besides that, the toxicity and flammability of organic solvents require careful handling. To satisfy broad operating temperature range and high energy density synchronously, the performance of organic electrolyte can be enhanced by using alternative solvents or binary salts. Lastly, improvement of electrochemical properties along with solvent and ion (size, type or concentration), interaction between the organic salts and the solvent, the desired levels of voltage window, and interaction between the electrolyte and the active materials require to be fully investigated.

List of Abbreviations

ACN- Acetonitrile

ADN-Adiponitrile

BP- Boiling Point

EC/DMC- Ethylene carbonate/Dimethyl carbonate

EMS- Ethyl Methyl Sulfones

$LiClO_4$- Lithium perchlorate

LICs-Lithium Ion Capacitors

MP-Melting Point

MWCNT- multiwalled carbon nanotube

PC- Propylene carbonate

rGO-reduced graphene oxide

SC-specific capacitance

$TEABF_4$-tetraethylammonium tetrafluoraborate

$Ti_3C_2T_x$- titanium carbide

V_2O_5- vanadium pentoxide

Acknowledgments

This research work was financially supported by the Impact-Oriented Interdisciplinary Research Grant (No. IIRG018A-2019) and Global Collaborative Programme - SATU Joint Research Scheme (No. ST012-2019).

References

[1] A. Balducci, Electrolytes for high voltage electrochemical double layer capacitors: A perspective article, J. Power Sources 326 (2016) 534-540. https://doi.org/10.1016/j.jpowsour.2016.05.029

[2] A. Brandt, S. Pohlmann, A. Varzi, A. Balducci, S. Passerini, Ionic liquids in supercapacitors, MRS bull. 38 (2013) 554-559. https://doi.org/10.1557/mrs.2013.151

[3] K. Chiba, T. Ueda, Y. Yamaguchi, Y. Oki, F. Shimodate, K. Naoi, Electrolyte systems for high withstand voltage and durability I. Linear sulfones for electric double-layer capacitors, J. Electrochem. Soc. 158 (2011) A872-A882. https://doi.org/10.1149/1.3593001

[4] X. Sun, C.A. Angell, Doped sulfone electrolytes for high voltage Li-ion cell applications, Electrochem.Commun.11 (2009) 1418-1421. https://doi.org/10.1016/j.elecom.2009.05.020

[5] K. Naoi, 'Nanohybrid capacitor': the next generation electrochemical capacitors, Fuel cells 10 (2010) 825-833. https://doi.org/10.1002/fuce.201000041

[6] A. Brandt, P. Isken, A. Lex-Balducci, A. Balducci, Adiponitrile-based electrochemical double layer capacitor, J. Power Sources 204 (2012) 213-219. https://doi.org/10.1016/j.jpowsour.2011.12.025

[7] H. Duncan, N. Salem, Y. Abu-Lebdeh, Electrolyte formulations based on dinitrile solvents for high voltage Li-ion batteries, J. Electrochem. Soc. 160 (2013) A838-A848. https://doi.org/10.1149/2.088306jes

[8] M. Nagahama, N. Hasegawa, S. Okada, High voltage performances of Li_2NiPO_4F cathode with dinitrile-based electrolytes, J. Electrochem. Soc.157 (2010) A748-A752. https://doi.org/10.1149/1.3417068

[9] R.-S. Kühnel, N. Böckenfeld, S. Passerini, M. Winter, A. Balducci, Mixtures of ionic liquid and organic carbonate as electrolyte with improved safety and performance for rechargeable lithium batteries, Electrochim. Acta 56 (2011) 4092-4099. https://doi.org/10.1016/j.electacta.2011.01.116

[10] S. Yoon, Y.-H. Lee, K.-H. Shin, S.B. Cho, W.J. Chung, Binary sulfone/ether-based electrolytes for rechargeable lithium-sulfur batteries, Electrochim. Acta 145 (2014) 170-176. https://doi.org/10.1016/j.electacta.2014.09.007

[11] P. Porion, Y.R. Dougassa, C. Tessier, L. El Ouatani, J. Jacquemin, M. Anouti, Comparative study on transport properties for LiFAP and $LiPF_6$ in alkyl-carbonates as electrolytes through conductivity, viscosity and NMR self-diffusion measurements, Electrochim. Acta 114 (2013) 95-104. https://doi.org/10.1016/j.electacta.2013.10.015

[12] M. Dahbi, F. Ghamouss, F. Tran-Van, D. Lemordant, M. Anouti, Comparative study of EC/DMC LiTFSI and $LiPF_6$ electrolytes for electrochemical storage, J. Power Sources 196 (2011) 9743-9750. https://doi.org/10.1016/j.jpowsour.2011.07.071

[13] G. Wang, L. Zhang, J. Zhang, A review of electrode materials for electrochemical supercapacitors, Chem. Soc. Rev. 41 (2012) 797-828. https://doi.org/10.1039/C1CS15060J

[14] C. Zhong, Y. Deng, W. Hu, J. Qiao, L. Zhang, J. Zhang, A review of electrolyte materials and compositions for electrochemical supercapacitors, Chem. Soc. Rev. 44 (2015) 7484-7539. https://doi.org/10.1039/C5CS00303B

[15] J. Krummacher, C. Schütter, L. Hess, A. Balducci, Non-aqueous electrolytes for electrochemical capacitors, Curr. Opinion Electrochem. 9 (2018) 64-69. https://doi.org/10.1016/j.coelec.2018.03.036

[16] F. Cheng, X. Yu, J. Wang, Z. Shi, C. Wu, A novel supercapacitor electrolyte of spiro-(1, 1')-bipyrolidinium tetrafluoroborate in acetonitrile/dibutyl carbonate mixed solvents for ultra-low temperature applications, Electrochim. Acta 200 (2016) 106-114. https://doi.org/10.1016/j.electacta.2016.03.113

[17] S. Park, K. Kim, Tetramethylammonium tetrafluoroborate: The smallest quaternary ammonium tetrafluoroborate salt for use in electrochemical double layer capacitors, J. Power Sources 338 (2017) 129-135. https://doi.org/10.1016/j.jpowsour.2016.10.080

[18] J. Han, N. Yoshimoto, Y.M. Todorov, K. Fujii, M. Morita, Characteristics of the electric double-layer capacitors using organic electrolyte solutions containing different alkylammonium cations, Electrochim. Acta (2018). https://doi.org/10.1016/j.electacta.2018.06.012

[19] X. Yu, D. Ruan, C. Wu, J. Wang, Z. Shi, Spiro-(1, 1')-bipyrrolidinium tetrafluoroborate salt as high voltage electrolyte for electric double layer capacitors, J. Power Sources 265 (2014) 309-316. https://doi.org/10.1016/j.jpowsour.2014.04.144

[20] S. Nakata, Characteristic of an adiabatic charging reversible circuit with a Lithium ion capacitor as an energy storage device, Results Phys. 10 (2018) 964-966. https://doi.org/10.1016/j.rinp.2018.08.016

[21] L. Zhang, Y. Zhu, W. Zhao, L. Zhang, X. Ye, J.-J. Feng, Facile one-step synthesis of three-dimensional freestanding hierarchical porous carbon for high energy density supercapacitors in organic electrolyte, J. Electroanal. Chem. 818 (2018) 51-57. https://doi.org/10.1016/j.jelechem.2018.04.031

[22] K.Ö. Köse, B. Pişkin, M.K. Aydınol, Chemical and structural optimization of $ZnCl_2$ activated carbons via high temperature CO_2 treatment for EDLC applications, Int J Hydrogen Energy 43 (2018) 18607-18616. https://doi.org/10.1016/j.ijhydene.2018.03.222

[23] S.-M. Li, S.-Y. Yang, Y.-S. Wang, H.-P. Tsai, H.-W. Tien, S.-T. Hsiao, W.-H. Liao, C.-L. Chang, C.-C.M. Ma, C.-C. Hu, N-doped structures and surface functional groups of reduced graphene oxide and their effect on the electrochemical performance of supercapacitor with organic electrolyte, J. Power Sources 278 (2015) 218-229. https://doi.org/10.1016/j.jpowsour.2014.12.025

[24] P. Pazhamalai, K. Krishnamoorthy, S. Manoharan, S.-J. Kim, High energy symmetric supercapacitor based on mechanically delaminated few-layered MoS2 sheets in organic electrolyte, J. Alloys Compd. 771 (2019) 803-809. https://doi.org/10.1016/j.jallcom.2018.08.203

[25] Z. Liu, H. Zhang, Q. Yang, Y. Chen, Graphene/V_2O_5 hybrid electrode for an asymmetric supercapacitor with high energy density in an organic electrolyte, Electrochim. Acta 287 (2018) 149-157. https://doi.org/10.1016/j.electacta.2018.04.212

[26] A.M. Navarro-Suárez, K.L. Van Aken, T. Mathis, T. Makaryan, J. Yan, J. Carretero-González, T. Rojo, Y. Gogotsi, Development of asymmetric supercapacitors with titanium carbide-reduced graphene oxide couples as electrodes, Electrochim. Acta 259 (2018) 752-761. https://doi.org/10.1016/j.electacta.2017.10.125

Supercapacitor Technology: Materials, Processes and Architectures   Materials Research Forum LLC
Materials Research Foundations **61** (2019) 11-30   https://doi.org/10.21741/9781644900499-2

Chapter 2

Inorganic Electrolytes in Supercapacitor

P.E. Lokhande[*], U.S. Chavan

Department of Mechanical Engineering, Vishwkarma Institute of Technology, Pune, India-411037

*prasadlokhande2007@gmail.com

Abstract

Supercapacitors are considered promising energy storage systems due to their high power density, fast charge-discharge, long service lifetime, wide operating temperature range and excellent capacitance retention. The electrochemical performance of the supercapacitors depends upon numerous factors such as nature of electrode materials, type of electrolyte and separator thickness, etc. Among these factors, electrolyte used in supercapacitor plays an important role in deciding final characteristics of supercapacitors. In recent decades, tremendous research work has been on the development of novel electrolytes and electrode/electrolyte configurations. In this chapter, we aimed to focus on the role of inorganic electrolytes used in supercapacitors.

Keywords

Supercapacitor, Electrolyte, Metal Oxide, Carbon-Based Material

Contents

1.   Introduction

Presently, energy storage technology has drawn global attention because of environment problems such as decreasing fossil fuel, increasing pollution and global warming effect [1]. In that context, the search of renewable energy sources has become the need of society [2][3]. Two most successful storage systems are Li-batteries and supercapacitors. The battery and supercapacitor are differentiated on the basis of charge storage mechanism and their material/structures [4]. Supercapacitors are regarded one of the promising energy storage systems for storing renewable energy due to their fast charge/discharge capability, low cost, high power density, good cyclability, wide operating temperature range and environment-friendly nature [5]. In the history, first experimentation was done by General Electric Research Group in 1957 who found porous carbon to store charges electrostatically along with exceptionally high capacitance. Further in 1966, the research group of Standard Oil of Ohio working on fuel cell accidentally found the electric double layer effect. In 1978, the term "supercapacitor" was used the first time by Nippon Electric Company [6]. Supercapacitors also called as ultracapacitors utilize high surface electrode material and thin electrolytic dielectric. Supercapacitors are used in various applications like hybrid electric vehicles, memory backup, large industrial equipment and renewable energy power plant. In these days, smart portable and wearable electronic devices such as smartphone, laptop, smartwatch, camera, etc. requires improved energy storage systems like supercapacitor [4]. One of the recent applications of supercapacitor is in the emergency door of the Airbus 380 low emission hybrid electric vehicle [7].

Supercapacitors with unique characteristics are superior to batteries and advantageous to applications where power burst are required. Although higher power density and cyclability, lower storing capacity or energy density than battery limits its use in wide range of applications [8]. Many scientists are working on improving the energy density along with high power density by developing advanced electrode, electrolyte materials and electrode/electrolyte configuration [9,10]. The electrochemical properties of supercapacitors can be improved by choosing an appropriate electrolyte. Fig. 1 shows

Ragone plot, illustrating energy density with respect to power density for different energy storage technologies. From this it is clear that supercapacitor bridges the gap between a high energy density battery and high power density conventional capacitor [7].

In this chapter, we intended to provide a brief overview of inorganic electrolytes used in supercapacitors.

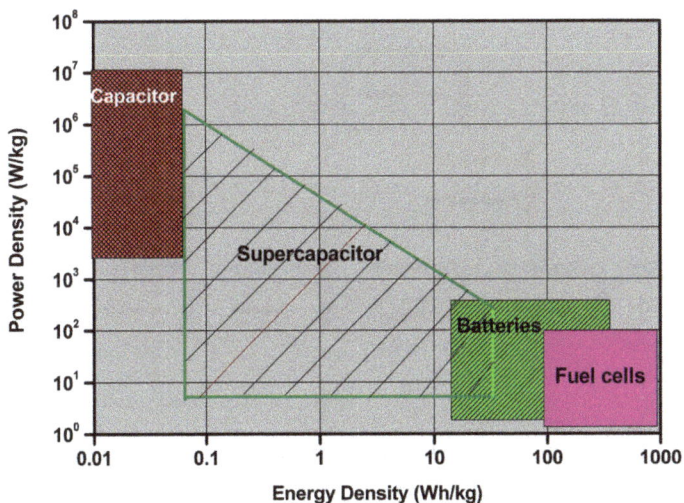

Figure 1. Ragone plot for various electrical energy storage devices (specific power against specific energy).

2.   Taxonomy of supercapacitor

Based on energy storage mechanism supercapacitors are classified into three groups: electrochemical double layer capacitors (EDLCs), pseudocapacitors and hybrid supercapacitors (Fig. 2) [11]. EDLCs store charge electrostatically using reversible ions of electrolyte which are reversibly adsorbed on active materials having chemically stable and high specific surface area as shown in Fig.3a [12]. The charge storage mechanism is non-Faradaic, where no oxidation-reduction reaction occurs. A very thin double layer is formed in between the interface of the electrode and the electrolyte and no transfer of charge from one electrode to the other occurred. Due to the physical charge transfer EDLCs have a relatively higher cycle life [13]. Pseudocapacitor uses fast, reversible redox reaction on the electrode surface and subsurface (Fig. 3b) [14]. The power density and cycle life for EDLC material are higher than that of pseudocapacitor while energy density and capacitance are higher for pseudocapacitor. To take advantages of both EDLC and the pseudocapacitor, composite are formed by using both type of material and are called hybrid capacitors [15]. The hybrid capacitors are further grouped into three categories as asymmetric supercapacitor, battery type and composite. Hybrid capacitor can achieve high power density and energy density along with good cycling stability [16].

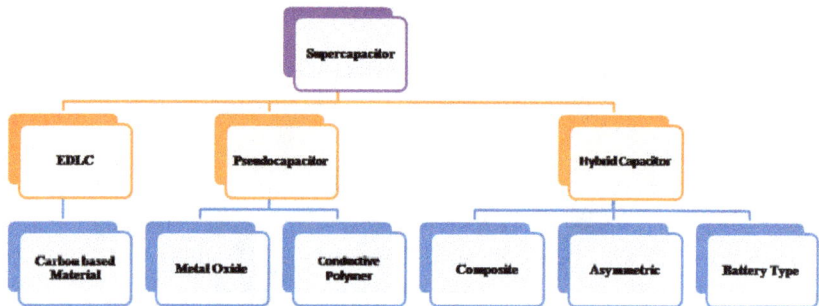

Figure 2. Classification of electrochemical supercapacitor.

Generally, a supercapacitor is composed of current collector, electrode, electrolyte and separator. Among these materials, electrode and electrolyte material are the key components of the supercapacitor. The properties like high specific surface area, good porosity, corrosion resistant, excellent electronic conductivity, high-temperature stability make materials attractive for electrode of supercapacitor [17]. In that context, various carbon-based materials, transition metal oxides/hydroxides, conducting polymers have been used as electrode for supercapacitors. The carbon-based materials such as activated carbon [18], porous carbon [19], carbon aerogels, hallow carbon [20], carbon nanotubes [21] and graphene [22], etc. have been utilized as electrode material in EDLC. The energy density obtained from EDLC is typically in the range of 3 to 5 Wh kg^{-1} and which is far lesser than battery (30-40 and 10-250 Wh kg^{-1} for lead acid and Li-ion batteries respectively) and is not enough to use such cell in practical application such as electric vehicles [23,24]. Hence, transition metal oxides/hydroxides have been investigated to achieve high energy density and specific capacitance. In case of pseudocapacitors transition metal oxides/hydroxides like NiO [6], RuO_2 [25], MnO_2 [26], Co_3O_4 [27],

$Ni(OH)_2$ [28], $Co(OH)_2$ [29], Fe_2O_3 [30], TiO_2 [31], etc. and the conducting polymers like PANI [32], PPy [33] and PTh [34]) have been widely used as electrode materials.

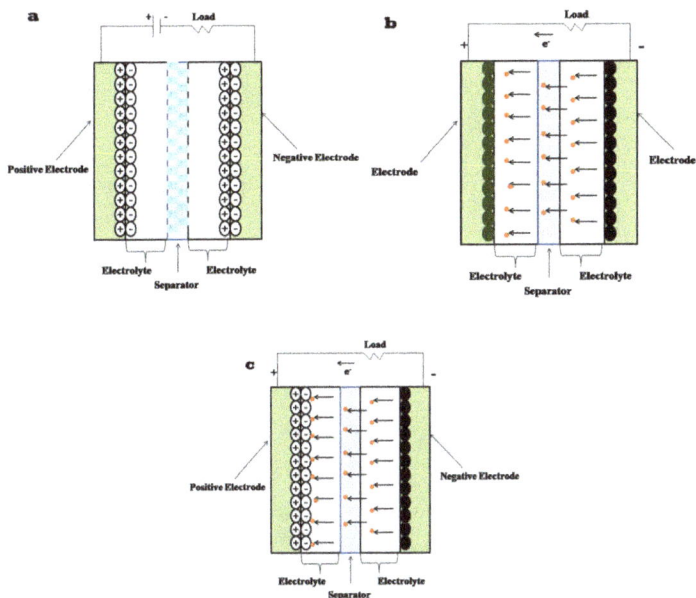

Figure 3. Schematic diagram of charge storage mechanism of various supercapacitors: (a) EDLC; (b) Pseudocapacitor (c) Hybrid capacitor[80].

3. Fundamentals of supercapacitor

The electrochemical performance of supercapacitor is measured in following terms 1. Capacitance value, 2. Capacity of energy storage, 3. Time required for charging, 4. Cyclability, 5. Charge-discharge capabilities, 6. Self-discharging and 7. Cost [7]. The parameters such as higher specific surface area of electrode, electrical conductivity of electrode material, appropriate porosity (easy access for electrolyte ions), crystallinity (for deep diffusion of electrolyte), and intrinsic properties of electrolyte, size and shape of redox active species affect the final electrochemical properties [35]. It is highly desirable to use appropriate combination of electrode material and electrolyte to achieve higher energy and power density. But still the need of versatile electrolyte material for use as electrode materials is felt [36,37].

Supercapacitor Technology: Materials, Processes and Architectures   Materials Research Forum LLC
Materials Research Foundations **61** (2019) 11-30   https://doi.org/10.21741/9781644900499-2

The specific capacitance of materials can be evaluated with the aid of two techniques cyclic voltammetry (CV) and galvanostatic charge-discharge (GCD). The equations used for measuring specific capacitance from CV curve is as follow and galvanostatic charge-discharge profile specific capacitance are as follows [38,39]

$$Cs = 1/(mv(v_f - v_i)) \int_{v_i}^{v_f} I\,(V)dV \tag{1}$$

$$Cs = \frac{I\Delta t}{m\Delta V} \tag{2}$$

Where, I (A), m(g), v (Vs^{-1}), (v_f - v_i) or ΔV and Δt represent the response current, electrode mass, scan rate, the potential window and discharge time respectively.

The energy density and power density are the two important parameters for evaluating electrochemical performance of supercapacitors. Following equations are used for measuring the maximum energy density and power density [40]:

$$E = {CV^2}/{2} \tag{3}$$

$$P = {V^2}/{4R} \tag{4}$$

Where V is the cells operating voltage in volt, C is the capacitance in faraday and R is the equivalent series resistance in ohm.

The cell voltage is dependent on thermodynamic stability of electrolyte and electroactive materials of electrode. Whereas the equivalent series resistance (ESR) is composed of following factors: 1. intrinsic resistance of active mass, 2. resistance between electroactive material and current collector, 3. ion diffusion resistance in electrode material and 4. separator and ionic resistance of electrolyte [7]. From the equations 3 & 4, it is clear that the electrochemical performance of supercapacitor can be enhanced by increasing specific capacitance, wide cell voltage while keeping minimum ESR. Recent days, major challenge in front of scientists is to enhance the energy density comparable to the battery [11]. In that context, parameters for increasing energy density have to be taken into consideration. The crucial parameters for enhancing energy density are shown in Fig. 4 [41]. Hence to increase energy density, this is the effective way for enhancing specific capacitance by increasing capacitance of both positive and negative electrolytes. To achieve this various advanced materials were developed, with improved properties such as higher specific surface area, optimized pore size, higher electric conductivity has been developed [42].

Second factor is the operating voltage which is very crucial because energy density is proportional to square of the operating voltage. Due to this various strategies have been applied to enhance operating voltage using organic electrolytes and ionic liquid to achieve higher operating voltage [15]. Another approach to enhance the operating voltage was developing of asymmetric supercapacitors. Asymmetric supercapacitors were prepared by using battery like electrode as energy source and capacitor electrode as a power source. It combines advantages of both battery and capacitors [43-44]. In the following section in-depth progress regarding electrode and electrolyte materials are discussed.

4.   Parameters affected by electrolyte

The electrochemical performance of supercapacitor in terms of energy and power densities are affected by electrolyte nature including type and size of ion, concentration of electrolyte, electrolyte and electrode interaction and operating potential window. In that context, extensive work has been devoted to get superior energy density, so that supercapacitors can be used in a variety of applications. From equation 4, it is clear that the energy density is directly proportional to the capacitance and square of voltage. Hence by developing advanced electrode materials or electrolytes with wider potential window can increase the energy density. The stable potential of electrolyte decides the cell voltage. From equation it is evident that widening the potential window of the electrolyte is more efficient than increasing capacitance because of the square of voltage proportion. Instead of finding a new electrode material, the development of new electrolyte with higher voltage range has been given primary priority to increase energy density. The electrolyte potential window affect other electrochemical properties like power density, equivalent series resistance (ESR) and operating temperature range etc. [45,46].

Apart from the potential window of the electrolyte, the interaction between electrode and electrolyte is another crucial factor for electrochemical performance. The specific capacitance is influenced by the degree of matching of electron-ion size and pore size of the electrode [47]. Also the pseudocapacitance values of the composite materials strongly depend on the nature and type of electrolyte [8]. The internal resistance of supercapacitor is mainly decided by ionic conductivity of electrolyte, specially ionic and organic electrolytes. The boiling point, viscosity, and freezing point of electrolytes decide the thermal stability and the operating temperature of supercapacitor. Also decomposition of the electrolyte leads to the aging and failure of supercapacitor. For high electrochemical performance, following properties of the electrolyte are required [45]: (i) wide potential window (ii) high chemical and electrochemical stabilities (iii) excellent ionic conductivity (iv) wide operating temperature (v) chemical inertness to supercapacitor component (vi) low volatility and flammability (vii) well matching of ions of electrolyte and pores of electrode and (viii) environmental friendliness. It is difficult to achieve all requirements and every electrolyte has its own advantages and shortcomings.

5.   Electrolyte

Electrolyte is composed of salt and solution and plays an important role in electrochemical performance of the supercapacitor. The electrolyte plays a key role in formation of the electrical double layer for EDLC and redox reaction in case of a pseudocapacitor. Generally three types of electrolytes are used in a supercapacitor: liquid-state electrolytes, solid-state electrolytes and quasi-solid state electrolytes. Among these the liquid electrolytes can be further categorized into aqueous, organic and ionic electrolytes, while solid and semi-solid electrolyte are broadly divided as organic and inorganic electrolytes [48]. Fig. 4 demonstrate the classification of electrolytes.

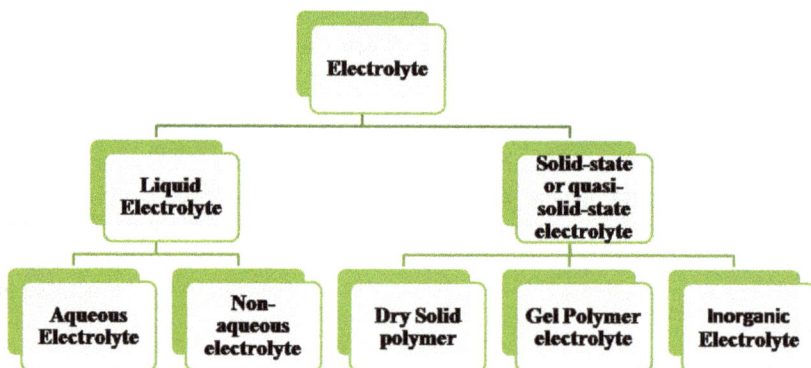

Figure 4. Classification of electrolytes for electrochemical supercapacitors

5.1   Liquid electrolyte

An liquid electrolyte is prepared by dissolving salt in polar solvent like water [6]. For the last two decades, researchers have been working on finding new materials for electrolytes having a wider electrochemical stability window and ionic conductivity. Liquid electrolyte are further classified into three categories 1. Aqueous electrolytes 2. Organic electrolytes and 3. Ionic Liquid.

5.1.1 Aqueous electrolytes

Generally, aqueous electrolytes (such as H_2SO_4, KOH, Na_2SO_4 and NH_4Cl aqueous solution) exhibit better conductivity as compared to organic and ionic liquid electrolytes [49]. For examples conductivity of 1M H_2SO_4 at 25 °C is 0.8 Scm^{-2}. Such higher conductivity of electrolyte reduces ESR and leads to enhancing power density of the supercapacitor. The size of the hydrated cations and mobility of their anions, decide the conductivity of aqueous electrolytes [44]. Even though aqueous electrolytes have good conductivity, lower resistance, small ionic size and ample supply of proton, aqueous electrolytes are not preferred for commercial use in supercapacitors due to low potential window (1V to 1.3 V) compared to other electrolytes [50]. Aqueous electrolytes are grouped into acidic, alkaline and neutral, of which H_2SO_4, KOH and Na_2SO_4 are most commonly used due to their excellent conductivity. The acidic and alkaline electrolytes restrict potential window up to 1.23 V while neutral electrolyte extends up to 1.6 V. Depending on pore size and specific surface area, proper selection of an aqueous electrolyte is important factor for obtaining higher electrochemical performance.

5.1.2 Organic electrolytes (OEs)

Organic electrolytes used in supercapacitors are prepared by dissolving salts having good conductivity in organic solvents. OEs nowadays dominate the market due to their higher operating voltage which is generally in between 2.5-2.8 V. With increase in cell operating voltage, energy and power densities are increased enormously. Also organic electrolytes allow the use of a cheap current collector and package materials to reduce total cost of the supercapacitor cell. Among organic electrolytes most commonly used solvents are acetonitrile (ACN) and propylene carbonate (PC). Acetonitrile is capable to dissolve more organic salts as compared to other solvents but its toxic nature limits its use. The organic electrolytes suffers from high internal resistance [51]. The salts used in preparation of organic electrolyte are tetraethylammonium tetrafluoroborate (TEABF$_4$)[52], tetraethylphosphonium tetrafluoroborate (TEPBF$_4$) and tetraethylmethyammonium tetrafluoroborate (TEMABF$_4$) [53]. Even though organic electrolytes possess higher operating cell voltage, they suffer from drawbacks like high cost, lower specific capacitance, less conductivity and most important safety concern related to flammability and toxicity [54].

5.1.3 Ionic electrolytes

Ionic liquids or low temperature molten salts (solvent free electrolyte at room temperature) attracted interest due to their stability and large voltage window [52]. Ionic liquids are salts generally composed of large asymmetric cations and inorganic or organic anions (ions) with melting points below 100 °C [55]. Hence by varying anions and cations, physical and chemical properties of electrolytes are easily tunable. Higher potential window (2 to 6 V) and excellent conductivity (~10 mS cm^{-1}) are the attractive properties of ionic liquids which have received significant interest as alternative electrolytes for supercapacitors [56]. Ionic liquids are nonflammable and thermally as well as chemically stable also have negligible vapor pressure [57]. Ionic liquids further classified based on composition into three categories aprotic, protic and zwitterionic [55]. The imidazolium-based ionic liquid electrolytes exhibited higher ionic conductivity while pyrrolidinium based electrolytes provide higher potential window [58,59]. The higher viscosities of ionic liquids at room temperature result in reduced electrochemical performance due to lower ionic conductivity. Baldacci et al. demonstrated the utility of hydrophobic   N-butyl-N-methylpyrrolidinium   bis(trifluromethansulfony)   imide (PYR$_{14}$TFSI) as ionic liquid. The electrode was prepared by using activated carbon which exhibited potential window up to 3.5V at 60 °C with superior stability [60]. The electrical conductivity of electrolyte was influenced by mainly two factors a. the mobility of dissociated ions in electrolyte and b. ability of salt dissociation. Hence more work needs to be carry out to develop ionic liquid electrolytes with ions having high dissociation constant, wide electrochemical stability window, and low molecular weight.

5.2   Solid or quasi-solid electrolytes

Solid electrolytes also called supersonic conductors have high ionic conductivity. The rapidly developing electronic field demands more power for portable devices, wearable electronic devices and printable electronics. Especially flexible electronic device have attracted attention for development of solid state electrolytes. Solid state electrolytes act not only as high ionic conductive electrolyte but also as separator. And they provide simplification in packaging and fabrication of the supercapacitor cell. Being light weight they exhibit excellent performance [61]. Solid state electrolytes are grouped into two major categories polymer electrolytes (solid state organic electrolyte) and inorganic solid material electrolytes.

5.2.1   Polymer electrolytes (Solid state organic electrolytes)

Major type of solid-state electrolytes developed for supercapacitors are based on polymer electrolyte. Further polymer electrolyte has been grouped into three categories: solid polymer electrolytes (SPEs), gel polymer electrolytes (GPEs) and polyelectrolyte. Solid polymer is also known as dry polymer because it is free from solvent where low lattice energy metal salt (e.g LiCl) is dissolved into the polar polymer matrix (e.g. PEO) having higher molecular weight with aprotic solvent to create ionic pathway. In other words, fast segmental movement of polymer matrix having Lewis-acid-base type interaction between cation and donor atoms responsible for ionic conduction in solid electrolyte. In solid polymer electrolyte following critical requirements should be considered during synthesis: 1. enhanced chemical and electrochemical stability, 2. superior thermal stability, 3. high ionic conductivity, 4. sufficient mechanical strength and, 5. dimensional stability. All three types of supercapacitors can be fabricated by using SPE type of electrolyte [62]. Majority of good solid polymer electrolytes reported till date are PEO and or PPO complexed/dissolved with different Li^+ ions [63].

In gel polymers some amount of solvent is present hence there are also called quasi-solid-state electrolytes where they are composed by polymer and aqueous electrolyte or conductive salt dissolved in a solvent. In case of GPEs, ions transport in the solvent present in the polymer matrix instead of the polymer base [64]. Among the three electrolytes, GPEs possess highest electronic conductivity hence the recent supercapacitor market prefers gel polymer electrolyte based supercapacitors [65]. When water is used as plasticizer then GPE is called hydrogel where water molecule is trapped into the polymer network mainly through surface tension [66]. For polyelectrolyte, a charged polymer chain contributes the ionic conductivity. GPE with various polymer hosts such as PEO [67], PVDF [68], PAN [69] and PMMA [70-71] have been synthesized with conductivity value 10^{-4} to 10^{-3} Scm^{-1} at room temperature.

5.2.2 Inorganic solid state electrolytes

Solid state inorganic electrolytes are mechanically robust and thermally stable. They have high lithium ionic conductivity below their melting points. The highest conductivity 2.5 * 10^{-2} Scm^{-1} reported for $Li_{9.54}Si_{1.74}P_{1.44}S_{11.7}C_{10.3}$ electrolyte is similar to the liquid electrolyte [72]. As compared to polymer electrolytes, inorganic solid state electrolytes have received less attention due to lack of bendability and almost non flexibility. Francisco et al. have developed Li_2-S-PS$_5$ solid-state inorganic electrolyte, which acted as separator as well as conductor of ions. The high Li-ion conductivity was found to this electrolyte and specific capacitance of 7.75 Fg^{-1} was observed in the devices constructed by nanostructured electrodes [73]. Ulihin et al. reported composite solid electrolyte $0.4LiClO_4$-$0.6Al_2O_3$ for both symmetric and asymmetric supercapacitors. The oxide electrode materials $LiMn_{1.5}Ni_{0.45}Mg_{0.05}O_4$, Mn_2O_3 and MnO_2 were used. The prepared cell showed specific capacitance of 29 Fg^{-1} at high temperature 150 °C[74]. Gu et al. prepared tetragonal perovskite $Li_{0.5}La_{0.5}TiO_3$ (LLTO) electrolyte by using solid-state reaction. Solid state electrolyte, $Li_{2/3-x}La_{3x}TiO_3$ (LLTO) ceramic exhibited attractive electrical conductivity in the rage of 10^{-5} to 10^{-3} Scm^{-1} at room temperature [75]. Hu et al. presented three-dimensional structured carbon filled porous/dense/porous layered ceramic electrolyte for solid state supercapacitor. Single phase $Li_{1.3}Al_{0.3}Ti_{1.7}P_3O_{12}$

(LATP) synthesized via one step solid state reaction was further upgraded by introducing LiMnPO$_4$. Prepared electrolyte exhibited a maximum capacitance of 0.13 Fcm^{-1} at low scan rate of 2 mVs^{-1} [76]. Liao et al. reported lithium aluminum titanium phosphate (LATP) (Li$_{1.4}$Al$_{0.4}$Ti$_{1.6}$(PO$_4$)$_3$) NASICON type structure as solid state electrolyte. By adding CNT in LATP, electrodes were prepared and two such similar electrodes separated by LATP were used as EDLC. The maximum specific capacitance of 0.52 mFcm^{-3} and 11.59 mFcm^{-3} when CNT percentage was increased up to 7.5%.

Graphene oxide (GO) has been also used as promising solid-state inorganic electrolyte material in supercapacitor [77]. GO acts as solid electrolyte due to water and ions mainly protons from hydrolysis of functional group within the GO layers [78]. Zhang et al. reported sandwich structured rGO/GO/rGO solid-state capacitor where GO was used as solid electrolyte. Prepared cell demonstrated fairly high specific capacitance of 0.86 mFcm^{-2} [79]. The GO film was prepared by using a modified Hummer's method while the sandwich film was prepared by using the lightscribing reduction method. Two-sided reduction of electrically conductive rGO (~30 Scm^{-1}, 12 atom % oxygen) and GO (30 atom % oxygen) interlayer was retained in the middle.

Conclusion

This chapter provides a comprehensive overview of electrolyte materials used in supercapacitor technology. Various types of electrolytes such as liquid electrolyte and solid state electrolytes have been used for supercapacitors. Among these recent trends concern solid-state electrolytes due to development of portable electronics. The combination of ionic conductivity with excellent mechanical strength, absence of intergrain resistance and manufacturing ability have made inorganic solid-state electrolytes promising materials for real applications. Inorganic solid state electrolytes have been used in supercapacitors due to their rigid nature. The properties of electrolytes such electronic conductivity, operating potential window and stability majorly affects the electrochemical performance of the supercapacitor such as energy density, power density which are reviewed in depth.

Acknowledgement

We acknowledge Mr. Ram Dayal for his help in collection of information.

References

[1]   Q. Du, L. Su, L. Hou, G. Sun, M. Feng, X. Yin, Z. Ma, G. Shao, W. Gao, Rationally designed ultrathin Ni-Al layered double hydroxide and graphene

heterostructure for high-performance asymmetric supercapacitor, J. Alloys Compd. 740 (2018) 1051–1059. https://doi.org/10.1016/j.jallcom.2018.01.069.

[2]   R.R. Salunkhe, Y. V. Kaneti, Y. Yamauchi, Metal-Organic Framework-Derived Nanoporous Metal Oxides toward Supercapacitor Applications: Progress and Prospects, ACS Nano. 11 (2017) 5293–5308. https://doi.org/10.1021/acsnano.7b02796.

[3]   G.Z. Chen, Supercapacitor and supercapattery as emerging electrochemical energy stores, Int. Mater. Rev. 62 (2017) 173–202. https://doi.org/10.1080/09506608.2016.1240914.

[4]   D.P. Dubal, N.R. Chodankar, D.-H. Kim, P. Gomez-Romero, Towards flexible solid-state supercapacitors for smart and wearable electronics, Chem. Soc. Rev. 47 (2018) 2065-2129. https://doi.org/10.1039/C7CS00505A.

[5]   Y. Li, H. Ye, J. Chen, N. Wang, R. Sun, C.P. Wong, Flexible β-Ni(OH)$_2$/graphene electrode with high areal capacitance enhanced by conductive interconnection, J. Alloys Compd. 737 (2018) 731–739. https://doi.org/10.1016/j.jallcom.2017.12.192.

[6]   R.S. Kate, S.A. Khalate, R.J. Deokate, Overview of nanostructured metal oxides and pure nickel oxide (NiO) electrodes for supercapacitors: A review, J. Alloys Compd. 734 (2018) 89–111. https://doi.org/10.1016/j.jallcom.2017.10.262.

[7]   L.L. Zhang, X.S. Zhao, Carbon-based materials as supercapacitor electrodes, Chem. Soc. Rev. 38 (2009) 2520-2531. https://doi.org/10.1039/b813846j.

[8]   A. González, E. Goikolea, J.A. Barrena, R. Mysyk, Review on supercapacitors: Technologies and materials, Renew. Sustain. Energy Rev. 58 (2016) 1189–1206. https://doi.org/10.1016/j.rser.2015.12.249.

[9]   P.E. Lokhande, U.S. Chavan, Surfactant-assisted cabbage rose-like CuO deposition on Cu foam by for supercapacitor applications, Inorg. Nano-Metal Chem. 0 (2019) 1–7. https://doi.org/10.1080/24701556.2019.1569685.

[10]   P.E. Lokhande, U.S. Chavan, Conventional chemical precipitation route to anchoring Ni(OH)$_2$for improving flame retardancy of PVA, Mater. Today Proc. 5 (2018) 16352–16357. https://doi.org/10.1016/j.matpr.2018.05.131.

[11]   Z.S. Iro, C. Subramani, S.S. Dash, A brief review on electrode materials for supercapacitor, Int. J. Electrochem. Sci. 11 (2016) 10628–10643. https://doi.org/10.20964/2016.12.50.

[12]   P.E. Lokhande, H.S. Panda, Synthesis and Characterization of Ni.Co(OH)$_2$
Material for Supercapacitor Application, IARJSET. 2 (2015) 10–13.
https://doi.org/10.17148/IARJSET.2015.2903.

[13]   M. Vangari, T. Pryor, L. Jiang, Supercapacitors: Review of Materials and
Fabrication Methods, J. Energy Eng. 139 (2012) 72–79.
https://doi.org/10.1061/(ASCE)EY.1943-7897.0000102.

[14]   P. Simon, Y. Gogotsi, Materials for electrochemical capacitors, Nat. Mater. 7
(2008) 845–854.

[15]   V.C. Lokhande, A.C. Lokhande, C.D. Lokhande, J.H. Kim, T. Ji, Supercapacitive
composite metal oxide electrodes formed with carbon, metal oxides and conducting
polymers, J. Alloys Compd. 682 (2016) 381–403.
https://doi.org/10.1016/j.jallcom.2016.04.242.

[16]   H.E. Density, R. Oxides, New materials and new configurations for advanced
electrochemical capacitors, Electrochem. Soc. Interface. (2008) 35.

[17]   C.D. Lokhande, D.P. Dubal, O.S. Joo, Metal oxide thin film based
supercapacitors, Curr. Appl. Phys. 11 (2011) 255–270.
https://doi.org/10.1016/j.cap.2010.12.001.

[18]   G. Moreno-Fernandez, J. Ibañez, J.M. Rojo, M. Kunowsky, Activated Carbon
Fiber Monoliths as Supercapacitor Electrodes, Adv. Mater. Sci. Eng. 2017 (2017) 1–8.
https://doi.org/10.1155/2017/3625414.

[19]   W. Lei, H. Liu, J. Xiao, Y. Wang, L. Lin, Moss-Derived Mesoporous Carbon as
Bi-Functional Electrode Materials for Lithium–Sulfur Batteries and Supercapacitors,
Nanomaterials. 9 (2019) 84. https://doi.org/10.3390/nano9010084.

[20]   P. Zhao, M. Yao, H. Ren, N. Wang, S. Komarneni, Nanocomposites of
hierarchical ultrathin MnO$_2$ nanosheets/hollow carbon nanofibers for high-
performance asymmetric supercapacitors, Appl. Surf. Sci. 463 (2018) 931-938. .
https://doi.org/10.1016/j.apsusc.2018.09.041.

[21]   H. Zheng, J. Wang, Y. Jia, C. Ma, In-situ synthetize multi-walled carbon
nanotubes@MnO$_2$ nanoflake core-shell structured materials for supercapacitors, J.
Power Sources. 216 (2012) 508–514. https://doi.org/10.1016/j.jpowsour.2012.06.047.

[22]   S.N. Alam, N. Sharma, L. Kumar, Synthesis of graphene oxide (GO) by modified
Hummers method and its thermal reduction to obtain reduced graphene oxide (rGO)*,
Graphene. 6 (2017) 1–18. https://doi.org/10.4236/graphene.2017.61001.

[23]   V.D. Patake, C.D. Lokhande, O.S. Joo, Electrodeposited ruthenium oxide thin films for supercapacitor: Effect of surface treatments, Appl. Surf. Sci. 255 (2009) 4192–4196. https://doi.org/10.1016/j.apsusc.2008.11.005.

[24]   M. Zhi, C. Xiang, J. Li, M. Li, N. Wu, Nanostructured carbon-metal oxide composite electrodes for supercapacitors: A review, Nanoscale. 5 (2013) 72–88. https://doi.org/10.1039/c2nr32040a.

[25]   J.C. Icaza, R.K. Guduru, Electrochemical characterization of nanocrystalline RuO_2 with aqueous multivalent (Be^{2+} and Al^{3+}) sulfate electrolytes for asymmetric supercapacitors, J. Alloys Compd. 735 (2018) 735–740. https://doi.org/10.1016/j.jallcom.2017.11.184.

[26]   A.J. Paleo, P. Staiti, A. Brigandì, F.N. Ferreira, A.M. Rocha, F. Lufrano, Supercapacitors based on AC/MnO_2 deposited onto dip-coated carbon nanofiber cotton fabric electrodes, Energy Storage Mater. 12 (2018) 204–215. https://doi.org/10.1016/j.ensm.2017.12.013.

[27]   Y. Tan, Y. Liu, L. Kong, L. Kang, F. Ran, Supercapacitor electrode of nano-Co_3O_4 decorated with gold nanoparticles via in-situ reduction method, J. Power Sources. 363 (2017) 1–8. https://doi.org/10.1016/j.jpowsour.2017.07.054.

[28]   P.E. Lokhande, U.S. Chavan, Nanoflower-like $Ni(OH)_2$ synthesis with chemical bath deposition method for high performance electrochemical applications, Mater. Lett. 218 (2018) 225–228. https://doi.org/10.1016/j.matlet.2018.02.012.

[29]   V. Gupta, T. Kusahara, H. Toyama, S. Gupta, N. Miura, Potentiostatically deposited nanostructured α-Co(OH)2: A high performance electrode material for redox-capacitors, Electrochem. Commun. 9 (2007) 2315–2319. https://doi.org/10.1016/j.elecom.2007.06.041.

[30]   J. Ma, X. Guo, Y. Yan, H. Xue, H. Pang, FeOx -Based Materials for Electrochemical Energy Storage, Adv. Sci. 5 (2018) 1700986. https://doi.org/10.1002/advs.201700986.

[31]   Y. Qian, J. Du, D.J. Kang, Enhanced electrochemical performance of porous Co-doped TiO_2 nanomaterials prepared by a solvothermal method, Microporous Mesoporous Mater. 273 (2019) 148–155. https://doi.org/10.1016/j.micromeso.2018.06.056.

[32]   Y. Zhou, Z.Y. Qin, L. Li, Y. Zhang, Y.L. Wei, L.F. Wang, M.F. Zhu, Polyaniline/multi-walled carbon nanotube composites with core-shell structures as

supercapacitor electrode materials, Electrochim. Acta. 55 (2010) 3904–3908.
https://doi.org/10.1016/j.electacta.2010.02.022.

[33]   A. Afzal, F.A. Abuilaiwi, A. Habib, M. Awais, S.B. Waje, M.A. Atieh,
Polypyrrole/carbon nanotube supercapacitorsTechnological advances and challenges,
J. Power Sources. 352 (2017) 174–186.
https://doi.org/10.1016/j.jpowsour.2017.03.128.

[34]   H. Zhang, Z. Hu, M. Li, L. Hu, S. Jiao, A high-performance supercapacitor based
on a polythiophene/multiwalled carbon nanotube composite by electropolymerization
in an ionic liquid microemulsion, J. Mater. Chem. A. 2 (2014) 17024–17030.
https://doi.org/10.1039/C4TA03369H.

[35]   A.G. Pandolfo, A.F. Hollenkamp, Carbon properties and their role in
supercapacitors, J. Power Sources. 157 (2006) 11–27.
https://doi.org/10.1016/j.jpowsour.2006.02.065.

[36]   P. Simon, Y. Gogotsi, Capacitive Energy storage in nanostructured carbon–
electrolyte systems, Acc. Chem. Res. 46 (2013) 1094–1103.
https://doi.org/10.1021/ar200306b.

[37]   W. Xia, A. Mahmood, R. Zou, Q. Xu, Metal-organic frameworks and their derived
nanostructures for electrochemical energy storage and conversion, Energy Environ.
Sci. 8 (2015) 1837–1866. https://doi.org/10.1039/c5ee00762c.

[38]   P.E. Lokhande, U.S. Chavan, Materials science for energy technologies
nanostructured $Ni(OH)_2$/rGO composite chemically deposited on Ni foam for high
performance of supercapacitor applications, Mater. Sci. Energy Technol. 2 (2019) 52–
56. https://doi.org/10.1016/j.mset.2018.10.003.

[39]   P.E. Lokhande, K. Pawar, U.S. Chavan, Chemically deposited ultrathin α-$Ni(OH)_2$
nanosheet using surfactant on Ni foam for high performance supercapacitor
application, Mater. Sci. Energy Technol. 1 (2018) 166–170.
https://doi.org/10.1016/j.mset.2018.07.001.

[40]   H. Pan, J. Li, Y.P. Feng, Carbon nanotubes for supercapacitor, Nanoscale Res.
Lett. 5 (2010) 654–668. https://doi.org/10.1007/s11671-009-9508-2.

[41]   J. Yan, Q. Wang, T. Wei, Z. Fan, Recent Advances in Design and Fabrication of
Electrochemical Supercapacitors with High Energy Densities, Adv. Energy Mater. 4
(2014) 1300816. https://doi.org/10.1002/aenm.201300816.

[42] W. Sun, X. Rui, M. Ulaganathan, S. Madhavi, Q. Yan, Few-layered Ni(OH)$_2$ nanosheets for high-performance supercapacitors, J. Power Sources. 295 (2015) 323–328. https://doi.org/10.1016/j.jpowsour.2015.07.024.

[43] Halper, S. Marin, J.C. Ellenbogen, Supercapacitors : A Brief Overview, Mitre Nanosyst. Gr. (2006).

[44] W. Raza, F. Ali, N. Raza, Y. Luo, K.H. Kim, J. Yang, S. Kumar, A. Mehmood, E.E. Kwon, Recent advancements in supercapacitor technology, Nano Energy. 52 (2018) 441–473. https://doi.org/10.1016/j.nanoen.2018.08.013.

[45] C. Zhong, Y. Deng, W. Hu, J. Qiao, L. Zhang, J. Zhang, A review of electrolyte materials and compositions for electrochemical supercapacitors, Chem. Soc. Rev. 44 (2015) 7484–7539. https://doi.org/10.1039/c5cs00303b.

[46] L. Soserov, T. Boyadzhieva, V. Koleva, C. Girginov, A. Stoyanova, R. Stoyanova, Effect of the electrolyte alkaline ions on the electrochemical performance of α-Ni(OH)$_2$/activated carbon composites in the hybrid supercapacitor cell, ChemistrySelect. 2 (2017) 6693–6698. https://doi.org/10.1002/slct.201701579.

[47] S. Schweizer, J. Landwehr, B.J.M. Etzold, R.H. Meißner, M. Amkreutz, P. Schiffels, J.-R. Hill, Combined computational and experimental study on the influence of surface chemistry of carbon-based electrodes on electrode–electrolyte interactions in supercapacitors, J. Phys. Chem. C. (2019) acs.jpcc.8b07617. https://doi.org/10.1021/acs.jpcc.8b07617.

[48] A. Lewandowski, A. Olejniczak, M. Galinski, I. Stepniak, Performance of carbon-carbon supercapacitors based on organic, aqueous and ionic liquid electrolytes, J. Power Sources. 195 (2010) 5814–5819. https://doi.org/10.1016/j.jpowsour.2010.03.082.

[49] G. Wang, L. Zhang, J. Zhang, A review of electrode materials for electrochemical supercapacitors, Chem. Soc. Rev. 41 (2012) 797–828. https://doi.org/10.1039/C1CS15060J.

[50] N. Blomquist, T. Wells, B. Andres, J. Bäckström, S. Forsberg, H. Olin, Metal-free supercapacitor with aqueous electrolyte and low-cost carbon materials, Sci. Rep. 7 (2017) 1–7. https://doi.org/10.1038/srep39836.

[51] X. Fang, D. Yao, An Overview of Solid-Like Electrolytes for Supercapacitors, in: Vol. 6A Energy, ASME, 2013: p. V06AT07A071. https://doi.org/10.1115/IMECE2013-64069.

[52]   X. Wang, Y. Li, F. Lou, M.E. Melandsø Buan, E. Sheridan, D. Chen, Enhancing capacitance of supercapacitor with both organic electrolyte and ionic liquid electrolyte on a biomass-derived carbon, RSC Adv. 7 (2017) 23859–23865. https://doi.org/10.1039/C7RA01630A.

[53]   F. Béguin, V. Presser, A. Balducci, E. Frackowiak, Carbons and electrolytes for advanced supercapacitors, Adv. Mater. 26 (2014) 2219–2251. https://doi.org/10.1002/adma.201304137.

[54]   E. Frackowiak, M. Meller, J. Menzel, D. Gastol, K. Fic, Redox-active electrolyte for supercapacitor application, Faraday Discuss. 172 (2014) 179–198. https://doi.org/10.1039/C4FD00052H.

[55]   M. Armand, F. Endres, D.R. MacFarlane, H. Ohno, B. Scrosati, Ionic-liquid materials for the electrochemical challenges of the future, Nat. Mater. 8 (2009) 621–629. https://doi.org/10.1038/nmat2448.

[56]   M.Y. Kiriukhin, K.D. Collins, Dynamic hydration numbers for biologically important ions, Biophys. Chem. 99 (2002) 155–168. https://doi.org/10.1016/S0301-4622(02) 00153-9.

[57]   A. Brandt, S. Pohlmann, A. Varzi, A. Balducci, S. Passerini, Ionic liquids in supercapacitors, MRS Bull. 38 (2013) 554–559. https://doi.org/10.1557/mrs.2013.151.

[58]   M. Galiński, A. Lewandowski, I. Stepniak, Ionic liquids as electrolytes, Electrochim. Acta. 51 (2006) 5567–5580. https://doi.org/10.1016/j.electacta.2006.03.016.

[59]   A. Lewandowski, M. Galinski, Practical and theoretical limits for electrochemical double-layer capacitors, J. Power Sources. 173 (2007) 822–828. https://doi.org/10.1016/j.jpowsour.2007.05.062.

[60]   A. Balducci, R. Dugas, P.L. Taberna, P. Simon, D. Plée, M. Mastragostino, S. Passerini, High temperature carbon-carbon supercapacitor using ionic liquid as electrolyte, J. Power Sources. 165 (2007) 922–927. https://doi.org/10.1016/j.jpowsour.2006.12.048.

[61]   R.C. Agrawal, G.P. Pandey, Solid polymer electrolytes: Materials designing and all-solid-state battery applications: An overview, J. Phys. D. Appl. Phys. 41 (2008) 223001. https://doi.org/10.1088/0022-3727/41/22/223001.

[62]   H. Gao, K. Lian, Proton-conducting polymer electrolytes and their applications in solid supercapacitors: A review, RSC Adv. 4 (2014) 33091–33113. https://doi.org/10.1039/c4ra05151c.

[63]   G.G. Cameron, Solid polymer electrolytes: Fundamentals and technological Applications. Fiona M. Gray. VCH Publishers Inc., New York 1991. pp. x + 245, price £44.00. ISBN 0–89573–772–8, Polym. Int. 32 (1993) 436–436. https://doi.org/10.1002/pi.4990320421.

[64]   L.-Q. Fan, J. Zhong, J.-H. Wu, J.-M. Lin, Y.-F. Huang, Improving the energy density of quasi-solid-state electric double-layer capacitors by introducing redox additives into gel polymer electrolytes, J. Mater. Chem. A. 2 (2014) 9011-9014.. https://doi.org/10.1039/c4ta01408a.

[65]   M.L. Verma, M. Minakshi, N.K. Singh, Synthesis and characterization of solid polymer electrolyte based on activated carbon for solid state capacitor, Electrochim. Acta. 137 (2014) 497–503. https://doi.org/10.1016/j.electacta.2014.06.039.

[66]   N.A. Choudhury, S. Sampath, A.K. Shukla, Hydrogel-polymer electrolytes for electrochemical capacitors: an overview, Energy Environ. Sci. 2 (2009) 55–67. https://doi.org/10.1039/B811217G.

[67]   S. Chintapalli, R. Frech, Effect of plasticizers on high molecular weight PEO-LiCF3SO3 complexes, Solid State Ionics. 86–88 (1996) 341–346. https://doi.org/10.1016/0167-2738(96)00144-0.

[68]   E. Tsuchida, H. Ohno, K. Tsunemi, Conduction of lithium ions in polyvinylidene fluoride and its derivatives-I, Electrochim. Acta. 28 (1983) 591–595. https://doi.org/10.1016/0013-4686(83)85049-X.

[69]   M. Watanabe, M. Kanba, K. Nagaoka, I. Shinohara, Ionic conductivity of hybrid films based on polyacrylonitrile and their battery application, J. Appl. Polym. Sci. 27 (1982) 4191–4198. https://doi.org/10.1002/app.1982.070271110.

[70]   G.B. Appetecchi, F. Croce, B. Scrosati, Kinetics and stability of the lithium electrode in poly(methylmethacrylate)-based gel electrolytes, Electrochim. Acta. 40 (1995) 991–997. https://doi.org/10.1016/0013-4686(94)00345-2.

[71]   X. Wen, T. Dong, A. Liu, S. Zheng, S. Chen, Y. Han, S. Zhang, A new solid-state electrolyte based on polymeric ionic liquid for high-performance supercapacitor, Ionics (Kiel). (2018) 1–11. https://doi.org/10.1007/s11581-018-2582-7.

[72]   Y. Kato, S. Hori, T. Saito, K. Suzuki, M. Hirayama, A. Mitsui, M. Yonemura, H. Iba, R. Kanno, High-power all-solid-state batteries using sulfide superionic conductors, Nat. Energy. 1 (2016) 16030. https://doi.org/10.1038/nenergy.2016.30.

[73]   B.E. Francisco, C.M. Jones, S.-H. Lee, C.R. Stoldt, Nanostructured all-solid-state supercapacitor based on Li 2 S-P 2 S 5 glass-ceramic electrolyte, Appl. Phys. Lett. 100 (2012) 103902. https://doi.org/10.1063/1.3693521.

[74]   A.S. Ulihin, Y.G. Mateyshina, N.F. Uvarov, All-solid-state asymmetric supercapacitors with solid composite electrolytes, Solid State Ionics. 251 (2013) 62–65. https://doi.org/10.1016/j.ssi.2013.03.014.

[75]   Y. Inaguma, C. Liquan, M. Itoh, T. Nakamura, T. Uchida, H. Ikuta, M. Wakihara, High ionic conductivity in lithium lanthanum titanate, Solid State Commun. 86 (1993) 689–693. https://doi.org/10.1016/0038-1098(93)90841-A.

[76]   X. Hu, Y. Chen, Z. Hu, Y. Li, Z. Ling, All-Solid-State Supercapacitors Based on a Carbon-Filled Porous/Dense/Porous Layered Ceramic Electrolyte, J. Electrochem. Soc. 165 (2018) A1269–A1274. https://doi.org/10.1149/2.0481807jes.

[77]   C. Ogata, R. Kurogi, K. Awaya, K. Hatakeyama, T. Taniguchi, M. Koinuma, Y. Matsumoto, All-Graphene Oxide Flexible Solid-State Supercapacitors with Enhanced Electrochemical Performance, ACS Appl. Mater. Interfaces. 9 (2017) 26151–26160. https://doi.org/10.1021/acsami.7b04180.

[78]   W. Gao, N. Singh, L. Song, Z. Liu, A.L.M. Reddy, L. Ci, R. Vajtai, Q. Zhang, B. Wei, P.M. Ajayan, Direct laser writing of micro-supercapacitors on hydrated graphite oxide films, Nat. Nanotechnol. 6 (2011) 496–500. https://doi.org/10.1038/nnano.2011.110.

[79]   Q. Zhang, K. Scrafford, M. Li, Z. Cao, Z. Xia, P.M. Ajayan, B. Wei, Anomalous capacitive behaviors of graphene oxide based solid-state supercapacitors, Nano Lett. 14 (2014) 1938–1943. https://doi.org/10.1021/nl4047784.

[80]   M. Vangari, T. Pryor, L. Jiang, Supercapacitors : Review of materials and fabrication methods, J. Energy Eng. 139 (2013) 72–79. https://doi.org/10.1061/(ASCE)EY.1943-7897.0000102.

Supercapacitor Technology: Materials, Processes and Architectures   Materials Research Forum LLC
Materials Research Foundations **61** (2019) 31-44   https://doi.org/10.21741/9781644900499-3

Chapter 3

Gel Polymer Electrolytes for Supercapacitor Applications

K.K. Purushothaman[1], B. Saravanakumar[2*], S. Vadivel[3], N. Krishna Chandar[4], Mohd Imran Ahamed[5]

[2] Department of Physics, Aringar Anna Government Arts and Science College, Karaikal-609605, Puduchery, India

[1] Department of Physics, Dr. Mahalingam College of Engineering and Technology, Pollachi, 642003, Tamilnadu, India

[3] Department of Chemistry, PSG College of Technology, Coimbatore-641004, Tamilnadu, India

[4] Department of Physics, School of Advanced Sciences, Vellore Institute of Technology, Vellore-632 014, Tamilnadu, India

[5] Department of Chemistry, Faculty of Science, Aligarh Muslim University, Aligarh-202 002, India

*saravanakumar123@gmail.com

Abstract

Flexible supercapacitors are emerging as potential energy storage devices for advanced mobile electronic devices. The gel polymer electrolytes are considered as the potential electrolyte for future advanced flexible electrical energy storage systems owing to their elastic polymeric nature with higher redox capacitance. The use of these gel polymer electrolytes in a supercapacitor system facilitates the reduction in size, reliability, weight, better flexibility and extended operating window with high range of operating temperature. The use of gel polymer electrolytes holds a greater hope for fabrication of new generation mobile electronics. In this chapter, we intend to discuss the advancement in design and types of gel-based polymer electrolyte towards supercapacitor device applications in detail.

Keywords

Supercapacitor, Flexible Electrode, Composite, Electrolyte, Gel Polymer

Contents

1. Introduction

The rapid growth of the world population is stressing the need for developing innovative electrical energy production technologies. At present, significant efforts have been made to develop new technologies for efficient utilization of renewable energies including solar, geothermal, tidal and ocean thermal energies [1, 2]. All these methods of producing electrical energy mostly depend on the nature of the environment. Here, the disparity in the regional renewable resources is considered as a critical concern for continuous supply. The development of stable, efficient and environmental benign electrochemical energy storage devices is important for effective utilization of renewable energy [3, 4].

Lithium batteries and electrochemical capacitors are the two important types of electrical energy storage devices, utilized for various energy storage needs. Among these, supercapacitors (SCs), also known as ultracapacitors, have generated significant interest due to their unparalleled physiochemical features including higher cyclic stability, zero maintenance, safety and very high power density. Due to these attractive features, the SCs can be used in power backup systems, hybrid electric vehicles, wind turbine blade pitching systems and different mobile electronics applications.

Basically the SCs are categorized, as electrical double-layer capacitors (EDLCs) and pseudocapacitors based on the charge storage principles. The EDLCs save the charges at the interface between electrolyte and electrode. Here, the separation of charges is a physical reaction not utilizing Faradic reactions. Normally, the carbon materials have been utilized as electrode materials owing to their higher surface area and chemically inertness. In contrast, pseudocapacitors store the charges through fast Faradic reactions. They involve metal oxides (MOs) and conducting polymers (CPs) and exhibit higher electrochemical performance.

The growth of flexible electronic devices and hybrid electric transportation systems need the fabrication of superior and stable electrical energy storage systems. At present, flexible electrical energy storage devices have captured reasonable attention for mobile electronics applications. The flexible energy storage devices possess advantageous features such as low cost, reliability, easy handling and wide operating temperatures [5, 6]. Mainly there are three different types of electrolytes used for supercapacitor applications which include water-soluble electrolytes, organic solutions and ionic liquids [7, 8]. All these electrolytes have their own advantages and disadvantages. But more specifically, these involve most expensive packaging materials and methods to reduce the issue of leakage of the electrolytes. Further, it is very tedious to design flexible energy storage devices using liquid electrolytes.

The flexible SC system is made of flexible electrode, separator and solid-state electrolyte as presented in the Figure 1.

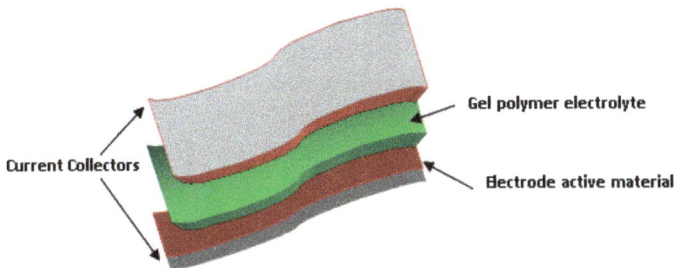

Figure 1. Fabrication schematic of a gel polymer based SC device

It is essentially required that the electrolyte and electrode should have some flexibility to design the flexible SC system. Most of the reports concentrate on the performance improvements in flexible electrodes. Here, the solid-state electrolyte is also a key component to determine the rate performance and cyclic stability of a SC device. Among various electrolytes, the polymer gel based electrolytes are considered as better candidate for flexible SCs. The development of polymer-based flexible electrolytes was first reported by Fenton et al. in the year of 1973 [9]. Since then, the development of polymer-based flexible electrolytes has received greater interest due to their significant usefulness in energy storage and conversion systems [10, 11].

2.   Gel polymer electrolytes

The electrolyte is a critical component of the SC device. Among different types of electrolytes, the polymer gel electrolytes are extending greater advantages like flexibility, ease of handling and good range of operating temperature. Especially, this kind of electrolyte significantly reduces the leakage and corrosion problems. It may cut the device packaging expenses drastically. Generally, a good gel polymer electrolyte should possess nontoxicity, better stability, wide potential window and higher ionic conductivity. This type of electrolytes possesses enhanced conductivity than dry solid polymer electrolytes.

Polymer gel electrolytes possess a host of polymeric material, a solvent (aqueous or organic) and plasticizer with additional electrolyte salt. Poly(vinyl alcohol) (PVA), poly (methyl methacrylate) (PMMA), polyacrylonitrile (PAN), poly(amine-ester) (PAE), poly (vinylidene fluoride) (PVdF), and poly (polyacrylate) (PAA), are the polymeric materials utilized as polymer host for the preparation of gel polymer based electrolytes. Further, ethylene carbonate, propylene carbonate, ethyl methyl carbonate, dimethyl carbonate and diethyl carbonate are the common solvents used for the synthesis of gel polymer based electrolytes. For providing mobile ions in electrolyte, electrolytic salt is added during preparation of the electrolyte. Based on the salt present in the electrolyte, the gel polymer electrolytes are categorized into four different types (Figure 2). (1) Lithium-based ion gel polymer electrolytes, (2) Proton conducting gel polymer electrolytes, (3) Alkaline based gel polymer electrolytes, and (4) Ionic liquid based gel polymer electrolytes.

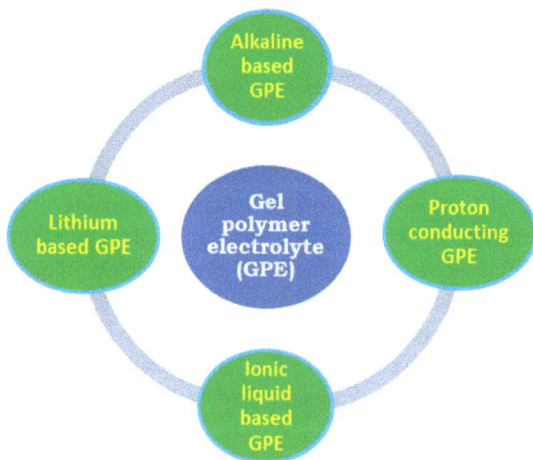

Figure 2. Classification of gel polymer electrolyte

2.1   Li-ion based gel polymer electrolytes

Li salt based gel polymer electrolytes were tested for energy storage applications including SC and Li-ion batteries. The Li-ion polymer gel electrolytes are synthesized by blending a polymer host with a salt of lithium dissolved with an organic solvent at a temperature of 80-170°C. During the last few decades, many polymer hosts have been developed and investigated that include poly (acrylonitrile) (PAN), poly (propylene oxide) (PPO), poly (ethylene oxide) (PEO), poly (vinylidene fluoride) (PVdF), poly (vinyl chloride) (PVC), poly (vinylidene fluoride-hexafluoropropylene) (PVdF-HFP) and poly (methyl methacrylate) (PMMA), etc. Most importantly, combinations like PVA/LiClO$_4$, PMMA/LiClO$_4$, PAN/LiClO$_4$ and PAE/PVA/SiO$_2$/LiClO$_4$ have been studied extensively for energy storage applications [12].

For instance, Chodankar et al. [13] designed the MnO$_2$ based flexible solid state SC device using PVA/LiClO$_4$GPE. This electrolyte demonstrates better electrochemical properties including a higher potential window with the capacitance of 112 Fg^{-1} with an energy density of 15 Whkg^{-1}. In addition, this device showed the cyclic stability up to 2500 CV cycles and exhibited the negligible decrease in electrochemical features even after 20 days (calendar life).

The poly (vinyl alcohol) potassium borate (PVAPB) based GPEs were electrodeposited on the surface of activated carbon electrodes and the resulting electrodes were used for fabrication of flexible SC device by Jiang and co-workers [14]. It is reasonable to mention here that PVAPB GPE served separator and also electrolyte for this SC device which showed the capacitance of 65.9 F g^{-1} at a current rate of 0.1 Ag^{-1}. Further, it retained 95% of initial specific capacitance after 2000 charge-discharge cycles. Hashmi et al. electrochemically fabricated MnO$_2$-polypyrrole based flexible SC devices using different GPEs including PVA, PVA-H$_3$PO$_4$, LiClO$_4$-PC and PMMA-EC-PC-NaClO$_4$ [15]. Interestingly, the SC cell using PVA-H$_3$PO$_4$ electrolyte displayed a non-capacitive nature behavior due to reversible reaction of water and MnO$_2$. In contrast, the MnO$_2$-PPy combination exhibited better performance with non-aqueous electrolytes and displaying capacitance ranges from 10–18 mF cm^{-2} at different current rates.

Deng et al. [16] designed a flexible symmetrical SC device utilizing Mn oxide electrodes and following different techniques. The authors utilized LiClO$_4$/PVA gel polymer electrolyte and explored its electrochemical features. This attractive device exhibited specific capacitance of 960 Fg^{-1} with superior retention of capacitance around 85% after 5000 cycles. Further, the supercapacitive features of this device measured in the temperature range 27–110 °C showed highest capacitance at 1100 Fg^{-1} at 90 °C. In addition, the authors studied the oxidation-state changes of the material by in-situ X-ray absorption spectroscopy.

Vanadium oxide nanowires SC electrodes were prepared by Wang and co-workers and tested their potential for SC device applications [17]. The authors addressed the issue of capacitance fade during charging/discharging cycling and material pulverization by using LiCl/ PVA polymer gel electrolyte. This device exhibited better retention of capacitance around 85% after 5000 charge- discharge cycles without losing its performance.

A composite of ZnO/amorphous ZnO/ MnO_2 nanostruture deposited on a cloth of carbon by Yang et al. The material exhibited specific capacitance of 1260.9 Fg^{-1} [18]. The authors assembled a flexible all-solid-state SC utilizing PVA/ LiCl GPE as electrolyte. This attractive SC showed areal capacitance of 26 $mFcm^{-2}$ and retained 87.5% of initial capacitance up to 10000 charge-discharge cycles. A supercapacitor electrode containing nanowires of PEDOT- MnO_2-core materials has been synthesized and utilized for fabrication of symmetric and asymmetric supercapacitor devices by Duay et al. [19]. The authors investigated electrochemical features and reported the energy density of 9.8 $Whkg^{-1}$ with a power density of 850 W kg^{-1}in a Li based gel polymer electrolyte. This highly flexible device retained 86% of its energy density at flexed state with the specific capacitance of 0.26 F at a potential window of 1.7 V.

2.2   Proton conducting gel polymer electrolyte

Proton-conducting gel polymer based electrolytes primarily utilized in fuel cells have exhibited superior ionic conductivity compared to all other kinds of polymer electrolytes. However, their narrow potential window restricted their use. The proton-conducting polymer electrolytes can be synthesized at room temperature. For achieving higher conductivity, the presence of water is needed. The retention of better hydration under ambient conditions is still a big challenge. This type of electrolytes has been extensively investigated for electrochemical energy storage applications. Mainly pure ionomers having negatively charged functional sites ($-SO_3H$, $-PO_4H_2$, $-COOH$) are concentrated owing to their better proton conductivity at lower temperature (below 100 °C). These types of electrolytes have multiphase structures having hydrophobic and hydrophilic areas. Among different proton-conducting gel polymer electrolytes, perfluorosulphonic acid (PFSA) membranes have been highly attractive for low-temperature applications. In this context, Nafion has been considered as a most standard proton-conducting polymer [20,21]. Staiti et al. have demonstrated the usefulness of the Nafion polymer electrolyte for EDLCs. The authors developed a many-cell device by combining five individual devices. This device exhibited the capacitance of 1.5 F with 5V of working window [21].

Interestingly, Choi et al. synthesized solution of Nafion and slowly transferred onto the thin film electrode to facilitate the ion migration at the electrode-electrolyte interface for EDLC [22]. It resulted in enhancement in the capacitance compared to EDLC devices without Nafion blending. The Nafion blended gel polymer electrolyte showed specific capacitance of 118.5 Fg^{-1}. Sarangapani et al. fabricated first Nafian based pseudo capacitor utilizing RuO_2-Nafion composite electrode [23]. Further, the mass loadings of Nafion isomers were optimized by Park et al. and the resulted device reached the maximum capacitance of 200 Fg^{-1} with a superior cyclic stability of 10000 cycles [24].

However, the preparation of Nafion is costly and not environmental friendly due to its fluorine-based production technologies. Due to this important reason, non-perfluorinated alternatives such as poly (ether sulfones) (PES) and poly (ether ketones) (PEK), poly (ether ether ketone) (PEEK), polyesters, poly (ether ether ketone ketone) (PEEKK) and poly(arylene ethers) have been considered. Additionally, these materials have better chemical resistance due to high C-H bond strength. The different chemical structures of these materials through doping have shown significantly improved proton conductivity [25]. Sivaraman et al. fabricated a SC device using PANI as the electrode active material and a poly (vinyl sulfonic acid) as the proton-exchanging gel polymer. [26]. This capacitor demonstrated a capacitance of 98 Fg^{-1} and 80% retention of capacitance after 1500 cycles.

On the other hand, acid- polymer blends such as mixture of strong acids (H_2SO_4 or H_3PO_4) with polymers have been studied extensively as supercapacitor electrolytes. This kind of electrolytes provided good conductivity, easy preparation techniques and high stability [27]. The polymers like PVA, PEO, PAAM, poly(4-vinylpyridine) (P4VP), poly (vinyl pyrrolidone) (PVP), and poly (2-vinyl pyridine) (P2VP) have been used for synthesis of this type of electrolytes. Among these polymers, PVA has been extensively investigated due to its good film-forming capability.

Nano-crystalline RuO_2 films were prepared on a stainless steel current collector by utilizing chemical deposition by Dubal et al. [28]. The authors extensively reported the formation of different types of films at varying coating temperatures. The H_2SO_4/PVA polymer electrolyte was used as a separator and electrolyte. This SC system displayed capacitance of 234 Fg^{-1} with a lower value of charge transfer resistance (ESR) of 0.63. This lower value of ESR is beneficial to perform charge-discharge activity of SC at higher current densities. Wu et al. fabricated *a* plane micro-supercapacitor utilizing graphene-based materials as electrode active material. The combination of H_2SO_4 and PVA was used as the electrolyte. Which exhibited the areal capacitance of 80.7 μFcm^{-2} [29]. H_2SO_4/PVA polymer gel electrolyte was used for assembling of PANI-based supercapacitors. The attractive device showed specific capacitance as 360 Fg^{-1} [30].

2.3 Alkaline based gel polymer electrolytes

Researchers are paying great attention on the synthesis of different varieties of alkaline polymer electrolytes due to their advantageous features. Some of the significant alkaline polymer gel electrolytes are P(ECH-co-EO)/KOH/ H_2O (poly (epichlorohydrin-co-ethylene oxide), (PAAK)/KOH/H_2O (potassium poly(acrylate), PVA/KOH/ H_2O and PEO/KOH/H_2O. For instance, Yuvan et al. assembled an SC using NiO with activated carbon based electrodes. The authors used the alkaline PVA/KOH/H_2O as electrolyte [31]. Interestingly, the authors reported the performance dependence of SC on temperature ranges between - 20 to 40 °C. The rise in the temperature significantly improved the conductivity of polymer gel electrolyte and the reduction in the charge-transfer resistance provided good interfacing between electrode and electrolyte. This attractive SC exhibited capacitance of 73 Fg^{-1} and an energy density of 26.1 $Whkg^{-1}$ at the current rate of 0.1 Ag^{-1}. An SC device was assembled by Kalpana et al. utilizing carbon aero gel as electrode and the alkaline PVA-KOH-PEO gel polymer electrolyte as electrolyte [32]. The authors reported the thickness of 3 mm with ionic conductivity of 10^{-2} Scm^{-1} at room temperature. This device displayed a capacitance of 9 F g^{-1}.

The alkaline polymer gel PVA/ KOH/ $K_3[Fe(CN)_6]$ based SC device was assembled by Ma et al. using activated carbon-based positive and negative electrodes [33]. Here the gel polymer electrode was utilized as separator as well. The use of alkaline and polymer electrolyte facilitated higher ionic conductivity, flexible and larger potential window. This device exhibited superior electrochemical features including specific capacitance of 430.95 Fg^{-1} with cyclic stability of 1000 cycles at a current rate of 1 Ag^{-1}. In addition, this device showed energy and power densities of 57.94 $Whkg^{-1}$ and 59.84 $kW kg^{-1}$ respectively. Fadakar et al. assembled an SC device using nanosilver as electrodes [34]. The authors sandwiched the electrodes using the alkaline-based gel polymer containing polyvinyl alcohol and potassium hydroxide (PVA–KOH) with a redox mediator potassium iodide (KI). In addition, they reported the performance of this device without addition of KI.

A semi quasi-solid-state SC device was fabricated utilizing activated carbon as electrodes and PVA–KOH–KI as gel polymer electrolyte [35]. The authors detailed the increasing of ionic conductivity by mixing of KI. It led the enhancement in electrochemical properties including improved capacitance, rate performance, and energy/ power densities. This device exhibited the capacitance of 236.90 F g^{-1} with the energy density of 7.80 Wh kg^{-1}. In addition, it exhibited power density of 15.34 kW kg^{-1}. It is interesting to mention here that the introduction of redox mediator KI resulted in increasing the capacitance value to the level of 74.8% in comparison to the PVA–KOH based device at the same current rate.

2.4   Ionic liquid based gel polymer based electrolyte

Since the first report in 1914, the ionic liquids (ILs) have been utilized for many significant applications [36]. Some of the unique advantages of ILs include higher ionic conductivity (10^{-3} to 10^{-2} SCm^{-1}), larger potential window (4-6V), nonflammability and nonvolatility. In addition, the higher chemical and environmental stabilities led their use in supercapacitor systems. For instance, Muchakayala et al. developed a novel ionic polymer gel electrolyte with a combination of 1- methyl -1 -propylpyrrolidinium bis (trifluoromethyl sulfonyl) imide ([PMpyr][NTf$_2$])/ poly (vinylidene fluoride-hexafluoropropylene) (PVdF-HFP) and tested its usefulness for SC electrode applications [37]. The authors reported the morphology, structure, thermal and electrochemical properties of this polymer gel electrolyte using different analytical techniques. The transparent materials showed good amorphicity and better thermal stability. The SC device assembled using MWCNT-added AC as electrodes and ionic liquid gel polymer electrolyte, exhibited specific capacitance of 156.64 F g^{-1} with a specific energy of 30.69 Wh kg^{-1}. Further, this SC device retained initial capacitance up to 2000 charge/discharge cycles.

The activated carbon-based flexible SC devices have used the combination of 1-butyl-3-methylimidazolium iodide (BMIMI), neutral Li$_2$SO$_4$ and PVA [38]. The authors reported that the PVA-Li$_2$SO$_4$-BMIMI ionic liquid based gel polymer electrolyte exhibited superior break up elongation up to 1200%. This flexible SC provided the energy density of 29.3 Whkg^{-1} due to the higher potential window and better salvation Li$^+$ cations and SO$_4^{2-}$ anions. In addition, this device showed good durability and mechanical performance.

Tiruye et al. fabricated SC devices using activated carbon electrode and ionic liquid-based polymer electrolyte (IL-*b*-PE) [39]. The electrolyte consisted of a (pDADMATFSI), (poly (diallyldimethylammonium) bis (trifluoromethanesulfonyl) imide) and PYR$_{14}$TFSI (N- butyl-N-methylpyrrolidinium bis (trifluoromethylsulfonyl) imide) in the ratio of 40:60. This device exhibited capacitance of 100 Fg^{-1} and the energy density of 32 Whkg^{-1}. The authors reported the energy density of the device was 42 Whkg^{-1} at a current density of 1mAcm^{-2}. The use of ionic liquid blended gel polymer electrolyte in the device enhanced the potential window as high as 3.5 V.

The all solid state SC was assembled utilizing functionalised MWCNT as electrode material and the ionic liquid based gel polymer electrolyte [40]. The 1-ethyl-3-methylimidazolium tris (pentafluoroethyl) trifluorophosphate, (EMImFAP) was used as electrolyte with LiPF$_6$. The authors reported the effect of adding Li salt in the electrolyte. The poly(vinylidine fluoride-co-hexafluoropropylene) (PVdF-HFP) was used as a polymeric host in the electrolyte. This film electrolyte exhibited a higher potential window as 4 Volts between -2.0 to 2.0 V with good ionic conductivity ($\sim 2.6 \times 10^{-3}$ S cm^{-1} at 20 °C). This device showed capacitance of 127 Fg^{-1} for IL-LiPF$_6$ based electrolyte and 76 F g^{-1} for without Li-salt. In addition, this device exhibited good cyclic stability.

Obeidat et al. prepared a ionic liquid based polymer gel electrolyte using 1-butyl-3methylimidazolium tetra fluoroborate (BMIBF$_4$) with poly (vinylidene fluoride hexa fluoro propylene) P(VdF-HFP) [41]. The polypyrrole coated graphite sheets prepared by pulsed polymerization technique were used as SC electrodes. The individual electrode showed areal capacitance values as 403.23 mF cm^{-2} and 336.08 mF cm^{-2} in H$_2$SO$_4$ and LiClO$_4$ electrolytes respectively. The very high ionic conductivities and extended potential operating windows of ionic liquid based GPE'S notably improved the electrochemical performance of the SC devices.

Conclusion

The commercialization of an all solid-state flexible supercapacitor device is not realized in spite of much notable advancement. The research on this technology still needs further refinement to achieve this goal. There are many researchers working on the chemistry of materials for designing innovative electrolytes to realize significant results. But comprehensive efforts are much needed on system integration with emphasis on active components such as electrodes, electrolyte, separator and packaging. The research on flexible electrolytes is much needed for the development of advanced energy storage systems. The critical constraints of polymer gel based electrolytes are lower ionic conductivity and limited potential window with the presence of hydroxyl ions. Therefore, further intensive research on electrolytes is largely appreciated to improve the conductivity and potential window.

References

[1] A. K. Akella, R. P. Saini, M. P. Sharma, Social, economical and environmental impacts of renewable energy systems, Renew. Energy. 34 (2009) 390-396. https://doi.org/10.1016/j.renene.2008.05.002

[2] B. V. Mathiesen, H. Lund, K. Karlsson, 100% Renewable energy systems, climate mitigation and economic growth, Appl. Energy. 88 (2011) 488-501. https://doi.org/10.1016/j.apenergy.2010.03.001

[3] A. N. Menegaki, A social marketing mix for renewable energy in europe based on consumer stated preference surveys, Renew. Energy. 39 (2012)30-39. https://doi.org/10.1016/j.renene.2011.08.042

[4] M. Cao, Z. H. Li, J. Wang, W. Ge, T. Yue, R. Li, V. L. Colvin, W. W.Yu, Food related applications of magnetic iron oxide nanoparticles: enzyme immobilization, protein purification, and food analysis, Mater. Sci. and Eng. Tech. 27 (2012) 47-56. https://doi.org/10.1016/j.tifs.2012.04.003

[5] S. Kim, H. J. Kwon, S. Lee, H. Shim, Y. Chun, W. Choi, J. Kwack, D. Han, M. Song, S. Kim, S. Mohammadi, I. Kee and S. Y. Lee, Low-power flexible organic light-emitting diode display device, Adv. Mater. 23 (2011) 3511-3516. https://doi.org/10.1002/adma.201101066

[6] M. Koo, K. I. Park, S. H. Lee, M. Suh, D. Y. Jeon, J. W. Choi, K. Kang, K. J. Lee, Bendable inorganic thin-film battery for fully flexible electronic systems. Nano Lett. 12 (2012) 4810-4816. https://doi.org/10.1021/nl302254v

[7] Y. G. Wang, Z. D. Wang, Y. Y. Xia, An asymmetric supercapacitor using RuO2/ TiO2 nanotube composite and activated carbon electrodes, Electrochim. Acta. 50 (2005) 5641-5646. https://doi.org/10.1016/j.electacta.2005.03.042

[8] A. E. Fischer, K. A. Pettigrew, D. R. Rolison, R. M. Stroud, J. W. Long, Incorporation of homogeneous, nanoscale MnO_2 within ultraporous carbon structures via self-limiting electroless deposition: implications for electrochemical capacitors, Nano Lett. 7 (2007) 281-286. https://doi.org/10.1021/nl062263i

[9] D.E. Fenton, J.M. Parker, P.V. Wright, Complexes of alkali metal ions with poly(ethylene oxide), Polymer. 14 (1973) 589–589. https://doi.org/10.1016/0032-3861(73)90146-8

[10] S. Ramesh, C. W. Liew, Investigation on the effects of addition of SiO_2 nanoparticles on ionic conductivity, FTIR, and thermal properties of nanocomposite $PMMA\text{-}LiCF_3SO_3\text{-}SiO_2$, Ionics. 16 (2010) 255–262. https://doi.org/10.1007/s11581-009-0388-3

[11] D. F. Shriver, P. G. Bruce, Solid State Electrochemistry, second ed., Cambridge University Press, Cambridge, New York, 1995.

[12] A. M. Stephan, Review on gel polymer electrolytes for lithium batteries, Eur. Polym. J. 42 (2006) 21–42. https://doi.org/10.1016/j.eurpolymj.2005.09.017

[13] N. R. Chodankar, D. P. Dubal, A. C. Lokhande, C. D. Lokhande, Ionically conducting $PVA\text{-}LiClO_4$ gel electrolyte for high performance flexible solid-state

supercapacitors, J. Colloid. Interf. Sci. 460 (2015)370-376.
https://doi.org/10.1016/j.jcis.2015.08.046

[14] M. Jiang, J. Zhu, C. Chen, Y. Lu, Y. Ge, X. Zhang, Poly (vinyl Alcohol) borate gel
polymer electrolytes prepared by electrodeposition and their application in
electrochemical supercapacitors, ACS Appl. Mater. Interfaces. 8 (5) (2016) 3473–
3481. https://doi.org/10.1021/acsami.5b11984

[15] S. A. Hashmi, H. M. Upadhyaya, MnO_2-polypyrrole conducting polymer composite
electrodes for electrochemical redox supercapacitors, Ionics. 8, (2002) 272-277.
https://doi.org/10.1007/BF02376079

[16] M. J. Deng, K. W. Chen, Y. C. Che, I. J. Wang, C. M. Lin, J. M. Chen, K. T. Lu, Y.
F. Liao, H. Ishii, Cheap, High-performance, and wearable Mn oxide supercapacitors
with urea-$LiClO_4$ based gel electrolytes, ACS Appl. Mater. Interfaces 9 (1) (2017)
479–486. https://doi.org/10.1021/acsami.6b13575

[17] G. M. Wang, X. H. Lu, Y. C. Ling, T. Zhai, H. Y. Wang, Y. X. Tong, Y. Li, LiCl/
PVA gel electrolyte stabilizes vanadium oxide nanowire electrodes for
pseudocapacitors, ACS Nano. 6 (2012) 10296-10302.
https://doi.org/10.1021/nn304178b

[18] P. Yang, X. Xiao, Y. Li, Y. Ding, P. Qiang, X. Tan, W. Mai, Z. Lin, W. Wu, T. Li,
H. Jin, P. Liu, J. Zhou, C. P. Wong, Z. L. Wang, Hydrogenated ZnO core–shell
nanocables for flexible supercapacitors and self-powered systems, ACS Nano 7(3)
(2013) 2617-2626. https://doi.org/10.1021/nn306044d

[19] J. Duay, E. Gillette, R. Liu, S. B. Lee, Highly flexible pseudocapacitor based on
freestanding heterogeneous MnO_2/conductive polymer nanowire arrays, Phys. Chem.
Chem. Phys. 14 (2012) 3329-3337. https://doi.org/10.1039/c2cp00019a

[20] S. Slade, S. Campbell, T. Ralph, Ionic conductivity of an extruded nafion 1100 EW
series of membranes. J. Electrochem. Soc. 149 (2002) A1556- A1564.
https://doi.org/10.1149/1.1517281

[21] P. Staiti, F. Lufrano. Design, fabrication, and evaluation of a 1.5 F and 5 V prototype
of solid-state electrochemical supercapacitor. J. Electrochem. Soc. 152 (2005) A617-
A621. https://doi.org/10.1149/1.1859614

[22] B. G. Choi, J. Hong, W. H. Hong, P. T. Hammond, H. Park, Facilitated ion transport
in all-solid-state flexible supercapacitors. ACS Nano 5 (2011) 7205-7213.
https://doi.org/10.1021/nn202020w

[23] S. Sarangapani, P. Lessner, J. Forchione, A. Griffith, A. B. Laconti, Advanced double layer capacitors, J. Power Sources 29(1990) 355-364. https://doi.org/10.1016/0378-7753(90)85010-A

[24] K.W. Park, H. J. Ahn and Y. E. Sung, All-solid-state supercapacitor using a Nafion®polymer membrane and its hybridization with a direct methanol fuel cell. J. Power Sources 109 (2002) 500-506. https://doi.org/10.1016/S0378-7753(02)00165-9

[25] J. Sumner, S. Creager, J. Ma, D. D. Marteau, Proton Conductivity in nafion® 117 and in a novel bis [(perfluoroalkyl) sulfonyl] imide ionomer membrane, J. Electrochem. Soc.145 (1998) 107-110. https://doi.org/10.1149/1.1838220

[26] P. Sivaraman, S. K. Rath, V. R. Hande, A. P. Thakur, M. Patri, A. B. Samui, All-solid-supercapacitor based on polyaniline and sulfonated polymers, Synth. Mat.156 (2006) 1057-1064. https://doi.org/10.1016/j.synthmet.2006.06.017

[27] P. Colomban, Proton Conductors: Solids, membranes and gels-materials and devices, first ed., Cambridge University Press, Cambridge, USA, 1992. https://doi.org/10.1017/CBO9780511524806

[28] D. P. Dubal, G. S. Gund, R. Holze, H. S. Jadhav, C. D. Lokhande, C.J. Park, Solution-based binder-free synthetic approach of RuO2 thin films for all solid state supercapacitors, Electrochim. Acta. 103 (2013) 103-109. https://doi.org/10.1016/j.electacta.2013.04.055

[29] Z. S. Wu, K. Parvez, X. Feng, K. Mullen, Graphene-based in-plane micro-supercapacitors with high power and energy densities. Nat. Commun. 4 (2013) 1-8. https://doi.org/10.1038/ncomms3487

[30] C. Meng, C. Liu, L. Chen, C. Hu, S. Fan, Highly flexible and all-solid-state paper like polymer supercapacitors, Nano Lett.10 (2010) 4025-4031. https://doi.org/10.1021/nl1019672

[31] C. Yuan, X. Zhang, Q. Wu, B. Gao, Effect of temperature on the hybrid supercapacitor based on NiO and activated carbon with alkaline polymer gel electrolyte, Solid-State Ionics 177 (2006) 1237-1242. https://doi.org/10.1016/j.ssi.2006.04.052

[32] D. Kalpana, N. G. Renganathan, S. Pitchumani, A new class of alkaline polymer gel electrolyte for carbon aerogel supercapacitors, J. Power Sources. 157 (2006) 621–623. https://doi.org/10.1016/j.jpowsour.2005.07.057

[33] G. Ma, J. Li, K. Sun, H. Peng, J. Mu, Z. Lei, High-performance solid-state supercapacitor with PVA–KOH–K$_3$ [Fe(CN)$_6$] gel polymer as electrolyte and

separator, J. Power Sources 256 (2014) 281-287.
https://doi.org/10.1016/j.jpowsour.2014.01.062

[34] Z. Fadakar, N. Nasirizadeh, S. M. Bidoki, Z. Shekari, V. Mottaghitalab, Fabrication of a supercapacitor with a PVA–KOH–KI electrolyte and nano silver flexible electrodes, Microelectronic Eng. 140 (2015) 29-32.
https://doi.org/10.1016/j.mee.2015.05.004

[35] H. Yu, J. Wu, L. Fan, K. Xu, X. Zhong, Y. Lin, J. Lin, Improvement of the performance for quasi-solid-state supercapacitor by using PVA–KOH–KI polymer gel electrolyte, Electrochim. Acta. 56 (20) (2011) 6881-6886.
https://doi.org/10.1016/j.electacta.2011.06.039

[36] P. Walden, On the molecular size and electrical conductivity of some molten salts, Bull. Acad. Sci. 8(6) (1914) 405–422.

[37] R. Muchakayala, S. Song, J. Wang, Y. Fan, M. Bengeppagari, J. Chen, M. Tan, Development and supercapacitor application of ionic liquid-incorporated gel polymer electrolyte films, J. Ind. Engg. Chem, 59 (2018) 79-89.
https://doi.org/10.1016/j.jiec.2017.10.009

[38] Q. M. Tu, L.Q. Fan, F. Pan, J. L. Haung , Y. Gu, J. M. Lin, M. L Huang, Y. F. Huang, J. H. Wu, Design of a novel redox-active gel polymer electrolyte with a dual-role ionic liquid for flexible supercapacitors, Electrochim. Acta. 268 (2018) 562-568.
https://doi.org/10.1016/j.electacta.2018.02.008

[39] G. A. Tiruye, D. M. Torrero, J. Palma, M. Anderson, R. Marcilla, All-solid-state supercapacitors operating at 3.5 V by using ionic liquid-based polymer electrolytes, J. Power Sources. 279 (2015) 472-480. https://doi.org/10.1016/j.jpowsour.2015.01.039

[40] G. P. Pandey, S. A. Hashmi, Solid-state supercapacitors with ionic liquid based gel polymer electrolyte: Effect of lithium salt addition, J. Power Sources 243 (2013) 211-218. https://doi.org/10.1016/j.jpowsour.2013.05.183

[41] A. M. Obeidat, M. A. Gharaibeh, M. Obaidat, Solid-state supercapacitors with ionic liquid gel polymer electrolyte and polypyrrole electrodes for electrical energy storage J. Energy Storage 13 (2017) 123-128. https://doi.org/10.1016/j.est.2017.07.010

Supercapacitor Technology: Materials, Processes and Architectures   Materials Research Forum LLC
Materials Research Foundations **61** (2019) 45-94   https://doi.org/10.21741/9781644900499-4

Chapter 4

Redox Electrolytes/Mediators for Supercapacitors

X. Zhou*, Y. Wang, C. Zhang, and X. Qiao

[1]Department of Mechanical and Aerospace Engineering, University of Miami, USA

*xzhou@miami.edu

Abstract

This chapter is intended to summarize research progress of relatively new supercapacitors, redox electrolyte/mediator supercapacitors using other types of supercapacitors as references. The chapter will articulate the most important milestones of the research and development. Much of the efforts have been on important fundamental issues including the mechanisms, electron shuttling effect, simulation approaches, and metrics for evaluating the specific capacitance, energy, and power. The aim of this chapter is to view the mediator supercapacitor as a hybrid of electric double layer capacitor or EDLC and rechargeable battery. In the end, the authors' visions for future research and development are provided.

Keywords

Redox Electrolyte, Mediator, Supercapacitor, Simulation, Nanomaterials

Contents

1. Introduction

The R&D of electrochemical energy storage devices including rechargeable batteries and supercapacitors (SCs) has gained an unprecedented momentum [1-4]. In a renewable or sustainable economy that largely relies on renewable energy power sources including solar energy and wind energy, the electrochemical energy storage devices will play an indispensable role in coping with the intermittent nature of the renewable energy sources. Timely implementation of the electrochemical energy storage devices has become one of the main goals of technological developments of the major industrial countries. Rechargeable batteries and SCs sharply differ in their operational properties [5, 6]. As shown in Figure 1, while rechargeable batteries have high charge capacity and specific energy but low charge/discharge rate, low specific power, and low cycleability, SCs have low charge capacity and specific energy but high charge/discharge rate, high specific power, and high cycleability. A competition between rechargeable battery technology (mainly lithium batteries) and SC technology in the arena of electrochemical energy storage is undergoing. The performance boundaries for both are advancing to previously unattainable regions quickly. Although, very rarely a SC can be used as a standalone power source, the high charge rate and high cycleablity are two merits that keep SCs in the competition. Indeed, there is a high stake on increasing the specific energy to level of the rechargeable batteries.

Most research efforts on electrochemical energy storage devices are primarily focused on electrode materials [7-9]. For example, SCs are often categorized by the electronically conducting materials such as carbon materials, metal oxides, and conducting polymers. Recently, the use of nanoscale 1-D (nanotubes or nanofibers) and 2-D (graphene and graphene oxide) materials has become the main stream of the research of SCs [10-13]. The underline logic is that the number of ion accessible sites on the surface of the electronic conducting phases dictates the specific capacitance and energy of the device. Probably in 2009-2011, researchers started to publish their results on redox electrolyte/mediator SCs [14-16]. The redox species or mediators dissolved or partially dissolved in either water [14, 15] or polymers [16] provided an additional charge storage capacitance or pseudocapacitance via electrochemical redox reactions. The redox species or mediators dissolved in an electrolyte can also promote the conductivity of the electrolyte [16]. Many kinds of mediators can release ions during the charging process thereby mitigating the problem of ion depletion in the electrolyte [17]. Because of these benefits, the redox electrolyte or mediator SC has drawn worldwide attention [18]. However, because this is a relatively new area of research, understanding of the mechanisms and even a unified metrics for evaluation of the performance of mediator/redox electrolyte SCs have not been well established. This leads to confusions, overestimations, invalid claims, and hindrances to advancement of the scientific and engineering goals of the research. This chapter is intended to summarize the major advancements, current status, and visions to initiate discussions on the mechanisms and some important fundamental issues. Specifically, the objectives of this chapter are to

1) survey the types, the operating mechanisms, and properties of the SCs;

2) summarize the advent and development of redox electrolyte/mediator SC;

3) understand the mechanisms and properties of the redox electrolyte/mediator SCs and its advantages and disadvantages versus other types of supercapacitors;

4) report the state-of-the-art solid-state mediator supercapacitors;

5) review the progress on nanomaterials based mediator supercapacitors.

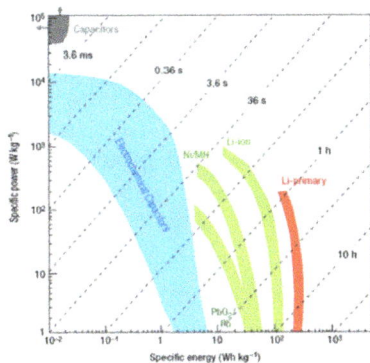

Figure 1. Ragone plots for power sources (Simon, P., and Gogotsi, Y. (2008). Materials for electrochemical capacitors. Nat. Mater. 7, 845–854.doi:10.1038/nmat2297).

2.   Types of Supercapacitors

2.1   Carbon EDLC

SCs can be categorized into five different types [6-9]: 1) carbon materials based electrical double layer capacitor or EDLCs; 2) metal oxide SCs; 3) conducting polymer SCs; 4) hybrid SCs; and 5) redox electrolyte/mediator SCs. The EDLCs were first commercialized SCs. The high energy charged state is characterized with a physical separation of positive and negative charges across the interface between electronic and ionic conducting phases. In an ideal situation, the positive and negative charges are attracted by the electrostatic force that is modulated by the dielectric species in-between. This is an analog of dielectric capacitors. However, due to the high interface area and short distance between the separated charges, the specific energy of EDLCs ranges from 3 to 15 Wh/kg much greater than that of an dielectric capacitor and specific power ranging from 0.8-3 kW/kg [6, 9]. The EDLCs show an ideal capacitive behavior including: rectangular cyclic voltammetry (CV) at low and high scan rates, high charge/discharge energy efficiency (>90%), and high cycle life up to 10^6. Typical specific capacitance is 120-150 F/g, which represents roughly 18% electronic charge (1.6×10^{-19} C) per carbon atom on the electrode/electrolyte interface at 1 V [6]. The relatively low utilization of the surface area is presumably due to the low accessibility of the counter ions to the deep portion of pores through tortuous paths.

2.2   Metal oxide supercapacitors

Many metal oxides can undergo redox reactions at a high reaction rate [6, 19]. These metal oxides are often considered as potential materials offering a high pseudocapacitance. Ruthenium oxide (RuO_2) is probably the first metal oxide material for making pseudocapacitors due to its ideal capacitor-like behaviors: high charge/discharge rate, nearly perfect rectangle-shaped CV curves, and high reversibility (i.e. high charge and energy efficiencies). In acidic aqueous electrolytes, a fast reversible electron transfer reaction and an electro-adsorption of protons to the RuO_2 electrolyte interface enable storage of charges at a voltage:

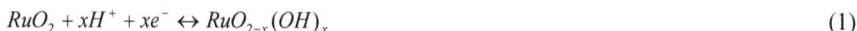

$$RuO_2 + xH^+ + xe^- \leftrightarrow RuO_{2-x}(OH)_x \tag{1}$$

Thus, the high energy charged state is realized via coupling/decoupling of electrons and ions (protons) from the electrolyte. It is also suggested that proton insertion into the 3-D bulk of RuO_2 and accordingly a valence transitions between Ru(IV), Ru(III), and Ru(II) could give rise to a very high specific capacitance of 600 F/g in H_2SO_4 aqueous electrolyte. Hydrous ruthenium oxide ($RuO_2 \cdot 0.5H_2O$) provides accessibility to a larger portion (70%) of RuO_2 by protons to form positive/negative charge couples. The specific capacitance could reach 900 F/g. According to the high capacitance of this material, it is estimated that the theoretical specific energy should be at least 40 Wh/kg with a voltage window of 1 V. However, prototypes made using RuO_2 materials only demonstrated a far lower value of 8.5 Wh/kg. The maximum specific power was reported to be 6 kW/kg by Zheng et al. [19]. Given the high cost of ruthenium, mass production and high market occupation for this type of supercapacitor are highly unlikely. Another issue with this type of SC is the aqueous electrolytes (acid or base solution) which offer a very limited voltage window (1.2 V).

Other extensively studied materials are manganese oxides which are much cheaper than RuO_2 [20-24]. Several redox couples, Mn(II)/Mn(III), Mn(III)/Mn(IV), and Mn(IV)/Mn(VI), for manganese oxides may function to provide pseudocapacitance via the following equation,

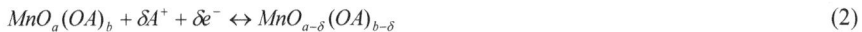

$$MnO_a(OA)_b + \delta A^+ + \delta e^- \leftrightarrow MnO_{a-\delta}(OA)_{b-\delta} \tag{2}$$

where A is H or alkali metals (Li, Na, K, etc). The specific capacitance of manganese oxides is generally lower than that of ruthenium oxide, usually 200-350 F/g though Mn has more functioning valences than Ru and is much lighter than Ru. The ideality, reversibility, and stability, specific power and energy of manganese oxides based SCs are not comparable to those of ruthenium oxides based SCs. However, Mn based electrode materials are considered favorable to Ru counterparts because they are much cheaper.

Many other metal oxides based SCs have been studied including Ni, Co, V, and Fe [25-35]. Although, efforts have been made in order to improve the specific surface area and accessibility using a variety of materials synthesis and processing approaches, and in a few cases, quite large specific capacitance have been achieved, the charge/discharge efficiency, reversibility, and ideality are not comparable with those of ruthenium oxide based SCs.

2.3   Conducting polymer supercapacitors

Conducting polymers are referred to polymers that possess π bonds giving rise to an electronic conductivity in the level of carbon materials (10^3 S/cm) [36, 37]. The commonly used conductive polymers are polyaniline (PANI)[38], polypyrrole (PPy)[39], polythiophene (PTh)[40] polyimides (PIs), and their derivatives. Conductive polymers are also coated on carbon materials such as graphene[41-45], graphene oxides (GO)[46], and carbon nanotube (CNT) [47, 48] to make composite materials. The conducting polymer SCs often do not show rectangular shaped CVs, fast charge/discharge rate, and good reversibility. In most of cases, the conducting polymer SCs are rather battery-like than capacitor-like. The maximum specific capacitance is 400-500 F/g. The specific power and energy are much lower than those for the metal-oxide supercapacitors in general.

2.4   Redox electrolyte supercapacitors

Some redox materials or mediators readily dissolve into electrolytes including aqueous, organic, polymer, superionic, and ceramic electrolytes. The redox materials dissolved in the electrolyte can be electrochemically oxidized or reduced at the interfaces between the electronic phases and electrolyte in the electrodes during the charging process. The products undergo a reverse redox reactions during the discharging process. In this way, the redox materials or mediators serve as charge storage or provide a pseudocapacitance. Thus, the charge storage mechanism of a redox electrolyte supercapacitor is similar to that of a flow battery [3].

2.5   Hybrid supercapacitors

The so-called hybrid SCs can be one of the following combinations [49]: 1) active carbon and redox materials (metal oxides or conducting polymers or redox electrolyte) in symmetrical electrodes; 2) active carbon in one electrode and redox materials in the other electrode; and 3) active carbon in one electrode and battery material in another electrodes. In fact, a pure EDLC does not exist because in an EDLC some species in the electrolyte may participate in a series of reversible chemisorption reactions. In other words, some redox reactions contribute to the total capacitance of an EDLC. In general, the combinations of active carbon and redox materials could enable higher specific capacitance and energy in comparison to carbon materials. However, these combinations often trade off the high cycleability, high charge and energy efficiency, and high charge/discharge rate. One of most recent developments has been so-called Li-ion SCs [50-56]. The commonly used configuration was Li-doped hard carbon as the negative electrode and a mixture of active carbon powder and a cathode material for Li-ion battery. At a higher charge/discharge rate, e.g. 60C (referred to a rate at which the full capacity can be charged or discharged in 1/60 hour), the behavior was capacitor-like and the specific energy was 12 Wh/kg for the cell. The charge efficiency was close to 100%. At the high charge/discharge rate, the cycleability can be more than 10000 with a capacity reduction of 6%. However, when the charge/discharge rate was low (e.g. 1C) the behavior was battery-like, the specific energy was 25 Wh/kg for the cell, the cycle life was only a few thousands similar to that of a typical Li-ion battery and the energy efficiency was 78%. Nevertheless, the Li-ion capacitors can achieve a high cycleability at a specific energy of more than 10 Wh/kg and specific power of ~1 kW/kg for a cell. The energy and charge efficiency were acceptable. This specific energy is at the high end of the range of that for the commercially available EDLCs and is mainly due to the greater voltage range (2.2-3.7V) than that of the EDLCs or 1.3V-2.7V.

2.6   Supercapacitor versus rechargeable battery

The energy of an ideal SC is 0.5qV, whereas the energy of an ideal rechargeable battery is 1.0qV where q is charge contained in each of the two electrodes and V is the voltage between the electrodes [6]. For non-ideal SCs and rechargeable batteries, the energy is between 0 and 1.0qV. The non-ideality of the SCs may result from sluggish transport of ionic species in the electrolytes or from irreversibility of the Faradaic processes which may result in loss of charges and energy or materials degradation. As discussed above, an EDLC is actually a combination of microscopic electrostatic capacitors and more or less charge storage capacity originated from redox reactions. In a series of publications [57-64], Conway summarized redox processes in a variety of systems and proposed a general concept and formula for evaluating the equilibrium potential for the redox reactions occurring at the interface between the electronic conducting and ionic conducting materials in electric energy storage devices as being given by,

$$E = E^0 + \frac{RT}{zF} \ln\left(\frac{X}{1-X}\right) \tag{3}$$

This is a typical Nernst equation, where X is occupancy fraction of redox states or sites. In the case of redox oxides,

$$X = \frac{[Ox]}{[Ox]+[Red]} \tag{4}$$

where [Ox] is the concentration of the oxidized form of a redox species and [Red] is the concentration of the reduced form.   In the case of underpotential deposition, X is the 2-dimentional site occupancy fraction, θ. In the case of Li^+ intercalation into electrode materials, X is the occupancy fraction of available layered lattice sites in the materials. For the pseudocapacitance originated from underpotential deposition the differential capacitance,

$$C_\emptyset = q_M \frac{d\theta_M}{dV} \tag{5}$$

can be evaluated using,

$$C_\emptyset = \frac{q_M F}{RT} \frac{K_1 C_{M^{z+}} exp[VF/RT]}{\left(1+K_1 C_{M^{z+}} exp[VF/RT]\right)^2} = \frac{q_M F}{RT} \frac{\theta_M(1-\theta_M)}{1+g\theta_M(1-\theta_M)} \tag{6}$$

Where K_1 is the equilibrium constant of an electrochemical Langmuir isothermal, V is the voltage, q_M is the charge required for completion of a mono-layer of atom/molecule M on electrode surface. $C_{M^{z+}}$ is the concentration of cation M^{z+}. g is the lateral interaction (repulsion) parameter.

Conway commented that when g=0, or there is no lateral repulsion between M on the electrode surface, $C_\emptyset$ as a function of V is characterized with a sharp peak or a battery-like behavior. When g is large, the function is characterized with a flat hump or a behavior more resembling to an EDLC. There exists a common idea that the difference between co-called capacitor-like and battery-like behaviors is reflection of an interface process versus a bulk process. In fact, the above theoretical analysis demonstrates that a pure interface process (adsorption) can result a battery-like behavior or capacitor-like behavior. Conway also pointed out that some redox oxides and conducting polymers could be battery-like and others are capacitor-like such as RuO_2. Some Li+ intercalation systems such as Li/TiS_2 are capacitor-like and others such as typical Li-ion battery materials are battery-like. These cases suggest that the difference between battery-like and capacitor-like is rather relative than absolute.

For the development of the SC technology, the main goal is how to increase the specific charge capacity/capacitance and specific energy while maintaining the high charge rate, high specific power, and high cycleability. Seemly there are two technical routes to achieve this goal: 1) to increase specific surface area and 2) to use or synthesize materials with a high intrinsic charge capacity. For many materials used in SCs their specific capacity increases with the specific area. In order to select an optimal material one has to know the possible maximum charge capacity that a material can reach by increasing the specific surface area. A fundamental way to measure the charge capacity of a material is to evaluate the charge that can be carried by one unit of the chemical formula, atom, or molecule. For example, the total charge is averaged onto each RuO_2 molecule for the RuO_2 electrode, to each LiC_6 for the negative electrode of a Li-ion battery, or to each C atom for an EDLC electrode. A microscopic quantity proportional to the specific surface area is the ratio of the atoms or molecules on the surface to the total atoms/molecules (RMST). In this way, it is easy for us to compare the average charge capacity per molecule (ACCPM) of a material with the valence of the material or number of electrons that a molecule can give off or accept in certain electrochemical potential ranges. The present authors collected information on specific capacitance versus the specific area. The ACCPM was calculated by

$$ACCPM = \frac{C \Delta V M_w}{F} \tag{7}$$

where C is the specific capacitance, ΔV is the voltage window, M_w is molecular mass, and F is Faraday constant. The RMST can be estimated by assuming that each molecule or atom can be approximated as a cube:

$$RMST = \frac{s}{\left(\frac{M_w}{d N_A}\right)^{\frac{2}{3}}} / \left(\frac{N_A}{M_w}\right) \tag{8}$$

where s is the specific area, d is the density, N_A is Avogadro constant.

The data of ACCPM vs. RMST are summarized in Figure 2. The data for metal oxides were taken from Refs. [19-35]. The data for carbon materials were taken from Refs. [5-13]. The data for graphite materials were collected from website of MTI Corp [65]. A set of reliable data for mediators were collected from Refs. [16, 17, 66-69]. The reason for choosing these papers but excluding other papers for mediators was that in those papers the mass of mediators or redox species was not accounted as the denominator for evaluating the mass specific capacitance.

As shown in Figure 2, it is clear that for most of metal oxides, it is true that increasing the specific surface results in an increased specific charge capacity. For most of the metal oxide materials, the specific charge capacity is enclosed between two limiting lines with a 45° angle to the axis of the coordinates. The data points for mediators are enclosed in the two lines but generally represent much greater ACCPM. This is because the mediators completely mix up with the electrolyte and there is no interior portion for them. Thus, the RMST is unity, the maximum.

The data for the graphite materials used in Li-ion batteries are above upper line and those for carbon materials used in EDLCs are below the lower line. For graphene, because there is no interior, the RMST is unity. The same is true for the mediators because if mediator dissolved into an electrolyte, each molecule is surrounded by electrolyte. For the graphite materials used in the Li-ion batteries, the ratio is the least among all.

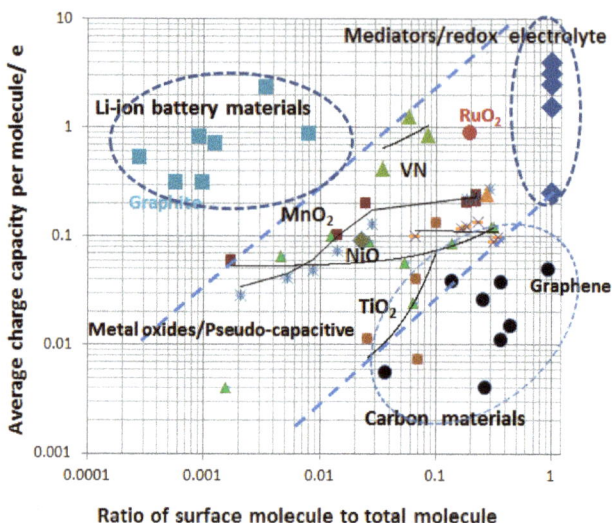

Figure 2. Average charge capacity per molecule versus ratio of the number of molecules on surface to the total number of molecules for a material.

It seems that there is no good correlation between the ACCPM and RMST for graphite and carbon materials. Further increasing the specific area for these two sets of materials does not seem to be effective. For the carbon materials in EDLCs, the specific surface area is at the high end among all other materials. Probably the gain of increasing the specific surface area is subsided by other adverse effects such as reduction of conductivity and increase of contact resistance. For the graphite in Li-ion battery, Li ions readily transport in the bulk. A high surface area is thus not that important.

The ACCPM of the graphite materials of the Li-ion batteries is in a range of 0.3-1 electron charge per LiC_6 while that of the carbon materials in the EDLCs is below 0.003-0.05 electron charge per C atom. However, if we average the charge onto each of the 6 C atoms in the LiC_6 for graphite, the ACCPM for the C materials in EDLCs can be in the same order of magnitude of that for the graphite materials. Thus, the apparent reasons for the difference between the specific energies for the EDLCs and Li-ion battery are 1) the cell voltage of the EDLCs is lower than that of the Li-ion batteries, i.e. 2.7V vs. 4.2V and 2) the charge/discharge behaviors are different which give rise to a difference of 2 folds.

For the metal oxide materials, though the ACCPM is greater than that of the carbon materials in EDLCs, the mass specific capacity may be lower than that of the carbon materials because of their higher densities. Interestingly, for almost reviewed data, the ACCPM was lower than one electron charge although it has been observed that RuO_2 can undergo multiple valence changes [6, 19]. Possibly, in these cases some of the Ru atoms underwent multiple valence change, some underwent single valence change, and some underwent no valence change within the window of the applied voltage. In other words, not all RuO_2 molecules were utilized or the charge density was not uniform in the electrodes. Presumably, due to finite electronic and ionic resistivities of the electrode materials, the materials close to the current collectors were first charged and the speed of the propagation of charged state was limited leaving some materials partially charged or uncharged during the charging cycle.

In order to further increase the ACCPM and utilization of active materials, one approach is to increase the voltage window to render higher valence changes, another one is to further increase RMST to 1.0, and yet another one is to increase the conductivities, in particular, the ionic conductivity. The redox materials/mediators dissolved in electrolyte can really reach RMST limit and improve the ionic conductivity. Figure 2 actually demonstrates one of the major benefits of mediator SCs: high intrinsic charge capacity.

2.7   The limitation of capacity by number of ions in the electrolyte

Li-ion batteries have a so-called "rock-chair" charge/discharge mechanism in which Li-ions released from the positive electrode during the charging process enter the bulk of the negative electrode to form LiC_6. In the case of an EDLC, during the charging process, positive ions migrate to negative electrode while negative ions migrate to the positive electrode. The charge capacity at certain voltage is limited by the total number of ions in the electrolyte or less than the total number of the ions in the electrolyte. In fact, the number of ions will not be utilized by 100% due to the slow transport when the ions are depleted. According to Zheng et al. [70] and Zheng [71] the capacitance of the device EDLC is,

$$\frac{1}{C_{tot}} = \frac{1}{C_{ed}} + \frac{1}{C_{el}} \tag{9}$$

Accordingly, the specific energy of an EDLC is not the commonly realized,

$$E_{ed} = \frac{1}{8} C_{ed} V^2 \tag{10}$$

but should be more properly expressed by

$$E_{tot} = \frac{1}{8} C_{ed} V^2 \frac{1}{1 + \frac{C_{ed}}{4\alpha C_{el} F}}$$
(11)

where C_{ed} is the capacitance of electrodes, C_{el} is the capacitance of electrolyte, V is voltage, α is the fraction of total ions removed from the electrolyte during charging, F is Faraday's constant. Equation (10) gives the specific energy based on the mass of the electrodes whereas Equation (11) gives the specific energy based on the total mass of the electrodes and electrolyte. Zheng et al. [70, 71] also reviewed published data and found that the specific energy based on the electrodes and electrolyte mass was less than 1/3 of that based on the mass of electrodes. If the separator, current collectors, and packaging materials are taken into account, the specific energy would be less than 1/6 of that based on the electrode mass. Equations (10) and (11) also suggest that if the issue of the depletion of ions in the electrolyte cannot be resolved, using high surface area nanomaterials and other low dimensional materials as the electrode materials, will not significantly improve the specific capacitance and energy. It will just make the total specific capacitance closer to the specific capacitance of the electrolyte, C_{el}. The depletion of ions during charging is also an issue for pseudocapacitors including ruthenium oxides and manganese oxides based SCs. For one valence change from III to IV for Ru in the positive electrode will release a proton which will be supplied to the negative electrode in which the valence of Ru changes from IV to III (Equation 1). However, valence change from II to IV will require more protons or cations in the electrolyte. The availability of cations in the electrolyte may be not sufficient. Some mediators including NaI and $K_4Fe(CN)_6$ [17], will release cations into the electrolyte during the charging process thereby mitigating the ion depletion problem.

2.8   Metrics for performance evaluation

In order to evaluate the operational properties of SCs including specific energy, power, and capacitance, four kinds of ways are found in literature which based on: 1) the active materials; 2) the electrode materials; 3) the electrodes and electrolyte; and 4) the full cell including electrodes, electrolyte, current collectors, separator/membrane, wiring, and packaging [54]. The differences among the specific capacitance, energy, and power evaluated using these ways can be easily one order of magnitude. However, this is not the only issue that leads to erroneous estimations and often overestimations of the performance. For example, the specific capacitance, power, and energy of EDLCs that were evaluated with different charge efficiency, energy efficiency, cycleability, reversibility, and self-discharge rate varied dramatically. Unfortunately, most of pseudocapacitors were not comparable with a typical EDLC with respect to the efficiencies, cycleability, and self-discharge rates. Thus, when one compares the specific capacitances, energies, and powers, the comparison may not be an apple-to-apple comparison.

In literature, there were many data that were evaluated using the three-electrode method where the counter electrode was supposed to have unlimited capacitance and the electrolyte was supposed to have unlimited amount of ions or capacitance. This could cause a strikingly high overestimation with respect to a two-electrode cell, not only because of the limited capacitances of the counter electrode and limited electrolyte capacitance but also because of a possible mismatch between the potential ranges for the redox reactions for the two electrodes.

Following the discussion above, firstly, evaluating the performance of a supercapacitor with and without considering the mass of electrolyte could lead a significant overestimation. Secondly, when specifying the specific energy, the charge/discharge rate (C-rate) or specific power and other parameters should be specified. A device made from pseudocapacitance materials may show very high specific energy but this may be achieved by trading off the other merits including the high charge/discharge rate and efficiencies. As discussed before, for a Li-ion hybrid supercapacitor, a relatively high specific energy (~30 Wh/kg) was obtained at a low charge rate and low cycleability whereas at a high charge rate and cycleability the specific energy was just at the high end of those for EDLCs (~10 Wh/kg). Thirdly, the self-discharge rate and durability should also be taken into account in the metrics. For Li-ion batteries, the open circuit voltage (OCV) could be constant for several weeks without external load. The OCV for EDLCs may last only for days. However, the open circuit voltage of pseudocapacitors often reduces much faster. In many cases, the OCV could reduce by 50% in less than 10 hours. For the commercial SCs, functioning well after a long-term storage at a certain temperature range is important. This property was not reported in most of research papers. Without implementation of a unified metrics, a rigorous and meaningful comparison between the operational properties of SCs is impossible.

3. Redox species or mediators in electrolytes

Mediators are referred to soluble or partially soluble compounds in electrolytes (ionic conductor) that can be reduced or oxidized at the electrodes of electrochemical cells. Application areas of mediators include electrochemical sensors and photovoltaic devices [72-74]. Mediators in the electrochemical cells and photovoltaic devices transport redox chemical states/charges between electrodes through electrolytes. Mediators in electrolytes not only increase electronic conductivity but also promote ionic conductivity. Ruff and

Friedrich [75] proposed a general transfer diffusion theory as follows. Consider an exchange reaction between A and A* which are chemically identical particles but differentiate themselves by their locations in space, and X is an atom, radical, ion, electron or an energy quantum state,

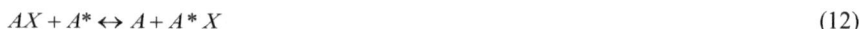

$$AX + A* \leftrightarrow A + A*X \tag{12}$$

The consequence of the reaction is that X transports from location A to location A*. Apparently, AX transports from the previous A position to A* position. If A particles are arrayed in the space and a gradient of chemical potential with respect to X exist in the space, X will be delivered along the direction of gradient via the exchange reactions. Thus, in electrolyte containing mediators, two mechanisms are in operation during diffusion and conduction: 1) the conventional migration of the species (physical diffusion) and 2) the exchange reaction which causes an apparent displacement of AX in space. Good examples are I_2/I^- and $Fe(CN)_6^{3-}/Fe(CN)_6^{4-}$ couples [75-80]. Diffusion coefficients in the mediator containing electrolytes can be described by

$$D_{AX} = D_{AX}^{phy} + D_{AX}^{ex} = D_{AX}^{phy} + \frac{k\delta^2(C_{AX}+C_A)}{6} \tag{13}$$

where D_{AX}^{phy} and D_{AX}^{ex} are physical and exchange diffusion coefficients respectively, k is exchange reaction rate constant, δ is the center-to-center distance at exchange reaction, C_{AX} and C_A are concentrations of A and AX respectively. If X is electron, the exchange reaction promotes the ionic conductivity (Equation 12) and enables electron transport in the electrolyte. Experimental data have shown that on increasing the concentrations of I_2 and I^- from 0.01 to 1 M, the diffusion coefficient increased by two orders of magnitude and conductivity by 4 orders of magnitude [77].

4.   Redox electrolyte-mediator supercapacitors

In 1999, Conway [6] mentioned that a potential electrical energy storage method was charging/discharging soluble redox species in electrolytes. Apparently, in order to realize a redox electrolyte/mediator SC, one must resolve the problem of electron shuttling between two electrodes by mediators. There are four ways to mitigate this problem: 1) using an ion-exchange membrane to block diffusion of mediators; 2) adsorbing or partially enclosing mediators in pores of electrodes; 3) withholding charged mediators on electrodes by electrostatic force; and 4) using mediators that form precipitates on electrodes upon charging. The primary advantage of redox electrolyte/mediator SCs is that both electrode and electrolyte can store charges and energy. The second advantage is that because the active species are dissolved into the electrolyte, the accessibility issue for the other active materials (porous carbon, metal oxides, and conducting polymer) is greatly mitigated. The third advantage is that addition of mediators into the electrolyte in the electrode significantly increases the conductivity of electrolyte. In the context of the limitation of capacitance set by the limited ions in the electrolyte, charging mediators will release free ions mitigating the ion depletion problem. For example, if NaI is used as the mediator in the electrolyte, at the positive electrode, oxidation of Na^+I^-, will release Na^+ cations to replenish the concentration of ions in the electrodes and in the separator. Similarly, if $VO^{2+}SO_4^{2-}$ as mediator in negative electrode is reduced, SO_4^{2-} anions will be released into the electrolyte.

Lota et al. studied [14, 15] the systems of activated carbon electrodes and aqueous electrolytes (1 mol/L H_2SO_4) with iodides (1 mol/L LiI, NaI, KI, RbI, and CsI) using a Swagelok cell. The authors did not mention how they resolved the electron shuttling between the electrodes. The results published in 2009 mentioned symmetrical CV curves. The authors claimed that the specific capacitance was ~200 F/g but for the positive electrode the specific capacitance can be high as 2772 F/g. These results have been quoted for many times. However, the authors evaluated the specific capacitance based on the mass of the one electrode (carbon) without taking into account the mass of electrolyte and mediators contained in the electrolyte. Thus, the capacitance values were overestimations. The authors claimed a high reversibility of the system. Nevertheless, the results revealed a considerable enhancement of the specific capacitance in comparison to the reference systems without mediator. The best performance was found with iodide (KI) systems with 250-280 F/g at a specific current of 2 A/g. About 80% of capacitance was retained up to 10,000 cycles. The authors conducted *in situ* Raman analyses of the chemical reactions at the positive electrode. They found that the I_2 produced during the electrochemical reaction,

$$2I^- \rightarrow I_2 + 2e^- \tag{14}$$

interacted with I^- in the electrolyte to form I_n^-

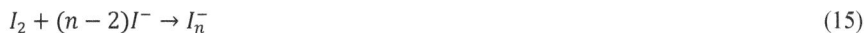

$$I_2 + (n-2)I^- \rightarrow I_n^- \tag{15}$$

Due to lack of the information regarding the geometrical configuration of the used cell and the type of membrane used as the separator, it is difficult to explain why the electron shuttling effect was not pronounced in this case. The authors suggested that the electrostatic attraction between the positive electrode and the anions, I_n^-, kept the oxidized mediators in the vicinity of the positive electrode mitigating the shuttling effect. It was also possible that because the body of the electrolyte was large, in the half cycle time (~100 seconds), the oxidized mediators could not arrive at the negative electrode.

Later, Frackowiak et al. [81] studied several other systems using a similar method. Their results showed that the specific capacitance of dihydroxybenzenes with varying –OH substitution (ortho, meta, para) as mediators was 283 F/g. Similar values were obtained with hydroquinone (HQ). All these values were obtained at a low charge/discharge rates (5 mV/s). The capacitance was stable up to 5000 cycles. After 5000 cycles, the capacitance decreased. The bromohydroxybenzene gave a high specific capacitance due to substitution of bromide in the benzene ring. The authors also conducted *in situ* Raman monitoring and confirmed formation of iodides.

Senthilkumar et al. [82. 83] also studied iodine and bromide species as mediators in acid electrolytes. They evaluated the specific capacitance, power, and energy with respect to the total mass of the active materials (active carbon) in both electrodes. The maximum specific capacitance (912 F/g) was obtained with electrolytes containing sulfuric acid and KI. However, again, since the total masses of the electrolyte and mediators were not used for the calculation and not specified, it was difficult to evaluate the contribution of the mediators to the total specific capacitance. The authors evaluated self-discharge for the supercapacitors. The open circuit voltage (OCV) reduced from the initial 0.8 to 0.4 V in about 20 minutes and was quite stable after 40 minutes. The galvanostatic charge and discharge (GCD) as well as CV curves showed appreciable asymmetry especially at charge/discharge rate of 10 mA/cm^2. These results confirmed a strong electron shuttling between the two electrodes.

Gao et al. [84] studied the effect of KI mediator in H_2SO_4 electrolyte on the specific capacitance of N-doped porous carbon. They found that adding 0.06 mol/L of KI into the electrolyte gave a specific capacitance of ~616 F/g or more than 100% increment in specific capacitance. They explained this increment with a catalytic effect of the N-doped porous carbon on the redox reactions. Unfortunately, the authors did not measure how much mediators was loaded in the SCs. Thus, there was no ground to judge what the contribution of mediator was and how much the "catalytic" effect was. In these previous studies, authors tended to consider the mediators as "promoters" or "catalysts" for the electrode materials. They did not really consider the mediators themselves as charge storage.

Frackowiak et al. [85] also investigated an asymmetrical $VOSO_4$/KI-H_2SO_4 system and received a specific capacitance of 550F/g. They proposed that $VOSO_4$ was reduced at the negative electrode while KI was oxidized at the positive electrode during the charging process.

Ferrocynide in combination of ferricynide ($(Fe(CN)_6]^{4-}/[Fe(CN)_6]^{3-}$) is a redox couple with fast electron exchange at electrodes. Several authors studied these mediators in aqueous solutions. Lee et al [86] used active carbon materials in electrodes and a PET (polyester) reinforced cation exchange membrane to suppress the electron shuttling effect. The specific capacitance, power and energy were evaluated based on the mass of the active materials in the electrodes. The authors emphasized the effect of excess volume of redox electrolyte which caused faster self-discharge. However, with the cation exchange membrane, the self-discharge rate was of $\sim$ 6 mV/hour with the initial voltage of 0.8 V. The maximum specific capacitance based on the active materials in electrodes (active carbon) was 597 F/g at a charge/discharge rate of 10 mA/g, which was a low rate for the standard of SCs. Within a voltage window of 0-1.2 V, the GCDs were quite symmetrical. However, when the voltage window was 0-1.8 V, the GDCs became asymmetrical and energy efficiency was only 57%. For charging/discharging between 0 and 1.8 V, the maximum specific energy is 28.3 Wh/kg at 20 W/kg. However, because the authors did not specify the total amount of mediators in the SCs and did not include this as the denominator for calculating the specific capacitance, the data cannot be used to assess the specific capacitance of the mediators and of the cell.

Su et al. [87] proposed the use of $Fe(CN)_6]^{4-}/[Fe(CN)_6]^{3-}$ redox couple in corporation of NiO in the electrodes. They obtained quite substantial promotion of specific capacitance up to 160 F/g on the basis of the active materials excluding the mediators. Still the reason why this worked so well was not quite clear. From the present authors' view, this was probably due to the promotion by the fast exchange diffusion/conduction of the mediators. According to Eq. (13), when the concentration of mediators is high, the total diffusivity and conductivity are dictated by the electron exchange mechanism. The electron exchange occurs via two paths: the reaction path and quantum tunneling path [80]. The latter is relatively insensitive to variation of temperature. Thus, not only the transport of mediators is promoted but due to electroosmotic effect, the transport of other ions pertinent to the redox reaction of NiO is also promoted. Su et al. [88] also studied the effect of $Fe(CN)_6]^{4-}/[Fe(CN)_6]^{3-}$ redox couple on the Co-Al layered double hydroxides based electrodes. Based on the active materials in the electrodes, the highest specific capacitance was 477 F/g. Chen et al. [89] studied the effect of $Fe(CN)_6]^{4-}/[Fe(CN)_6]^{3-}$ on $CoCl_2$, $CuCl_2$, $NiCl_2$ and $FeCl_3$ electroactive colloid electrodes (mixture of the salts, carbon black, and polyvinylidene floride). Addition of

$Fe(CN)_6]^{4-}/[Fe(CN)_6]^{3-}$ improved the charge capacity of the systems. However, the CV and GCD curves were asymmetrical and the energy efficiency of charge/discharge was 50% or less.

Chen et al. [68, 90, 91] published a series of papers on the use of chloride salts as mediators in aqueous electrolytes. The authors used a 3-electrode setup to conduct CV and GCD measurements. The specific capacitance based on the mediators was evaluated but excluding the mass of carbon materials which accounted to 1/8 of that of the mediators. The authors proposed a solvation/crystallization mechanism during the charging/discharging processes. However, the charge and energy efficiencies were very low especially at higher charge/discharge rates. Specific capacitances were 5442 F/g for $CuCl_2$, 1962 F/g for $CoCl_2$, or 2210 F/g for $YbCl_2$. Because these values were evaluated in a very narrow potential range (0-0.45 V), the specific energies were low according to Eq. (11).

In addition to charge/discharge kinetics and self-discharge, another issue with redox electrolyte SCs is the limited solubility of mediators or how much mediators can be carried by the electrolyte. One of the approaches to increase the quantity of mediator is adsorption or immobilization of mediators on the surface of carbon materials. However, this will reduce the accessibility of ions to the mediators. Another approach is to attach redox moieties to the molecules and ions of electrolytes with covalent bonds. Xie et al. [92] synthesized two ionic liquids by modifying cation or anion of [EMIm][NTf₂] with ferrocene, i.e. [FcEIm][NTf₂] and [EMIm][FcNTf]. The results indicated that [EMIm][FcNTf] was effective to promote the capacitance and energy but [FcEIm][NTf₂] was not. The IR drop in the GCDs was in the range of 400-500 mV which was too high for a typical SC. The specific energy based on carbon materials in electrode was 32.5 Wh/kg at a low power range of 150 W/kg. At a higher specific power of 840 W/kg, the specific energy, 7.5 Wh/kg was lower than that for the EDLC with non-active [EMIm][NTf₂] electrolyte. The symmetry of the GCDs was acceptable. During 1.25 hour, the OCV for the [EMIm][FcNTf] system dropped from 2 V to 1 V. Yamazaki et al. [93] also used an ionic liquid electrolyte, EMImBr in $EMImBF_4$, to promote the storage capacity of carbon electrodes. The specific capacitance was 59 F/g at a low discharge rate of 0.1 A/g. The GCD was very symmetric. 98% of the initial capacitance was retained up to 10^4 cycles. The self-discharge current was 0.03 mA indicating a low rate. The authors demonstarted the low self-discharge rate indicating the ability of the porous carbon to withhold the bromide species in the electrodes. Many other inorganic and organic redox species were examined including $VOSO_4$, indigo carmine, methylene blue lignosulfonates, and 1-ethy-3-methylmidazolium tetrafluoroborate+Cu^{2+}. Most of these were soluble in water. Wang et al. [94] used a chemical vapor deposition (CVD) prepared high surface carbon thin film as the electrode material and $FeBr_3$ as the mediator. They claimed that the specific capacitance was 885 F/g based on the active materials in the electrodes at a rate of 2A/g. Since the total amount of the mediator was not reported, this case was similar to the other studies[14, 15, 81]. It was impossible to judge what were the real specific capacitance, energy, and power of the SC . The capacitance of the SC did not show significant reduction within 10000 cycles. This may be explained by the fact that redox pair Br^-/ Br_3^- was maintained at the positive electrode and Fe^{3+}/Fe^{2+} was maintained at the negative electrode thereby mitigating the shuttling effect. The charge efficiency dropped dramatically from 100 to 30% when the charge/discharge current density increased from 2 to 20 A/g.

Navalpotro et al [95] fabricated SCs using para-benzoquinone (p-BQ) as mediator in ionic liquid, N-butyl-N-methylpyrrolidinium bis(-trifluoromethanesulfonyl) imide (PYR 14 TFSI) electrolyte. The electrodes were made using two types of active carbons, Vulcan with a specific surface area of 240 m^2/g and Pica with a specific surface area of 2400 m^2/g. On using Pica as the electrode material, addition of mediator resulted in a promotion of ~25% in the specific capacitance. However when Vulcan was used as the electrode material, the specific energy was promoted more than 2 folds by addition of the mediator. The absolute value of the 25% increment in the case of the Pica carbon, 40 F/g, was comparable to that of the 200% increment in the case of the Vulcan carbon, 45 F/g. The total capacitance can be considered as a sum of the capacitance of the carbon, i.e. EDLC and the capacitance of the mediator, i.e. pseudocapacitance. The charging processes of these two were competing. The charge rate for the mediator was a function of exchange current density and the potential of the electrode. It seems that at the carbon with the high specific surface area, the exchange current density is similar to that for the carbon with the low specific surface area. The potential of the electrode increased faster for the carbon with the low specific surface area or low specific capacitance than that for the carbon with the high specific surface area. Thus, the mediator was charged faster on Vulcan than on Pica. Since the charge time was longer for Pica than Vulcan, the final amounts of the charge on mediator for both cases were similar.

Zhang et al. [96] added p-phenylenediamine (PPD) into 6M KOH as the redox electrolyte. They used a 3-electrode setup to evaluate the specific capacitance as 501.5 F/g on the mass of the active carbon.

Komatsu et al. [97] used tetra-chlorohydroquinone (TCHQ) and anthraquinone (AQ) in H_2SO_4 electrolyte. The authors improved the electron conduction network in the electrode by adding a small amount of highly conductive carbons to the organic redox materials/activated carbon composites. By doing so, they claimed to have achieved both a high utilization rate (>70%) and high organic material loading ratio (70 wt% in the

Supercapacitor Technology: Materials, Processes and Architectures   Materials Research Forum LLC
Materials Research Foundations **61** (2019) 45-94   https://doi.org/10.21741/9781644900499-4

organic materials/ activated carbon composites). The resulting SCs exhibited a high specific energy of 30 Wh/kg. This result demonstrates that embedding the organic redox materials into the network of carbon is a viable approach.

Ren et al. [98] combined a conducting polymer, PANI with the active carbon to make electrodes. $FeSO_4$ and $Fe_2(SO_4)_3$ were used as mediators in H_2SO_4 electrolyte. The specific capacitance based on the active carbon and PANI was 1062 F/g. The authors claimed that the mediators promoted the charging process of the PANI. Similar results were obtained by Liu et al. [99] with $PANI/SnO_2$ electrodes. Zhu et al. [100] used hydroquinone (HQ) as the mediators to promote the performance of $PANI/SnO_2$ based SCs. The specific capacitance was 857 F/g. However, this was evaluated between -0.5 and 0.5 V. Since in practice a capacitor as an energy storage device operates in a range without switching polarity, this result is not meaningful from the application point of view. Nevertheless, with the mediator, the specific capacitance was increased. Following the similar strategy, Zhang et al. [101] used 4-oxo-2,2,6,6-tetramethylpiperidinooxy (4-oxo TEMPO) as the mediator to promote the performance of a Poly(3,4-ethylenedioxythiophene) (PEDOT) based SC. They obtained a specific capacitance of 363 F/g. Zhang et al. [104] used so-called dual redox species, phosphotungstic acid and potassium ferricyanide in the electrolyte to promote nanoporous carbon based SCs. The authors intended to have the mediators adsorbed onto the carbon materials. They obtained a specific capacitance of ~90 F/g.

Table 1. *The operational properties of redox electrolyte/mediator supercapacitors*

Redox/ electrolyte/Ref.	C/ΔV	E/P	Composition of mass	η_C/η_E	Cy	Sd
I_2 and I^- H_2SO_4 [14,15]	1840/ 0-1.2V	~/~	active carbon excluding mediator	H/~	5%/ 10000	~/~
I_2 and I^- H_2SO_4 [81]	283/ 0-1.2 V	~/~	active carbon excluding mediator	H/~	5%/ 5000	~/~
I_2 and I^- H_2SO_4 [82, 82]	912/ 0-1.2V	19/0.2	active carbon excluding mediator	L/L	5%/ 4000	50%/ 0.33h
I_2 and I^- H_2SO_4 [84]	616/ 0-1.2V	~/~	active carbon excluding mediator	~/~	5%/ 5000	~/~
$VOSO_4/KI-H_2SO_4$ [85]	550/ 0-0.8V	20/0.04	active materials excluding mediators	~/~	~/~	~/~
$Fe(CN)_6]^{4-}$ and $Fe(CN)_6]^{3-}$ [86]	597/ 0-1.2V	28.3/0.0 2	active carbon excluding mediator	65%/ 57%	25%/ 15000	25%/ 10 h

$Fe(CN)_6]^{4-}$ and $Fe(CN)_6]^{3-}$ [87]	160/ 0-1.2V	~/~	NiO excluding mediator	L/L	~	~/~
$Fe(CN)_6]^{4-}$ and $Fe(CN)_6]^{3-}$ [88]	477/ 0-1.2V	~/~	Co–Al layered double hydroxide	L/L	~	~/~
$CuCl_2$ [90]	5442/ 0-0.45V	~/~	only mediators	L/L	~	~/~
$CoCl_2$ [91]	1962/ 0-0.45V	~/~	only mediators	L/L	~	~/~
$YbCl_2$ [68]	2210/ 0-0.45V	~/~	only mediators	L/L	~	~/~
[FcEIm][NTf2] and [EMIm][FcNTf] RILs [92]	125/ 0-2.5V	32.5/0.15	only carbon electrodes excluding the mediator	L/L	5%/ 200	50%/ 1.25h
1-ethyl-3-methylimidazolium bro-mide dissolved in 1-ethyl-3-methylimidazolium tetrafluoroborate [93]	60/ 0-2.5V	~/~	active carbon excluding mediator	H/H	5%/ 10000	~/~
$FeBr_3$ [94]	885/ 0-1.6V	40/0.5	active carbon excluding mediator	85/~	5%/ 10000	~/~
methyl viologen (MV)/bromide [67]	32 / 0-1.4V	13.8/0.05	active carbon + electrolyte+ mediator	98.8/92	5%/ 20000	50%/ 2.5h
methyl viologen (MV)/bromide [67]	125/ 0-1.4V	51.0/0.2	active carbon excluding mediator	98.8/92	5%/ 20000	50%/ 2.5 h
para-Benzoquinone (p-BQ) in ionic liquid N-butyl-N-methylpyrrolidinium bis(-trifluoromethanesulfonyl) imide (PYR14 TFSI) [95]	70 / 0-3V	10.3/0.4	active carbon excluding mediator	~/~	50%/ 1000	~/~
p-phenylenediamine (PPD) in KOH [96]	501.4/ 0-1V	~/~	active carbon excluding mediator	L/L	15 %/ 5000	~/~
tetra-chlorohydroquinone (TCHQ) and anthraquinone (AQ) [97]	~/ 0-1.4V	32.5/~	unknown	~/~	50%/ 1000	~/~
$H_2SO_4 + Fe^{3+}/Fe^{2+}$ [98]	1062/ -0.2-0.6V	22.1/~	active carbon and PANI excluding mediator	L/L	7%/ 10000	~/~
Fe 3+ /Fe 2+ redox electrolyte for use in polyaniline/tin oxide (PANI/SnO₂) [99]	1172/ 0-0.5V	~/~	active materials excluding mediators	L/L	7% / 2000	~/~

hydroquinone (HQ) for PANI/SnO$_2$ [100]	857/ -0.5-0.5V	~/~	active materials excluding mediator	~/~	12.5% / 2000	~/~
Ferrocene (Fc) and 4-oxo-2,2,6,6-tetramethylpiperidinooxy (4-oxo TEMPO) on PEDOT [101]	363/ 0-1.5V	27.4/5	active materials excluding mediator	~/~	10%/ 300	~/~
p-phenylenediamine on nano-porous carbon [102]	557/ 0-1V	~/~	active materials excluding mediators	~/~	23%/ 500	~/~
hydroquinone-HQ on CNT+ PANI [103]	809.6/ 0-0.8V	72.5/4.9	active materials excluding mediators	~/~	25% / 2000	~/~
Phosphotungstic acid and potassium ferricyanide on nanoporous carbon [104]	90/ 0-1V	21.1/1.0	active materials excluding mediators	L/L	13%/ 5000	~/~
Sulfanilic acid azo chromotrop (SA) on GO [105]	1023/ 0-0.6V	80/1.0	active materials excluding mediators	L/L	15 %/ 10000	~/~
Hydroquinone and p-phenylenediamine [106]	116/ 0-1V	1.85/0.15	active materials excluding mediators	L/L	50%/ 20000	~/~
hydroquinone (HQ) in H$_2$SO$_4$ [107]	900/ 0-1V	~/~	active materials excluding mediators	L/L	50%/ 1000	~/~
HQ-H$_2$SO$_4$ PANI/MWCNTs [108]	7926/ 0-0.7V	113/~	active materials excluding mediators	30%/ 30%	4% / 1000	~/~

C: Specific capacitance (F/g); ΔV: Voltage window (V); E: Specific energy (Wh/kg); P: Specific power (kW/kg); η_C: Charge efficiency (%,);H(high) or L(low); η_E: Energy efficiency (%); Cy: Cycleability (Reduction%/cycle); Sd: Self-discharge (Reduction%/hour)

Generally speaking, a wide range of variations in specific capacitance and energy can be witnessed (Table 1), from 32 to 7926 F/g. The cycleability for the redox electrolyte/mediator SCs was far lower than that for typical EDLCs. In most cases the data for self-discharge were not available.

Akinwolemiwa et al. [18] reviewed and analyzed a number of papers. They correctly described how the mediators provided an additional charge storage as schematically shown in Figure 3. The oxidized mediator was reduced at a negative electrode during the charging process and the reduced mediator generated during the charging process released the electrons during the discharging process. These reversible processes can convert the mediators in the pores of the electrode and in the bulk. The authors proposed three possible routes for the charge/discharge processes: 1) charge storage enhanced by electro-sorption of anionic species on the electrode surface, e.g. oxidation of I^- at positive electrode; 2) charge storage enhanced by chemisorption of transition metal cations, e.g. the redox reactions of Cu^{2+}/Fe^{2+} in $CuSO_4$ -$FeSO_4$ -H_2SO_4 system; and 3) charge storage enhanced by electron transfer between the redox electrolyte and the electrode material, e.g. the electron exchange between the redox electrode materials NiO and $Fe(CN)_6^{3-}$ /$Fe(CN)_6^{4-}$ in electrolyte [87].

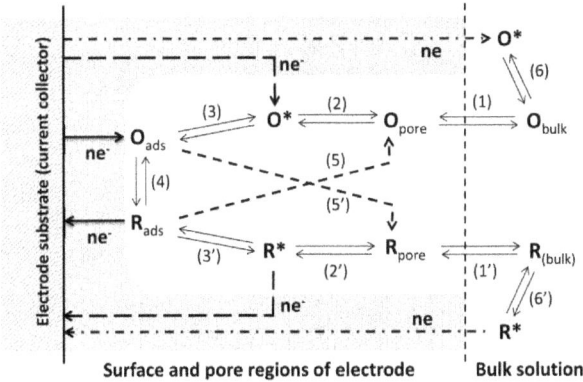

Figure 3. Illustration of charge storage mechanisms in the porous carbon electrode of a supercapacitor with a redox electrolyte.

The authors also pointed out a few "claims" that caused overestimation of specific capacitance and energy due to misuse of the following equations:

$$C = 4 \times \frac{\Delta Q}{\Delta V} \tag{16}$$

$$E = \frac{1}{2} C \Delta V^2 \tag{17}$$

where ΔV is the voltage range, ΔQ is the charge generated in the voltage range, C is the capacitance of **one electrode**, and E is the energy of the cell. Obviously, Eq. (17) is wrong and the correct way was to use Eq. (10) or Eq. (11).

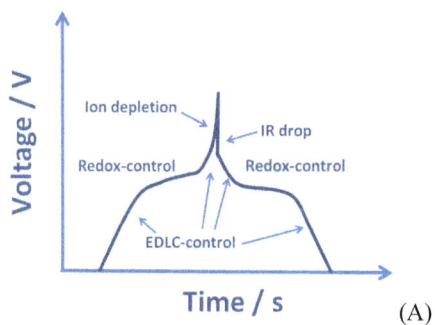

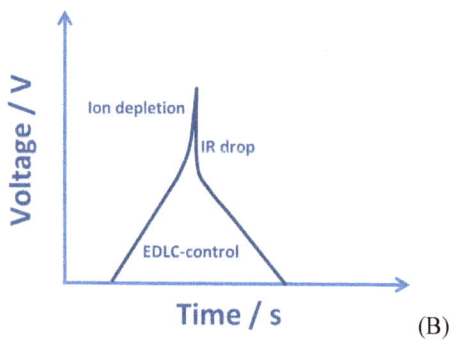

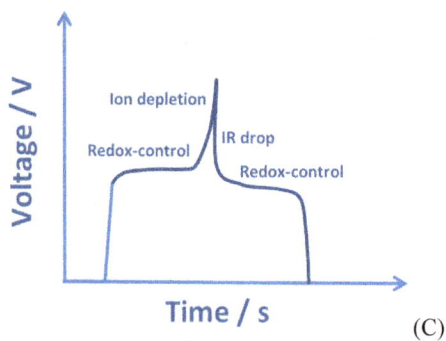

Figure 4. Schematics of typical GCDs of mediator SC (A), EDLC (B) and battery (C).

Supercapacitor Technology: Materials, Processes and Architectures   Materials Research Forum LLC
Materials Research Foundations **61** (2019) 45-94   https://doi.org/10.21741/9781644900499-4

In most of the cases listed in Table 1, active carbons or high surface area carbon materials were used. The charge/discharge processes were in a competition between charging of EDLs or EDLC, and redox reactions of mediators or pseudocapacitance. The current densities for these two processes can be written as [109]

$$i_{dl} = C_{dl}\frac{\partial(\phi_s-\phi_l)}{\partial t} \tag{18}$$

$$i_{redox} = i_{0c}(\frac{c_m}{c_m^*}\exp\left(\frac{\alpha_a F\eta}{RT}\right)) \tag{19}$$

where i_{dl} and i_{redox} are respectively current densities for charging the EDL and for the redox reaction, ϕ_s is the potential of the electrode, ϕ_l is the potential of the electrolyte, c_m is the concentration of the mediator, c_m^* is the reference concentration, and i_{0c} is the exchange current density. The overpotential is expressed by

$$\eta = \phi_s - \phi_l - E_{eq} \tag{20}$$

where E_{eq} is the equilibrium potential. In the case of a GCD test, the summation of i_{dl} and i_{redox} is a constant. For a typical mediator SC, the initial charging process is dominated by charging of the EDLs with linear increase of the voltage. If the concentration of mediator is high, when $\phi_s - \phi_l$ increases to a value fairly different from E_{eq}, i_{redox} increases exponentially and dominates the charging process, resulting in stagnation of $\phi_s - \phi_l$ or a voltage plateau as shown by case A in Figure 4. When the mediator is exhausted, i_{redox} approaches to zero and i_{dl} dominates the section of linear dependence. Further charging the SC exhausts the ions in the electrolyte, leading to exponential increase of the cell voltage with time.. During the discharging process, one often first witnesses a sharp IR drop followed by EDLC control, redox control, and EDLC control sections. If however the initial concentration of the mediators is low, charging of the EDLs will dominate for the entire charging /discharging process showing a capacitor-like behavior (case B in Figure 4). If the concentration of active carbon is low and that of the mediator is high, the behavior will be battery-like as shown in case C in Figure 4.

According to the data shown in Table 1, it was found that in most cases the authors just counted the mass of the active carbon as the denominator for evaluating the specific capacitance even when the mediators served as the major charge storage material. According to Zheng et al. [70] not only the mass of the mediators but also that of the electrolyte must be considered as the "active" materials and used as the denominator. Mediators may serve as promotors that increase the capacitance of carbon materials in electrodes via improving the transport of ions, interface between the electrodes and electrolytes, and intrinsic charge capacity as shown in Figure 2. However, these effects cannot be defined, evaluated, and verified without determination of the effect of mediators as charge storage materials or without precisely counting the mass of mediators in the supercapacitors for evaluating the specific capacitances. The very large specific capacitance values up to a few thousands of F/g reported might be a reflection of the effect of mediators as the major charge storage materials in the electrodes and in the separators.   The report by Chun et al. [67] serves as the perfect case for clarifying the reality of the data shown in Table 1. For KBr/MVCl$_2$, the "wet" specific energy was 13.8 Wh/kg at 0.05 kW/kg but the "dry" specific energy was 51.0 Wh/kg at 0.2 kW/kg. The "wet" specific capacitance was 32 F/g whereas the "dry" specific capacitance was 125 F/g. Thus, when the masses of the mediators and electrolyte were counted, the values of the properties were only 1/3 or 1/5 of those for the case when the masses of the mediator and electrolyte were not counted.

5.   Solid-state Redox electrolyte-mediator supercapacitors

Solid-state polymer electrolyte supercapacitors have appealing benefits of no-leakage, non-flammability, non-corrosion, high safety, and environment-benign. They can be developed into so-called structural supercapacitors that serve as an energy storage device and mechanical load bearing component. However, development of solid-state supercapacitor used to be very difficult due to the low ionic conductivity of solid electrolytes. In general, the ionic conductivity of polymer electrolytes is below 10^{-3} S/cm in comparison to 10^{-1} S/cm for aqueous electrolytes and 10^{-2} S/cm for organic liquid electrolytes. Previously, the specific power and energy of a solid-state supercapacitor were small fractions of those for liquid electrolyte supercapacitors. Availability of mediators or soluble redox species provides an excellent opportunity because addition of mediators in polymer electrolytes will significantly promote the electrical conductivity of polymer electrolytes in the electrodes.

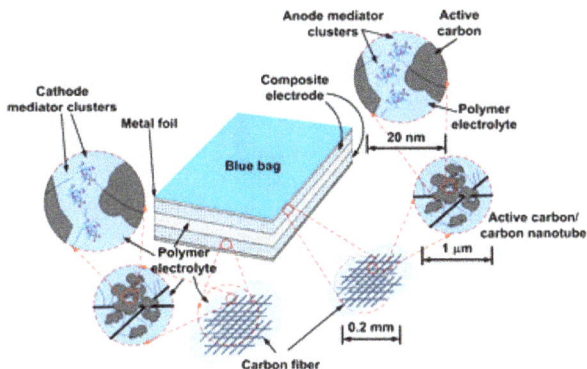

Figure 5. Schematic of the structure of solid-state mediator supercapacitor.

Zhou et al. [16] first proposed the structure of solid-state mediator supercapacitor shown in Figure 5. The multilayer structure consists of two current collectors (Al foil), two composite electrodes (active carbon, conductive carbon, polymer electrolyte, and mediators), and a polymer electrolyte membrane which serves as a barrier to the shuttling process and path for cations. There is no mediator in the membrane. In the composite electrodes, mediators present in the mixture of active carbon and polymer electrolyte. They may be adsorbed onto the surface of the porous active carbon or exist as solutes in the polymer electrolyte.

Zhou et al. [16] first studied a system with a Nafion membrane (a proton exchange membrane) as the separator. The polymer electrolyte contained in the electrodes was PEO/LiClO$_4$. The electrical conductivity of the polymer electrolyte was increased from 6.22×10^{-4} to 2.02×10^{-2} S/cm by adding 10 wt% mediators, NaI and I$_2$. Based on the mass of all active materials in two electrodes (carbon and mediators), the specific capacitance was 210 F/g. The specific energy was almost a constant of ~20 Wh/kg in a range of specific power from 0.4 to 1.8 kW/kg. The GCDs between 0-1 V were symmetrical in this range of the specific power. The specific energy initially increased, then decreased, and eventually became stable at 90% of retention up to 10^4 cycles. The OCV initially decreased and then became stable beyond 12 hours.

In order to further increase voltage window beyond 1.0 V and the specific energy, a PVDF/LiCF$_3$SO$_3$ (LiTFS) membrane was used to replace the Nafion membrane [17]. The authors reported that the conductivity of the PVDF/LiTFS membrane was 0.017 S/cm.

Two types of mediators, NaI/I_2 and $K_4Fe(CN)_6/K_3Fe(CN)_6$ were used in the electrodes. For NaI/I_2, the specific capacitance was 209 F/kg whereas for $K_4Fe(CN)_6/K_3Fe(CN)_6$ the specific capacitance was 138 F/kg based on the total mass of carbon materials and mediators. The degree of symmetry of the GCDs in the voltage window of 0-2.5V was acceptable. The specific energy was 49 Wh/kg at 1.5 kW/kg or 34 Wh/kg at 4 kW/kg for NaI/I_2 mediator supercapacitors whereas for $K_4Fe(CN)_6/K_3Fe(CN)_6$ it was 32 Wh/kg at 1.2 kW or 21 Wh/kg at 2.9 kW/kg. In both Nafion and PVDF/LiTFS membrane based SCs the contribution of carbon materials was ~20%. Because most of capacitance was provided by mediators and charging mediators does not need additional ions from the polymer electrolyte, the amount of polymer electrolyte can be reduced. Recently, we have reduced the polymer electrolyte to the level of 10% of the total electrode mass andsimilar or better performance has been obtained.

Zhou's group has recently studied Prussian blue derivatives or analogues (PBAs) as mediators in solid-state supercapacitors [69]. PBAs have fast reversible redox kinetics and high specific capacitance. Compared to other redox mediators, PBAs have three unique advantages:

1. PBAs are insoluble in the liquid electrolyte used for making the SCs. However, the PBAs were dispersed in the polymer electrolyte. For other redox mediators, such as $K_3Fe(CN)_6/K_4Fe(CN)_6$ and NaI/I_2, which are soluble in an aqueous electrolyte, an expensive selective membrane has to be employed to prevent the electron shuttling issue for liquid electrolyte SCs.

2. PBAs containing multi-valent transition metals, such as $K_2MnFe(CN)_6$, can provide greater specific capacitance than $K_3Fe(CN)_6$ and $K_4Fe(CN)_6$, because Mn can have oxidation states of 2+, 3+, and 4+ in a normal voltage range.

3. For EDLCs and pseudocapacitors, the ions are consumed to form the EDL or complete redox reactions. However, PBAs can provide extra ions to the electrolyte to maintain the concentration of ions, as shown in the following redox reaction regarding to a Mn containing PBAs:

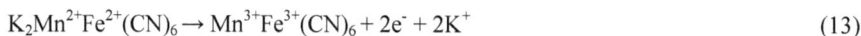

$$K_2Mn^{2+}Fe^{2+}(CN)_6 \rightarrow Mn^{3+}Fe^{3+}(CN)_6 + 2e^- + 2K^+ \tag{13}$$

The results shown in Figure 6 indicate that only 10wt% of mediators in the electrodes of the supercapacitors can provide a specific capacitance comparable to that provided by 90wt% of active carbon. The CVs and GCDs of the supercapacitors are all capacitor-like with a charge efficiency of above 95%. If it is assumed that the total capacitance is the summation of the capacitance of active carbon and that of mediators, the specific capacitance of the mediators in a voltage range of 0-2 V is 580 F/g, which corresponds to a specific charge capacity of 1160 C/g. For the Mn containing PBAs, the theoretical specific charge capacity is 296 C/g if one molecule charges/discharges 1 electron or 1180 C/g if charges/discharges 4 electrons. In other words, in order to realize the specific capacitance in 0-2 V, the Mn containing PBA should be rendered a valence transition to give off or accept 4 electrons. However, this is very unlikely because a transition from Mn^{2+} to Mn^{4+} will give off two electrons and at the same time a transition from Fe^{2+} to Fe^{3+} will give off additional one electron. The total that will give off is only 3 electrons. A reasonable explanation for this problem is that the total specific capacitance is the summation of the contribution of active carbon, the contribution of the mediators, and interaction or synergy between the active carbon and mediators, or

$$C = C_{active\_carbon} + C_{mediators} + C_{interaction} \qquad (21)$$

The third term is often wrongly attributed to the capacity of the active carbons resulting in an overestimation of the specific capacitance of electrode materials, especially when the amount of the electrode material is small.

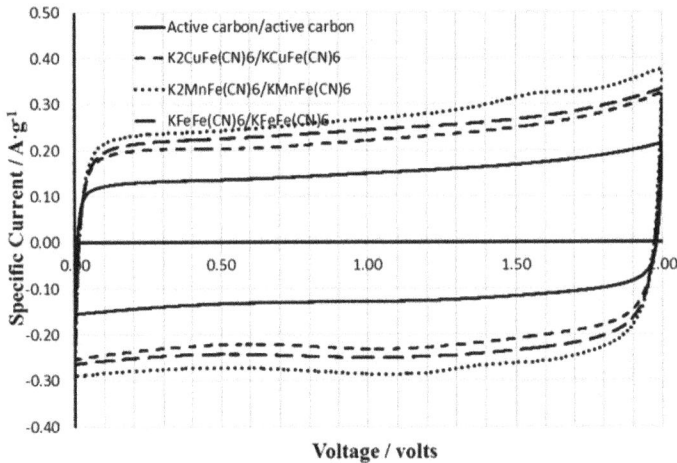

Figure 6. CVs for PBA mediator SCs.

When the PBAs get oxidized some free K+ ions will be released into the electrolyte, replenishing the reservoir of ions in the electrolyte during the charging process. This has been studied by Zhang [66]. As shown in **Figure 7**, three types of SCs were tested: 1) active carbon/active carbon; 2) $K_2Mn^{2+}Fe^{2+}(CN)_6$+active carbon/ $KMn^{2+}Fe^{3+}(CN)_6$+active carbon; and 3) $FeFe(CN)_6$+active carbon/$FeFe(CN)_6$+active carbon. All the GCDs shown have a linear section which corresponds to a normal charging process on EDLC and mediators and an exponential section. The exponential increase of voltage versus time is thought to be the sign of depletion of ions in the electrolyte which results in continuous reduction of conductivity or increase of the resistance. Interestingly, the depletion of the ions occurred at much later time for the second type 2) than those for the first 1) and third types 3). It was because $K_2Mn^{2+}Fe^{2+}(CN)_6$ released K+ ions during the charging process in contrast to the active carbon and $FeFe(CN)_6$.

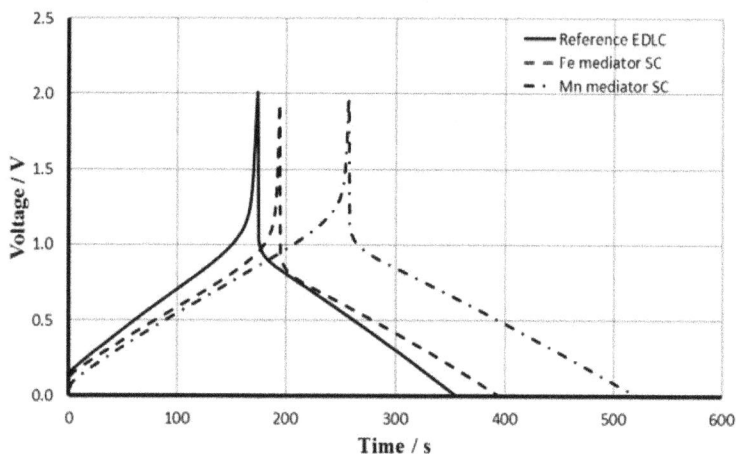

Figure 7. GCD results of three SCs with 0.1 M electrolyte concentration at charge/discharge current of 0.089A g^{-1}.

In order to elucidate the mechanisms of charge/discharge of the mediators in the solid-state mediator SCs, *in situ* X-ray absorption spectroscopy (XAS) was employed to detect the valence changes during charge/discharge. The setup is shown in Figure 8. The supercapacitor is an asymmetric SC with the working electrode similar to those electrodes given in Ref [110]. The counter electrode contains active carbon to provide a matching capacitance to that of the mediator containing working electrode. In this way, the signals relating to oxidation and reduction can only come from the working electrode not from the counter electrode which is at a different potential from the potential of the working electrode.

From the data shown in Figure 9, the initial mediator was Na^+I^-. Sweeping the cell voltage from -0.08 V to 0.80 V converted I^- to I_2. However, when the applied voltage was swept back to -0.08 V, I_2 did not go back to the original I^- but formed I_3^-. In later cycles, the chemical state of the iodide species switched between I^- and I_3^-. Between 0-1 V, the theoretical specific capacitance was 255 F/g. The bonding between carbon and iodide species was detected. This indicated that upon charging, almost all iodide species in the electrode contribute to reversible charge storage.

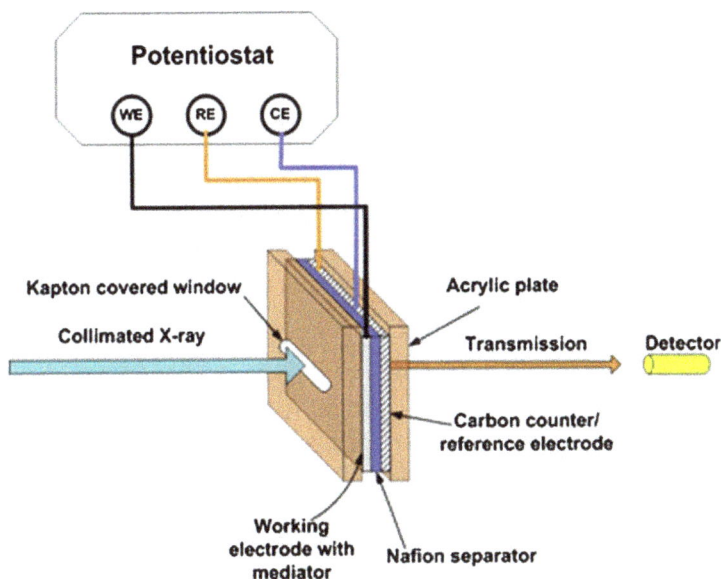

Figure 8. Schematic of the experimental setup for in situ XAS investigation of charge/discharge of mediators.

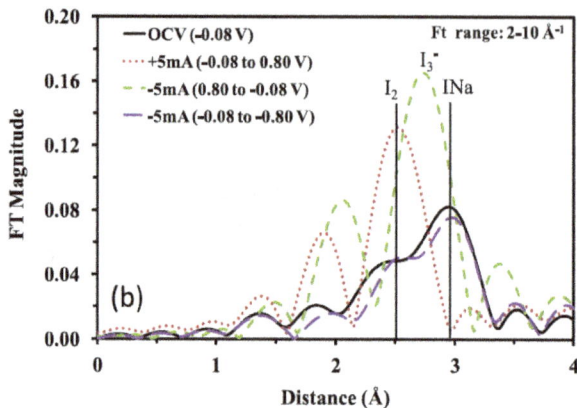

Figure 9. Phase uncorrected Fourier transforms of In situ I K-edge k2-weighted EXAFS spectra of a cell polarized to different potentials as indicated in the legend.

Fan et al. [111] explored a quasi-solid-state supercapacitor with a Nafion membrane separator. The polymer electrolyte was polyvinyl alcohol (PVC)-H_2SO_4 with $VOSO_4$ in negative electrode and with Na_2MoO_4 in the positive electrode. The conductivity of PVC-H_2SO_4-$VOSO_4$ was 0.018 S/cm and that of PVC-H_2SO_4-Na_2MoO_4 was 0.032 S/cm. The specific energy was 18 Wh/kg at 0.12 kW/kg or 6 Wh/kg at 2 kW/kg. The degree of symmetry of GCDs was acceptable. The advantage of this design was that when $VOSO_4$ was reduced, the redox part, VO^{2+}, was attracted by the negative electrode and when Na_2MoO_4 was oxidized, the redox part, MoO_4^{2-}, was attracted by the positive electrode. Thus, the shuttling processes were suppressed. Tripathi et al. [112] studied another polymer electrolyte, (PVDF-HFP)-PMMA-NaI and employed in supercapacitors. The conductivity was 0.016 S/cm. It was claimed that the specific energy was 44 Wh/kg at 4 kW/kg based on the mass of the conducting polymer electrodes excluding the mass of the mediators. Tu et al. [113] used an ionic liquid 1-butyl-3-methylimidazolium iodide (BMIMI) as plasticizer and redox additive, along with a neutral Li_2SO_4 aqueous solution to achieve wide operating voltage. A redox-active poly(vinyl alcohol) (PVA)-Li_2SO_4-BMIMI gel polymer electrolyte (GPE) was prepared and used for fabricating a flexible supercapacitor. Since the electrolyte (ionic liquid) acted both as a redox material and electrolyte, they referred this material as a dual-role material. The specific capacitance based on the active materials was 370 F/g and specific energy was 29 Wh/kg at ~0.1 kW/kg. The cycleability was reduced by ~20% at 10000 cycles. Self-discharge of the supercapacitors was ~25% reduction of the OCV within 6 hours. The charge and energy efficiency were both above 90%. However at higher rates, the cycles became very asymmetrical suggesting low efficiencies.

The charging and discharging processes of a mediator SC are complicated and include charging the EDL, electrochemical reactions at the interfaces, bulk reactions in the electrolyte, ion conduction, electroosmosis, and heat generation and transfer. Without comprehensive numerical modeling it is difficult to understand how mediators promote the performance and cause self-discharge. Ike et al. [114] studied self-discharge phenomenon of asymmetric/hybrid SCs with redox active electrolyte using a numerical simulation method. The 1-D numerical simulation results indicated that the mediator shuttle related self-discharge contributed to the majority of the self-discharge in an asymmetric/hybrid SCs.

Wang et al. [109] developed a 3-D numerical simulation model for mediator SCs. Figure 10 illustrates the simulation results for probing the effect of concentration of mediators at an intermediate charge rate of 20 mA/cm^2. Obviously, at a high concentration the behavior was battery-like with a long voltage plateau. The length of the plateau shrinked with decreasing the concentration and the behavior became capacitor-like. However, at the low concentration, the specific capacitance was greater than that of a pure EDLC. This explains the experimental results shown in Figures 6 and 7. Although the behavior was capacitor-like, the specific capacitance was promoted substantially by mediators. Figure 11 shows the concentration of the product of the electrochemical reaction of the mediator. The plateaus indicate that after some period of time the initial mediator have been completely converted into the product no matter how low the concentration is. This demonstrates that for the previous experiments listed in Table 1, given a long period of time, the mediators like I$^-$ would be completely converted.

Figure 12 shows the effect of diffusivity of mediators in a polymer electrolyte in electrodes. The results indicate that when the diffusivity is high the GCD is battery-like with a plateau and two linear sections. However, if the diffusivity is low, the behavior is capacitor-like. The critical point for a transition from the battery-like to capacitor-like behavior is 3×10^{-12} m^2/s. The self-discharge due to electron shuttling can also be simulated and analyzed using the model.

As shown in Figure 13, when the diffusivity of mediators in the separator/membrane is lower than 10^{-11} m^2/s, the GCD is normal presenting a plateau and two linear sections. However, if the diffusivity is greater than 10^{-10} m^2/s, the second linear section does not appear. In this case, the oxidized mediators in the cathode have time to transport to the anode and get reduced at anode. These reduced mediators are transported back to the cathode to get oxidized there. Because of the formation of this internal cycle in the cell, the supply of mediators is infinite and plateau is infinitely long. Thus, in order to fabricate a workable SC, one should use a polymer electrolyte with diffusivity greater than 3×10^{-12} m^2/s in the electrodes and at the same time a polymer electrolyte with a diffusivity less than 10^{-10} m^2/s be used to make a separator/membrane.

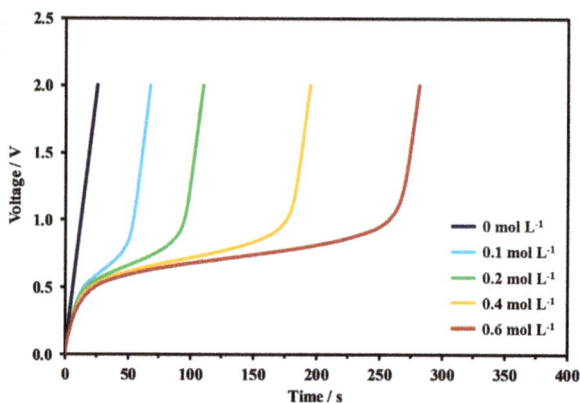

Figure 10. The galvanostatic charge curves with mediator initial concentrations (mol L^{-1}): 0.0, 0.1, 0.2, 0.4, and 0.6 at $i_{app} = 20$ mA cm^{-2}.

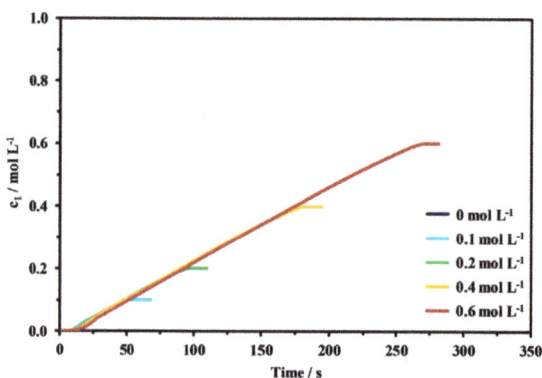

Figure 11. The concentration of the reduced mediator at the anode with mediator initial concentrations as 0, 0.1, 0.2, 0.4, and 0.6 mol L^{-1}.

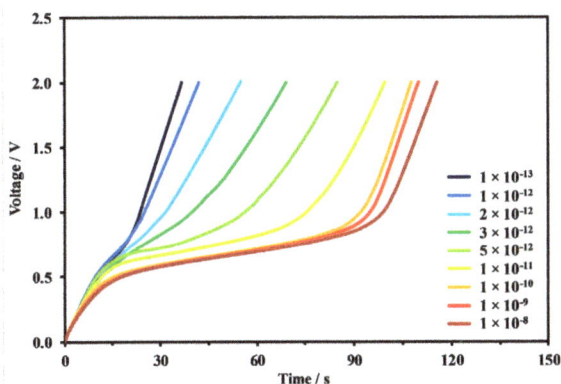

Figure 12. The galvanostatic charge curves with mediator diffusivity from 1×10^{-13} m^2 s^{-1}, to 1×10^{-8} m^2 s^{-1} at $i_{app} = 20$ mA cm^{-2}. The concentration of the mediators is 0.2 mol L^{-1} and the concentration of the ions in electrolyte is 1 mol L^{-1}.

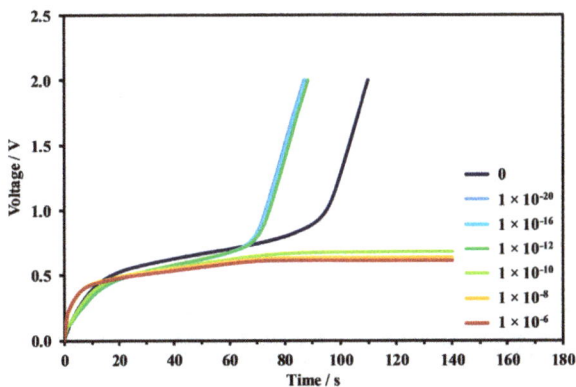

Figure 13. The galvanostatic charge curves with different diffusivities of the cathode mediators in the polymer electrolyte membrane domain at $i_{app} = 20$ mA cm^{-2}. The concentration of the reduced mediators is 0.2 mol L^{-1} and the concentration of the ions in electrolyte is 1 mol L^{-1}.

6.   Nanomaterials based redox electrolyte supercapacitors

The advent of nanomaterials including CNTs and graphene has opened the opportunities for development of SCs including redox electrolyte SCs. As discussed above, one of the major issues for redox electrolyte supercapacitors is the electron shuttling effect. Conway et al. [6] have proposed four methods to resolve this problem. Many researchers have used the nanomaterials in fabrication of the redox electrolyte/mediator supercapacitors to mitigate this problem.

Huang et al. [115] developed a method as illustrated in **Figure 14** to combine SiWLi mediators with graphene oxide (GO) and a PVA/Li$_2$SO$_4$ polymer electrolyte. They found that the mediators adsorbed onto GO could function as an additional charge storage capacity and could form "expressways" in the electrolyte or bridge the charge delivery in a supercapacitor or battery electrode. This can be characterized as a so-called *in situ* synthesis of nano-sized clusters of mediators anchored to GO pellets. When the concentration of mediators anchored by GO is sufficiently high, the electrons can be passed from/to the mediators in the bulk of the electrolyte in the electrode. The authors also claimed that the mediators anchored by GO pellets promoted the ionic conductivity.

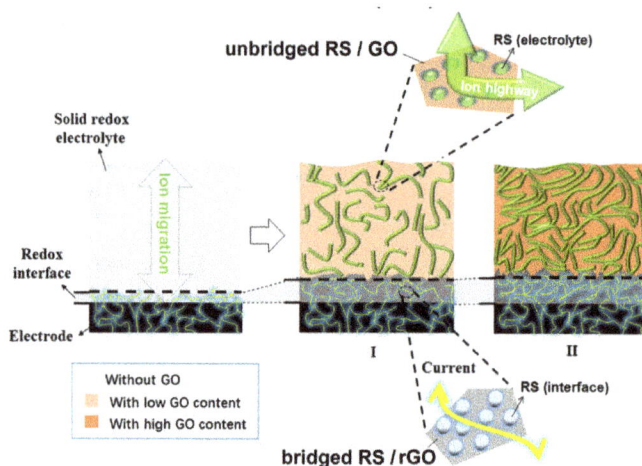

Figure 14 schematics of RS-coated reduced GO (RS/rGO) sheets as extended interface for extending battery life with increasing GO content in the solid electrolyte, as well as unbridged RS/GO sheets for facilitating ionic transportation.

Jana et al. [116] prepared electrolyte accessible layer-by-layer (LL) assembly of reduced graphene oxide (RGO) and sulfanilic acid azo-chromotrop (SA) as the redox additive. The materials were referred as LSARGO and NSARGO. The materials were evaluated using a three-electrode method in 6 M KOH. The observed specific capacitance on the basis of mass of active materials in one electrode was up to 1023 F/g at scan rate of ~10 m V/s in the working potential of ~0–0.6 V. The masses of mediators and electrolyte were not counted in this calculation. The asymmetrical GCDs suggested a low energy efficiency. However, 80% of the specific capacitance could be retained at 10,000 cycles.

Potphode et al. [103] synthesized partially exfoliated multi-walled carbon nanotubes (Px-MWCNT)/Polyaniline (PANI) nanocomposites. The nanocomposite was added into a gel polymer electrolyte (PVA/H_2SO_4) with a redox-active material, hydroquinone to form a redox electrolyte membrane. The gel membrane was sandwiched between two Px-MWCNT)/Polyaniline (PANI) electrodes. The authors observed a high specific capacitance of 809F/g on the basis of the active materials in the electrode. However, the mass of the active materials in the electrolyte or membrane was not counted. The self-discharge were not evaluated.

Conclusion

We conducted a literature review on the topic of the redox electrolyte/mediator SCs. The number the published articles has been increasing fast in recent years. The benefits of the mediators or redox species dissolved or partially dissolved in electrolyte can be summarized as follows. Firstly, the electrochemical reduction and oxidation of mediators at the electrode/electrolyte interfaces transfer charges from electrodes into electrolyte, enabling charge storage capacity in the electrolyte . Secondly, the presence of mediators in the electrolyte in the porous electrodes promotes transport processes including diffusion and conduction in the electrodes. Thirdly, some mediators release free ions, helping mitigate the problem of ion depletion during the charging process. Finally, there were evidences that a synergy between the electrodes and mediators existed or the capacitance of the electrodes was promoted by mediators. However, further investigation is required for confirmation of the promotion effect of mediators. It is pertinent to establish and implement unified metrics for evaluating the specific capacitance, energy, and power. In most reports reviewed, the mass of the mediators was not counted as part of the denominator for calculating the specific capacitance, energy, and power. This leads to confusions and overestimations. The specific capacitance of a redox electrolyte/mediator SC is generally greater than that of an EDLC. However, with respect to the charge and energy efficiencies, cycleability, and self-discharge, the redox electrolyte/mediator SC has been inferior to the EDLC. Electron shuttling by the transport of mediators between the electrode is one of major sources for the lower efficiencies, lower cycleability, and higher self-discharge rate. Future research may advance in the following directions: 1) identifying redox materials with multiple valences or high molecular charge capacity; 2) designing systems that have electrodes and electrolyte with matching potential ranges and capacitances; 3) developing ion-exchange polymer electrolyte membrane and using graphene, GO, CNT, or other nanomaterials to anchor the mediators to mitigate the electron shuttling effect; 4) utilizing the novel polymer electrolytes and mediators to develop high performance solid-state redox electrolyte/mediator SCs; 5) conducting numerical and atomistic modeling and simulations in order to probe the mechanisms and to provide guidance for cell design; 6) conducting in situ physicochemical investigation technology to monitor the electrochemical redox reactions of mediators.

Acknowledgement

This work was partially supported by Emil Buehler Foundation and Office of Naval Research (Grant number: N00014-17-1-2362).

List of abbreviations

EDLC: Electrical double layer capacitor
SC: Supercapacitor
CV: Cyclic voltammetry
PANI: Polyaniline
PPy: Polypyrrole
PTh: Polythiophene
PIs: Polyimides
GO: Graphene oxide
CNT: Carbon nanotube
RMST: Ratio of the atoms/molecules on the surface to the total atoms/molecules
ACCPM: Average charge capacity per molecule
OCV: Open circuit voltage
HQ: Hydroquinone
GCD: Galvanostatic charge and discharge
PET: Polyester
CVD: Chemical vapor deposition
PYR 14 TFSI: N-butyl-N-methylpyrrolidinium bis(-trifluoromethanesulfonyl) imide
p-BQ: Para-benzoquinone
PPD: P-phenylenediamine
TCHQ: Tetra-chlorohydroquinone

AQ: Anthraquinone

PEDOT: Poly(3,4-ethylenedioxythiophene)

MV: Methyl viologen

SA: Azo chromotrop

PVDF: Polyvinylidene fluoride

LiTFS: $LiCF_3SO_3$

PBAs: Prussian blue derivatives or analogues

XAS: X-ray absorption spectroscopy

PVC: Polyvinyl alcohol

HFP: Hexafluoropropylene

PMMA: Poly(methyl methacrylate)

BMIMI: 1-butyl-3-methylimidazolium iodide

GPE: Gel polymer electrolyte

LL: Layer-by-layer

RGO: Reduced graphene oxide

Px-MWCNT: Exfoliated multi-walled carbon nanotubes

RS: Redox

References

[1] L. Wang, Y. Han, X. Feng, J. Zhou, P. Qi, B. Wang, Metal–organic frameworks for energy storage: Batteries and supercapacitors, Coord. Chem. Rev. 307 (2016) 361–381. https://doi.org/10.1016/j.ccr.2015.09.002

[2] Mustafa Inci, Omer Türksoy, Review of fuel cells to grid interface: Configurations, technical challenges and trends, J. Clean. Prod. 213 (2019) 1353-1370. https://doi.org/10.1016/j.jclepro.2018.12.281

[3] P. Alotto, M. Guarnieri, F. Moro, Redox flow batteries for the storage of renewable energy: A review, Renew. Sust. Energ. Rev. 29 (2014) 325–335. https://doi.org/10.1016/j.rser.2013.08.001

[4] Mi. Horn, J. MacLeod, M. Liu, J.Webb, Nunzio Motta, Supercapacitors: A new source of power for electric cars? Econ. Anal. Pol. 61 (2019) 93–103. https://doi.org/10.1016/j.eap.2018.08.003

[5] P. Simon, Y. Gogotsi, Materials for electrochemical capacitors. Nat. Mater. 7 (2008) 845–854. https://doi.org/10.1038/nmat2297

[6] B.E. Conway, Electrochemical Supercapacitors: Scientific Fundamentals and Technological Applications, Kluwer Academic/Plenum Publishers, New York, 1999.

[7] M. Wohlfahrt-Mehrens, J. Schenk, P.M. Wilde, E. Abdelmula, P. Axmann, J. Garche, New materials for supercapacitors, J. Power Sources 105 (2002) 182–188. https://doi.org/10.1016/S0378-7753(01)00937-5

[8] A. Muzaffara, M. Basheer Ahamed, K. Deshmukh, J. Thirumalai, A review on recent advances in hybrid supercapacitors: Design, fabrication and applications, Renew. Sust. Energ. Rev 101 (2019) 123–145. https://doi.org/10.1016/j.rser.2018.10.026

[9] A. González, E. Goikolea, J. Barrena, R. Mysyk, Review on supercapacitors: Technologies and materials, Renew. Sust. Energ. Rev 58 (2016) 1189–1206. https://doi.org/10.1016/j.rser.2015.12.249

[10] W. Yang, M. Ni, X. Ren, Y. Tian, N. Li, Y. Su, X. Zhang, Graphene in Supercapacitor Applications, Curr. Opin. Colloid Interface Sci. 20 (2015) 416–428. https://doi.org/10.1016/j.cocis.2015.10.009

[11] Qingqing Ke, John Wang, Graphene-based materials for supercapacitor electrodes: A review, J. Materiomics 2 (2016) 37-54. https://doi.org/10.1016/j.jmat.2016.01.001

[12] C. Portet, P.L. Taberna, P. Simon, E. Flahaut, Influence of carbon nanotubes addition on carbon–carbon supercapacitor performances in organic electrolyte, J. Power Sources 139 (2005) 371–378. https://doi.org/10.1016/j.jpowsour.2004.07.015

[13] C. Portet, P.L. Taberna, P. Simon, E. Flahaut, C. Laberty-Robert, High power density electrodes for Carbon supercapacitor applications, Electrochim. Acta 50 (2005) 4174–4181. https://doi.org/10.1016/j.electacta.2005.01.038

[14] G. Lota, E. Frackowiak, Striking capacitance of carbon/iodide interface, Electrochem. Commun. 11 (2009) 87–90. https://doi.org/10.1016/j.elecom.2008.10.026

[15] G. Lota, K. Fic, E. Frackowiak, Alkali metal iodide/carbon interface as a source of pseudocapacitance, Electrochem. Commun. 12 (2011) 38–41. https://doi.org/10.1016/j.elecom.2010.11.007

[16] Juanjuan Zhou, Yijing Yin, Azzam N. Mansour, and Xiangyang Zhou, Experimental studies of mediator-enhanced polymer electrolyte supercapacitors, Electrochem. Solid-State Lett. 14 (2011) A25-A28. https://doi.org/10.1149/1.3526094

[17] X. Zhou, X. Qiao, C. Zhang, Y. Wang, A. N. Mansour, G. H. Waller, C.A. Martin, "The Effects of Potassium Ferrocyanide/Potassium Ferricyanide and their Derivatives on the Performance of Solid-State Supercapacitor", 48th Power Sources Conference, Denver, Colorado, June 11-14, 2018, 292-295.

[18] B. Akinwolemiwa, C. Peng, G. Z. Chen, Redox electrolytes in supercapacitors, J. Electrochem. Soc. 162 (2015) A5054-A5059. https://doi.org/10.1149/2.0111505jes

[19] J. P. Zheng, P. J. Cygan, and T. R. Jow, Hydrous ruthenium oxide as an electrode material for electrochemical capacitors, J. Electrochem. Soc.142 (1995) 2699-2703. https://doi.org/10.1149/1.2050077

[20] W. Wei, X. Cui, W. Chen, D.G. Ivey, Manganese oxide-based materials as electrochemical supercapacitor electrodes, Chem. Soc. Rev. 40 (2011) 1697–1721. https://doi.org/10.1039/C0CS00127A

[21] M. Sawangphruk, P. Srimuk, P. Chiochan, A. Krittayavathananon, S. Luanwuthi, J. Limtrakul, High-performance supercapacitor of manganese oxide/reduced graphene oxide nanocomposite coated on flexible carbon fiber paper, Carbon 60 (2013) 109-116. https://doi.org/10.1016/j.carbon.2013.03.062

[22] S. Hassan, M. Suzuki, A. Abd El-Moneim, Capacitive Behavior of Manganese Dioxide/Stainless Steel Electrodes at Different Deposition Currents, Am. J. Mater. Sc. 2 (2012) 11-14. https://doi.org/10.5923/j.materials.20120202.03

[23] S. Devaraj, N. Munichandraiah, Electrochemical supercapacitor studies of nanostructured/MnO_2 synthesized by microemulsion method and the effect of annealing, J Electrochem. Soc. 154 (2007) A80-A88. https://doi.org/10.1149/1.2404775

[24] P. Ragupathy, D. H. Park, G.Campet, H. N. Vasan, S.J. Hwang, J.H. Choy, N. Munichandraiah , Remarkable capacity retention of nanostructured manganese oxide upon cycling as an electrode material for supercapacitor, J. Phys. Chem. C, 113 (2009) 6303–6309. https://doi.org/10.1021/jp811407q

[25] J. Cheng, B. Zhao, W. Zhang, F. Shi, G. Zheng, D. Zhang, J. Yang, High-performance supercapacitor applications of NiO-nanoparticle-decorated millimeter-long vertically aligned carbon nanotube arrays via an effective supercritical CO_2-assisted method, Adv. Funct. Mater. 25 (2015) 7381–7391. https://doi.org/10.1002/adfm.201502711

[26] Y. Zhu, C. Cao, S.Tao, W. Chu, Z. Wu, Y. Li, Ultrathin nickel hydroxide and oxide nanosheets: Synthesis, Characterizations and excellent supercapacitor performances, Sci. Rep. 4 (2014) 1-7. https://doi.org/10.1038/srep05787

[27] C. Xiang, M. Li, M. Zhi, A. Manivannan. N. Wu, Reduced graphene oxide/titanium dioxide composites for supercapacitor electrodes: shape and coupling effects, J. Mater. Chem. 22 (2012) 19161–19167. https://doi.org/10.1039/c2jm33177b

[28] M. Fukuhara, T. Kuroda, F. Hasegawa, Amorphous titanium-oxide supercapacitors, Sci. Rep. 6(35870) (2016) 1-5. https://doi.org/10.1038/srep35870

[29] D. Choi, G. E. Blomgren, P.N. Kumta, Fast and reversible surface redox reaction in nanocrystalline vanadium nitride supercapacitors, Adv. Mater. 18 (2016) 1178–1182. https://doi.org/10.1002/adma.200502471

[30] S. Boukhalfa, K.Evanoff, G. Yushin, Atomic layer deposition of vanadium oxide on carbon nanotubes for high-power supercapacitor electrodes, Energy Environ. Sci. 5 (2012) 6872–6879. https://doi.org/10.1039/c2ee21110f

[31] M. Yu, Y. Zeng, Y. Han, X. Cheng, W. Zhao, C. Liang, Y. Tong, H. Tang, X. Lu, Valence-optimized vanadium oxide supercapacitor electrodes exhibit ultrahigh capacitance and super-long cyclic durability of 100 000 cycles, Adv. Funct. Mater. 1 (2015) 1-5. https://doi.org/10.1002/adfm.201501342

[32] R. Thangappan, S. Kalaiselvam, A. Elayaperumal, R. Jayavel, Synthesis of graphene oxide/vanadium pentoxide composite nanofibers by electrospinning for supercapacitor applications, Solid State Ionics 268 (2014) 321–325. https://doi.org/10.1016/j.ssi.2014.10.025

[33] D. Shu, C. Lv, F. Cheng, C. He, K. Yang, J. Nan, Lu Long, Enhanced capacitance and rate capability of nanocrystalline VN as electrode materials for supercapacitors, Int. J. Electrochem. Sci. 8 (2013) 1209 – 1225.

[34] C. Guan, J. Liu, Y. Wang, L. Mao, Z. Fan, Z. Shen, H. Zhang, J. Wang, Iron oxide-decorated carbon for supercapacitor anodes with ultrahigh energy density and outstanding cycling stability, ACS Nano 9 (2015) 5198–5207. https://doi.org/10.1021/acsnano.5b00582

[35] Q. Xia, M. Xu, H. Xia, J. Xie, Nanostructured iron oxide/hydroxide-based electrode materials for supercapacitors, ChemNanoMat 2 (2016) 588–600. https://doi.org/10.1002/cnma.201600110

[36] G.A. Snook, P. Kao, A.S. Best, Conducting-polymer-based supercapacitor devices and electrodes, J Power Sources 196 (2001) 1–12. https://doi.org/10.1016/j.jpowsour.2010.06.084

[37] Y. Liu Y, K. Ai, L. Lu, Polydopamine and its derivative materials: Synthesis and promising applications in energy, environmental, and biomedical fields, Chem. Rev. 114 (2014) 5057–5115. https://doi.org/10.1021/cr400407a

[38] G. Ciric-Marjanovic, Recent advances in polyaniline research: Polymerization mechanisms, structural aspects, properties and applications, Synth. Met. 177 (2013) 1–47. https://doi.org/10.1016/j.synthmet.2013.06.004

[39] H. Tang, J. Wang, H. Yin, Growth of polypyrrole ultrathin films on MoS_2 monolayers as high-performance supercapacitor electrodes, Adv. Mater. 27 (2015) 1117–1123. https://doi.org/10.1002/adma.201404622

[40] A. Laforgue, P. Simon, C. Sarrazin, J. Fauvarque, Polythiophene-based supercapacitors, J. Power Sources 80 (1999)142–148. https://doi.org/10.1016/S0378-7753(98)00258-4

[41] H-P Cong, X-C Ren, P. Wang, S-H Yu, Flexible graphene-polyaniline composite paper for high-performance supercapacitor, Energy Environ. Sci. 6 (2013) 1185–1191. https://doi.org/10.1039/c2ee24203f

[42] Y. Meng, K. Wang, Y. Zhang, Z. Wei, Hierarchical porous graphene/polyaniline composite film with superior rate performance for flexible supercapacitors, Adv. Mater. 25 (2013) 6985–6990. https://doi.org/10.1002/adma.201303529

[43] J. Yan, T. Wei, Z. Fan, Preparation of graphene nanosheet/carbon nanotube/polyaniline composite as electrode material for supercapacitors, J. Power Sources 195 (2010) 3041–3045. https://doi.org/10.1016/j.jpowsour.2009.11.028

[44] Q. Hao, X. Xia, W. Lei, Facile synthesis of sandwich-like polyaniline/boron-doped graphene nano hybrid for supercapacitors, Carbon 81 (2015) 552–563. https://doi.org/10.1016/j.carbon.2014.09.090

[45] K. Zhang, LL Zhang, XS Zhao, J. Wu, Graphene/polyaniline nanofiber composites as supercapacitor electrodes, Chem. Mater. 22 (2010) 1392–1401. https://doi.org/10.1021/cm902876u

[46] H. Wang, Q. Hao, X.Yang, Graphene oxide doped polyaniline for supercapacitors, Electrochem. Commun. 11 (2009) 1158–1161. https://doi.org/10.1016/j.elecom.2009.03.036

[47] K. Wang, J. Huang, Z. Wei, Conducting polyaniline nanowire arrays for high performance supercapacitors, J. Phys. Chem. C 114 (2010) 8062–8067. https://doi.org/10.1021/jp9113255

[48] K. Wang, Q. Meng, Y. Zhang, High-performance two-ply yarn supercapacitors based on carbon nanotubes and polyaniline nanowire arrays, Adv. Mater. 25 (2013) 1494–1498. https://doi.org/10.1002/adma.201204598

[49] A. Muzaffar, M. Basheer Ahamed, K. Deshmukh, J. Thirumalai, A review on recent advances in hybrid supercapacitors: Design, fabrication and applications, Renew. Sust. Energy Rev. 101 (2019) 123–145. https://doi.org/10.1016/j.rser.2018.10.026

[50] S.R. Sivakkumar, A.G. Pandolfo, Evaluation of lithium-ion capacitors assembled with pre-lithiated graphite anode and activated carbon cathode, Electrochim. Acta 65 (2012) 280– 287. https://doi.org/10.1016/j.electacta.2012.01.076

[51] Sheng S. Zhang, Eliminating pre-lithiation step for making high energy density hybrid Li-ion capacitor, J. Power Sources 343 (2017) 322-328. https://doi.org/10.1016/j.jpowsour.2017.01.061

[52] A. Shellikeri, S. Yturriaga, J.S. Zheng, W. Cao, M. Hagen, J.A. Read, T.R. Jow, J.P. Zheng, Hybrid lithium-ion capacitor with $LiFePO_4/AC$ composite cathode – Long term cycle life study, rate effect and charge sharing analysis, J. Power Sources 392 (2018) 285–295. https://doi.org/10.1016/j.jpowsour.2018.05.002

[53] L. Jin, J. Zheng, Q. Wu, A. Shellikeri, S. Yturriaga, R. Gong, J. Huang, Jim P. Zheng, Exploiting a hybrid lithium ion power source with a high energy density over 30 Wh/kg, Mater. Today Energy 7 (2018) 51-57. https://doi.org/10.1016/j.mtener.2017.12.003

[54] J. Zhang, J. Wang, Z. Shi, Z. Xu, Electrochemical behavior of lithium ion capacitor under low temperature, J Electroanal. Chem. 817 (2018) 195–201. https://doi.org/10.1016/j.jelechem.2018.04.014

[55] X. Zhang, C. Lu, H. Peng, X. Wang, Y. Zhang, Z. Wang, Y. Zhong, G. Wang, Influence of sintering temperature and graphene additives on the electrochemical performance of porous $Li_4Ti_5O_{12}$ anode for lithium ion capacitor, Electrochim. Acta 246 (2017) 1237-1247. https://doi.org/10.1016/j.electacta.2017.07.014

[56] N.W. Li, X. Du, J. L. Shi, X. Zhang, W. Fan, J. Wang, S. Zhao, Y. Liu, W. Xu, M. Li, Y.G. Guo, C. Li, Graphene@hierarchical meso-/microporous carbon for ultrahigh energy density lithium-ion capacitors, Electrochim. Acta 281 (2018) 459-465. https://doi.org/10.1016/j.electacta.2018.05.147

[57] B. E. Conway, Two-dimensional and quasi-two-dimensional isothermals for Li intercalation and UPD processes at surface, Electrochim. Acta 38 (1993) 1249-1258. https://doi.org/10.1016/0013-4686(93)80055-5

[58] B. E. Conway, Transition from "Supercapacitor" to "Battery" behavior in electrochemical energy storage, J. Electrochem. Soc. 38 (1991) 1537-1547.

[59] H. Angerstein-Kozlowska, B.E. Conway, Evaluation of rate constants and reversibility parameters for surface reactions by the potential-sweep method, J. Electroanal. Chem. 95 (1979) 1-28. https://doi.org/10.1016/0368-1874(79)80001-5

[60] B.E. Conway, E. Gileadi, Kinetic theory of pseudo-capacitance and electrode reactions at appreciable surface coverage, Transaction of The Faraday Society 58 (1963) 2493-2508. https://doi.org/10.1039/tf9625802493

[61] S. Hadzi-Jordanov, H. Angerstein-Kozlowska, M. Vukoviff, B. E. Conway, Reversibility and growth behavior of surface oxide films at ruthenium electrodes, J. Electrochem. Soc. 125 (1978) 1471-1480. https://doi.org/10.1149/1.2131698

[62] R. P. Simpraga, B. E. Conway, The real-area scaling factor in electrocatalysisand in charge storage by supercapacitors, Electrochim. Acta 43 (1998) 3045–3058. https://doi.org/10.1016/S0013-4686(98)00045-0

[63] W. G. Pell, B. E. Conway, Peculiarities and requirements of asymmetric capacitor devices based on combination of capacitor and battery-type electrodes, J. Power Sources 136 (2004) 334–345. https://doi.org/10.1016/j.jpowsour.2004.03.021

[64] B.E. Conway, V. Birss, J. Wojtowicz, The role and utilization of pseudocapacitance for energy storage by supercapacitors, J. Power Sources 66 (1997) 1-14. https://doi.org/10.1016/S0378-7753(96)02474-3

[65] MTI Corp. https://www.mtixtl.com/

[66] C. Zhang, PhD Dissertation, University of Miami, 2018.

[67] S.E. Chun, B. Evanko, X. Wang, D. Vonlanthen, X. Ji, G. D. Stucky, S. W. Boettcher, Design of aqueous redox-enhanced electrochemical capacitors with high specific energies and slow self-discharge, Nature Commun. 6 (2015) 7818-7828. https://doi.org/10.1038/ncomms8818

[68] K. Chen, D. Xue, $YbCl_3$ electrode in alkaline aqueous electrolyte with high pseudocapacitance, J. Colloid Inter. Sci. 424 (2014) 84–89. https://doi.org/10.1016/j.jcis.2014.03.022

[69] J. Zhou, J. Cai, S. Cai, X. Zhou, A. N. Mansour, Development of all-solid-state mediator-enhanced supercapacitors with polyvinylidene fluoride/lithium trifluoromethanesulfonate separators, J. Power Sources 196 (2011) 10479– 10483. https://doi.org/10.1016/j.jpowsour.2011.08.051

[70] J.P. Zheng, J.Huang, R. Jow, The Limitations of Energy Density for Electrochemical Capacitors, J. Electrochem. Soc. 144 (1997) 2027-2031. https://doi.org/10.1149/1.1837738

[71] J. P. Zheng, Theoretical energy density for electrochemical capacitors with intercalation electrodes, J. Electrochem. Soc. 152 (2005) A1864-A1869. https://doi.org/10.1149/1.1997152

[72] N. Papageorgiou, P. Liska, A. Kay, and M. Grätzel, Mediator transport in multilayer nanocrystalline photoelectrochemical cell configurations, J. Electrochem. Soc. 146 (1999) 898-907. https://doi.org/10.1149/1.1391698

[73] J. Gong, K. Sumathy, Q. Qiao, Z. Zhou, Review on dye-sensitized solar cells (DSSCs): Advanced techniques and research trends, Renew. Sust. Energy Rev. 68 (2017) 234–246. https://doi.org/10.1016/j.rser.2016.09.097

[74] B. Kannan, D. E. Williams, M. A. Booth, J. Travas-Sejdic, High-sensitivity, label-free DNA sensors using electrochemically active conducting polymers, Anal. Chem. 83 (2011) 3415–3421. https://doi.org/10.1021/ac1033243

[75] I. Ruff, V. J. Friedrich, Transfer Diffusion. I. Theoretical, J. Phys. Chem. 75 (1971) 3297-3301. https://doi.org/10.1021/j100690a016

[76] I. Ruff, V. J. Friedrich, K. Demeter, K. Csillag, Transfer diffusion. II. Kinetics of electron exchange reaction between ferrocene and ferricinium ion in alcohols, J. Phys. Chem. 75 (1971) 3303-3309. https://doi.org/10.1021/j100690a017

[77] P. Wang, S.M. Zakeeruddin, P. Comte, I. Exnar, M. Gratzel, Gelation of ionic liquid-based electrolytes with silica nanoparticles for quasi-solid-state dye-sensitized solar cells, J. Am. Chem. Soc. 125 (2003) 1166-1167. https://doi.org/10.1021/ja029294+

[78] R. Kawano, M. Watanabe, Equilibrium potentials and charge transport of an I_2/I_3^- redox couple in an ionic liquid, Chem. Commun. 3 (2003) 330–331. https://doi.org/10.1039/b208388b

[79] H. Dahms, Electronic conduction in aqueous solution, J. Phys.Chem. 78 (1968) 362-364. https://doi.org/10.1021/j100847a073

[80] E. Frackowiak, M. Meller, J. Menzel, D. Gastol, K. Fic, Redox-active electrolyte for supercapacitor application, Faraday Discuss. 172 (2014) 179- 198. https://doi.org/10.1039/C4FD00052H

[81] S. T. Senthilkumar, R. Kalai Selvan, Y. S. Lee, J. S. Melo, Electric double layer capacitor and its improved specific capacitance using redox additive electrolyte, J. Mater. Chem. A, 1 (2013) 1086–1095. https://doi.org/10.1039/C2TA00210H

[82] S. T. Senthilkumar, R. Kalai Selvan, J. S. Melo, Redox additive/active electrolytes: a novel approach to enhance the performance of supercapacitors, J. Mater. Chem. A 1 (2013) 12386–12394. https://doi.org/10.1039/c3ta11959a

[83] Z. Gao, L. Zhang, J. Chang, Z. Wang, D. Wu, F. Xu, Y. Guo, K. Jiang, Catalytic electrode-redox electrolyte supercapacitor system with enhanced capacitive

performance, Chem. Eng. J. 335 (2018) 590–599.
https://doi.org/10.1016/j.cej.2017.11.037

[84] E. Frackowiak, K. Fic, M. Meller, G. Lota, electrochemistry serving people and
nature: high-energy supercapacitors based on redox-active electrolytes, ChemSusChem
5 (2012) 1181-1185. https://doi.org/10.1002/cssc.201200227

[85] J. Lee, S. Choudhury, D. Weingarth, D. Kim, V. Presser, High performance hybrid
energy storage with potassium ferricyanide redox electrolyte, ACS Appl. Mater.
Interfaces, 8 (2016) 23676−23687. https://doi.org/10.1021/acsami.6b06264

[86] L. Su, L. Gong, Y. Zhao, A new strategy to enhance low-temperature capacitance:
combination of two charge-storage mechanisms, Phys. Chem. Chem. Phys. 16 (2014)
681-684. https://doi.org/10.1039/C3CP53747A

[87] L.H. Su, X.G. Zhang, C.H. Mi, B. Gao, Y. Liu, Improvement of the capacitive
performances for Co–Al layered double hydroxide by adding hexacyanoferrate into the
electrolyte, Phys. Chem. Chem. Phys. 11 (2009) 2195–2202.
https://doi.org/10.1039/b814844a

[88] K. Chen, S. Song, D. Xue, An ionic aqueous pseudocapacitor system: electroactive
ions in both a salt electrode and redox electrolyte, RSC Adv., 4 (2014) 23338–23343.
https://doi.org/10.1039/c4ra03037k

[89] K. Chen, S. Song, K. Li, D. Xue, Water-soluble inorganic salts with ultrahigh specific
capacitance: crystallization transformation investigation of $CuCl_2$ electrodes,
CrystEngComm. 15 (2013)10367–10373. https://doi.org/10.1039/c3ce41802b

[90] K. Chen, Y. Yang, K. Li, Z. Ma, Y. Zhou, D. Xue, $CoCl_2$ Designed as excellent
pseudocapacitor electrode materials, ACS Sustainable Chem. Eng. 2 (2014) 440−444.
https://doi.org/10.1021/sc400338c

[91] H. J. Xie, B. Gélinas, D. Rochefort, Redox-active electrolyte supercapacitors using
electroactive ionic liquids, Electrochem. Commun. 66 (2016) 42–45.
https://doi.org/10.1016/j.elecom.2016.02.019

[92] S.Yamazaki, T. Ito, M. Yamagata, M. Ishikawa, Non-aqueous electrochemical
capacitor utilizing electrolytic redox reactions of bromide species in ionic liquid,
Electrochim. Acta 86 (2012) 294–297. https://doi.org/10.1016/j.electacta.2012.01.031

[93] Y. Wang, Z. Chang, M. Qian, Z. Zhang, J. Lin, F. Huang, Enhanced specific
capacitance by a new dual redox-active electrolyte in activated carbon-based
supercapacitors, Carbon 143 (2019) 300-308.
https://doi.org/10.1016/j.carbon.2018.11.033

[94] P. Navalpotro, J. Palma, M. Anderson, R, Marcilla, High performance hybrid supercapacitors by using para-Benzoquinone ionic liquid redox electrolyte, J. Power Sources 306 (2016) 711-717. https://doi.org/10.1016/j.jpowsour.2015.12.103

[95] Z.J. Zhang, Y. Q. Zhu, X.Y. Chen, Y. Cao, Pronounced improvement of supercapacitor capacitance by using redox active electrolyte of p-phenylenediamine, Electrochim. Acta 176 (2015) 941-948. https://doi.org/10.1016/j.electacta.2015.07.136

[96] D. Komatsu, T. Tomai, I. Honma, Enhancement of energy density in organic redox capacitor by improvement of electric conduction network, J. Power Sources 274 (2015) 412-416. https://doi.org/10.1016/j.jpowsour.2014.10.069

[97] L. Ren, G. Zhang, Z. Yan, L. Kang, H. Xu, F. Shi, Z. Lei, Z.H. Liu , High capacitive property for supercapacitor using Fe^{3+}/Fe^{2+} redox couple additive electrolyte, Electrochim. Acta 231 (2017) 705–712. https://doi.org/10.1016/j.electacta.2017.02.056

[98] T.T. Liu, Y.H. Zhu, E.H. Liu, Z.Y. Luo, T.T. Hu, Z.P. Li, R. Ding, Fe^{3+}/Fe^{2+} redox electrolyte for high-performance polyaniline/SnO_2 supercapacitors, Trans. Nonferrous Met. Soc. China 25 (2015) 2661–2665. https://doi.org/10.1016/S1003-6326(15)63889-4

[99] Y. Zhu, E. Liu, Z. Luo, T.T. Hu, T.T. Liu, Z.P. Li, Q.L. Zhao, A hydroquinone redox electrolyte for polyaniline/SnO_2 supercapacitors, Electrochim. Acta 118 (2014) 106–111. https://doi.org/10.1016/j.electacta.2013.12.015

[100]H.H. Zhang, J. Li, C. Gu, M. Yao, B. Yang, P. Lu, Y. Ma, High performance, flexible, poly(3,4-ethylenedioxythiophene) supercapacitors achieved by doping redox mediators in organogel electrolytes, J. Power Sources 332 (2016) 413-419. https://doi.org/10.1016/j.jpowsour.2016.09.137

[101]L. Liu, R. Feng, Y. Pan, X.P. Zheng, L. Bai, Nanoporous carbons derived from poplar catkins for high performance supercapacitors with a redox active electrolyte of p-phenylenediamine, J. Alloys Compd. 748 (2018) 473-480. https://doi.org/10.1016/j.jallcom.2018.03.073

[102]D. D. Potphode, L. Sinha, P. M. Shirage, Redox additive enhanced capacitance: Multi-walled carbon nanotubes/polyaniline nanocomposite based symmetric supercapacitors for rapid charge storage, Appl. Surf. Sci. 469 (2019) 162–172. https://doi.org/10.1016/j.apsusc.2018.10.277

[103]Z.J. Zhang, J.X. Li, T.T. Huang, M.R. Liu, X.Y. Chen, Large performance improvement of carbon-based supercapacitors using dual-redox additives phosphotungstic acid and potassium ferricyanide, J. Alloys Compd. 768 (2018) 756-765. https://doi.org/10.1016/j.jallcom.2018.07.316

[104] M. Jana, P. Samanta, N. C. Murmu, T. Kuila, Surface modification of reduced graphene oxide through successive ionic layer adsorption and reaction method for redox dominant supercapacitor electrodes, Chem. Eng. J. 330 (2017) 914–925. https://doi.org/10.1016/j.cej.2017.08.046

[105] Y.C. Chen, L.Y. Lin, Investigating the redox behavior of activated carbon supercapacitors with hydroquinone and *p*-phenylenediamine dual redox additives in the electrolyte, J. Colloid Interface Sci. 537 (2019) 295–305. https://doi.org/10.1016/j.jcis.2018.11.026

[106] S. Roldn, C. Blanco, M. Granda, R. Menndez, R. Santamar, Towards a further generation of high-energy carbon-based capacitors by using redox-active electrolytes, Angew. Chem. Int. Ed. 50 (2011) 1699–1701. https://doi.org/10.1002/anie.201006811

[107] Y. Zhang, X. Cui, L. Zu, X. Cai, Y. Liu, X. Wang, H. Lian, New supercapacitors based on the synergetic redox effect between electrode and electrolyte, Materials 734 (2016) 1-13. https://doi.org/10.3390/ma9090734

[108] Y. Wang, C. Zhang, X. Qiao, A. N. Mansour, X. Zhou, Three-dimensional modeling of mediator-enhanced solid-state supercapacitors, J. Power Sources 423 (2019) 18-25. https://doi.org/10.1016/j.jpowsour.2019.03.012

[109] A. N. Mansour, J.J. Zhou, X.Y. Zhou, X-ray absorption spectroscopic study of sodium iodide and iodine mediators in a solid-state supercapacitor, J. Power Sources 245 (2014) 270-276. https://doi.org/10.1016/j.jpowsour.2013.06.129

[110] LQ Fan, J. Zhong, J. Wu, J. Lin, Y. Huang, Improving the energy density of quasi-solid-state electric double-layer capacitors by introducing redox additives into gel polymer electrolytes, J. Mater. Chem. A 2 (2014) 9011−9014. https://doi.org/10.1039/c4ta01408a

[111] S.K. Tripathi, A. Jain, A. Gupta, M. Kumari, Studies on redox supercapacitor using electrochemically synthesized polypyrrole as electrode material using blend polymer gel electrolyte, Indian J. Pure Ap. Phys. 50 (2013) 315-319.

[112] Q.M. Tu, L.Q.Fan, F. Pan, J.L. Huang, Y. Gu, J.M. Lin, M.L. Huang, Y.F. Huang, J.H. Wu, Design of a novel redox-active gel polymer electrolyte with a dual-role ionic liquid for flexible supercapacitors, Electrochim. Acta 268 (2018) 562-568. https://doi.org/10.1016/j.electacta.2018.02.008

[113] I. Ike I. Sigalas, S. Iyuke, The effects of self-discharge on the performance of asymmetric/hybrid electrochemical capacitors with redox-active electrolytes: Insights

from modeling and simulation, J. Electron. Mater. 47 (2018) 470-492.
https://doi.org/10.1007/s11664-017-5796-y

[114] Y.F. Huang, W. H. Ruan, D. L. Lin, M. Q. Zhang, Bridging redox species-coated
graphene oxide sheets to electrode for extending battery life using nanocomposite
electrolyte, ACS Appl. Mater. Interfaces, 9 (2017) 909−918.
https://doi.org/10.1021/acsami.6b13145

[115] M. Jana, P. Samanta, N. C. Murmu, T. Kuila, Surface modification of reduced
graphene oxide through successive ionic layer adsorption and reaction method for redox
dominant supercapacitor electrodes, Chem. Eng. J. 330 (2017) 914–925.
https://doi.org/10.1016/j.cej.2017.08.046

Supercapacitor Technology: Materials, Processes and Architectures   Materials Research Forum LLC
Materials Research Foundations **61** (2019) 95-120   https://doi.org/10.21741/9781644900499-5

Chapter 5

Separators for Supercapacitors

A. Amin Izazi[1,3], Chin-Wei Lai[1*,] Joon-Ching Juan[1,2], Siew-Moi Phang[3,4], Guan-Tin Pan[5], Thomas C-K. Yang[5]

[1]Nanotechnology and Catalysis Research Centre (NANOCAT), Institute for Advanced Studies, University of Malaya, 50603 Kuala Lumpur, Malaysia

[2]School of Science, Monash University, Sunway Campus, Jalan Lagoon Selatan, 47500 Bandar Sunway, Malaysia

[3]Institute of Ocean and Earth Science (IOES), Institute for Advanced Studies, University of Malaya, 50603 Kuala Lumpur, Malaysia

[4]Institute of Biological Sciences, Faculty of Science, University of Malaya, 50603, Kuala Lumpur, Malaysia

[5]Department of Chemical Engineering and Biotechnology, National Taipei University of Technology. Taipei 106, Taiwan

*cwlai@um.edu.my

Abstract

Separators being one of the important components in a supercapacitor are gaining interest and demand for the development of efficient, reliable, flexible and environmentally friendly supercapacitors. Many studies search of suitable materials for a separator that possess high porosity, high electrolyte wettability, high ionic conductivity, high mechanical stability and a lower price. This chapter addresses recent advances and summarizes the main characteristics of separators as used in emerging supercapacitors. Highlighted are the challenges related with the current state-of-the-art materials and methods that should be considered for future supercapacitor development with emphasize on the separator.

Keywords

Supercapacitor, Separator, Membrane, Electrode Contact, Wettability

Contents

1.   Introduction

Basically, the separator separates anode and cathode, and it is important in each electrochemical power devices [1]. Practically, supercapacitors (SCs) can function without a separator, whenever there is a distance between electrodes. However, in order to prevent a short circuit within the supercapacitors during the fabrication, separators are necessary. Due to the huge market, the separator of supercapacitors must be inexpensive [2].

The basic components of SCs include two electrodes which are immersed in electrolytes and a separator that prevents electrical contact in SCs [3]. The separator for supercapacitors is essentially an ion-permeable membrane that allows the transfer of ionic charge from electrolytes but not from the electron [4]; which at the same time prevents the electrical and physical contact between the two electrodes [5]. The separator must have high electrical resistance, fast ionic transfer and low thickness [6].

Electrodes are tightly pressed and separated from each other by the separator. Thus, the thickness, porosity, electrical and mechanical properties of the separator must provide an effective isolation of the electrode while simultaneously allowing for a free flow of ions in the electrolyte [7].

The selection of separator in a supercapacitor depends on the nature of electrolytes used. Solid-state electrolytes, organic or aqueous electrolytes are usually used, depending on the power requirement. For inorganic electrolytes; polymer and paper separator are used while in aqueous electrolytes, ceramic or glass fiber separators are used [6]. There have been numerous research activities on developing supercapacitor separator material viz polypropylene, eggshell membrane, polyvinyl alcohol (PVA), filter paper and Teflon ring [8].

Noticeably, the mesoporous separator parameters (chemical composition, thickness, wettability) influence the characteristic relaxation frequencies, specific energy and power density of supercapacitors. Thus, problems encountered by non-aqueous solvent and electrolyte properties at/inside the mesoporous separator, are the adsorption/desorption characteristics of solvent molecules at/inside the mesoporous separator (wettability problem), and mass transfer characteristics of ions i.e. molar conductivity of electrolytes inside the polymer matrix.

2. Properties of separators for supercapacitors

The main issue faced for the SCs has been the mechanical properties of the separator since its function is to avoid internal short circuit and maintaining safety [9]. The separator avoids the occurrence of electrical contact between anode and cathode, so that the choice of the separator is an important criterion in designing high-performance supercapacitors [10]. In addition, the separator is an inactive material [9] and it should be cost-effective [11]. A supercapacitor is a common type of the electrical double-layer capacitor (EDLC). EDLCs store the electrical charge at the metal-electrolyte interface. The electrolyte ions are diffused through the separator into the pores of both charged electrodes causing a recombination of ions as there is no back movement of the ions [4]. The separators must possess superior conductivity and affinity towards the electrolyte ions as well as good thermal stability [12].

At present, the separators can be categorized as macroporous and dense separators. The macroporous separators made of polyvinylidene fluoride homopolymer (PVdF) and poly (fluoride-co hexafluoropropylene) (PVdF-HFP) have been used in lithium-ion batteries (LIBs).

Dense separator materials like sulfonated poly (ether ether ketone) (SPEEK) and Nafion have been used in fuel cells as proton exchange polymer membranes because they have good mechanical properties and thermal stability. The SPEEK and Nafion separators have been fabricated by immersing them in a sulfuric acid solution.

A polymeric material separator can be categorized into; (i) electrolyte gel; an entrapped liquid electrolyte (H_2SO_4 or KOH) or organic solvent (acetonitrile or propylene carbonate) inside a highly porous (40-80% porosity) swollen separator to conduct ions and (ii) solid polymer electrolyte/separator (solvent free) where fixed ionic groups performed the ion conduction.

3.   Motivation

The performance of a supercapacitor is influenced by the choice of the active electrode material, that determines the capacitance of the supercapacitor; and the choice of electrolyte, that defines the operational voltage. The internal resistance (ESR) of the supercapacitor is influenced by the following phenomenon [11]:

- The ionic resistance of ions moving through the separator/electrolyte ion permeability [13]

- The ionic (diffusion) resistance of ions moving in small pores

The SCs internal resistance is due to electrolyte and not because of the separator used. Furthermore, the separator material and dimension texture should not influence the electric capacity of the supercapacitor, the energy and power density. The power capability of the SCs can be limited by high internal resistance.

Maximum power from a supercapacitor can be achieved by increasing its voltage and/or decreasing its equivalent series resistance (ESR). Ohmic barriers that are present in the electrical and electrochemical system will generate the ESR. The separator layer can contribute to the highest ohmic barrier. Typically, the power output of a supercapacitor is high, thus requires a faster transfer of ions from one electrode to the other. Moreover, substantial thermal stability is preferable to ease the stack drying step (organic electrolyte devices). The characteristic of the separator has to be high in porosity, and low in thickness with high thermal stability [14].

The capacitance of the supercapacitor depending on the electrolytes/electrode, indirectly influenced by the separator [15]. Many types of materials have been used as separator to achieve the targeted capacitance of a supercapacitor. The energy capacity of SCs is governed by its capacitance value, which inevitably depends on the surface area of the electrode and the electrolyte concentration [16]. The specific capacitance of SCs increased with electrolyte concentration. Thus, a high concentration of the aqueous solution can be employed as an electrolyte to achieve higher energy capacity of SCs at a relatively lower cost.

The use of a high concentrated electrolyte needs to couple with a corrosive-resistant separator to achieve a superior power and energy density SC. A conventional cellulose separator is incapable to withstand a high concentrated electrolyte. Regardless, most of the conventional separators like paper and cellulose are incapable to withstand the high concentration aqueous electrolyte [17].

High surface area electrodes and high efficient electrolytes have tremendously led the selection of separators as a key requirement in designing high-performance supercapacitor [10]. Commercially, separators have been synthesized from rubber, plastic, aqua-gel, resorcinol formaldehyde polymers, and polyolefin films [18]. However, the separator tends to dry out or rupture after a certain duration, thus exhibiting poor ionic conductivity. Inevitably, it is important that a highly porous separator can provide almost no resistance for ion movement and act as an electronic insulator between electrodes [19].

Numerous intensive researches have been done to improve cyclability of SCs by improving the energy as well as power density and designing porous electrode materials and electrolytes.

However, little work has been performed on developing new separator materials and their impact on the electrochemical stability of SCs as well as application for high-performance SCs [20]. Thus, this chapter will discuss different alternative separator materials that have been employed to achieve higher performance of supercapacitor. The electrochemical properties of the separator will be discussed as well.

The materials that have been used to produce solid separators are cellulose (paper), textiles and polymers [9]. Polymer separators have been considered promising for use in supercapacitor because of their high chemical and mechanical resistance, lower price, and ease of syntheses. Glass fiber membranes have been used as separator for high-temperature supercapacitors (HTSCs) because of their excellent thermal and electrochemical stabilities, thus HTSCs fabricated with glass fiber films exhibit outstanding cycling stability [21].

4.   Cellulose separator

Conventionally, fossil-fuel based polymers have been used as separators in a supercapacitor. Few cellulose-based biomaterials, such as hydroxyethyl cellulose, starch mixed with glycerol, and nano-fibrillated cellulose, have been employed as the separator for lithium ion batteries [22]. Recently, nonwoven cellulose has been widely investigated as separator for SCs. Many studies have showed that cellulose-derived materials are suitable for separator applications, because of their low cost and high porosity (45-90%). The properties of biomass-based separators including (1) favorable separator surface and electrolyte interface wettability, (2) high ionic conductivity, and (3) lower cost due to large-scale production [2].

Currently, many cellulose-based paper products are used as separator in commercially available SCs systems include one layer cellulose fiber materials, multilayer cellulosic and composite materials with various densities. These materials have good desirable properties like electrolyte absorption, electronic insulation, and physical strength [2]. Moreover, cellulose paper is usually suitable for use in organic electrolyte, but not in aqueous electrolytes due to the possibility of degradation of the cellulose in H_2SO_4 or KOH solution [23].

Double layer nitrocellulose (NC) separators i.e. nano fibrillate cellulose (NFC) and cellulose nanocrystal (CNC) on a flexible polymer substrate for SCs have been reported for the first time by Tuukkanen et al. [24]. NC is a robust separator material because of its nano-scale size and hence it is capable to form a strong entangled nanoporous network, as a pathway for ion conduction. NC layer has been deposited as a separator on CNT electrode through (a) thin CNC film as an adhesion promoter layer between thick electrode and separator films and (b) gel of NFC that deposited in the area enclosed via drop-casting and dehydration at room temperature to form the firmly attached thick films of NFC. The SEM micrograph showed the uniformity of the electrode after the separator deposition and dehydration, while also showing the electrode/separator films homogeneity and absence of gaps between the adjacent layers from cross-section analysis. The CV curve showed a box-like shape which indicated the good capacitive behavior of the supercapacitor. A relatively, high capacitance supercapacitor, 14.9 – 1605 mF, was obtained with active 1.8 cm^2, and specific capacitance of 7.4 – 9.1 mF/cm^2 and 2.4-2.9 F/g. The capacitance retained its performance during the 2000 times charge/discharge cycling.

5.   Filter paper as a separator

Nor et al. [25], reported filter paper as a nanoporous separator for carbon monolith-based SCs. It was found that the mobility of ion into the pore network related to the nanoporous structure of the separators, has led to the noteworthy value of specific capacitance, power, and energy of SCs. The filter paper separator composed of an interlaced fiber-like structure and nanoporous networks that provides a superhighway for promoting diffusion of ions into the electrode pore is illustrated in Figure 1 Moreover, the nanoporous structure is capable of retaining a high amount of electrolytes. In addition, the filter paper separator has a high electrolyte uptake (140%) compared to other separators and can retain electrolytes after four hours with only 11% loss and 100% loss after four days (Figure 2). These observations are consistent with the nanopores network characteristics of the filter paper as evident from the FESEM images. In Figure 3(a), the CV curves of the filter paper-based SCS (cell C) show a larger current window. The rectangular-like shape of nanoporous separators provides a superhighway for ion diffusions. The GCD curves (Figure 3(b)) exhibit a very low ohmic resistance at the potential switching point, indicating excellent electrochemical reversibility of the filter paper-based SCs with a typical symmetric charge-discharge performance. Electrochemical characterization of symmetrical SCs fabricated using H_2SO_4 electrolyte and oil palm empty fruit bunches (EFB)-based activated carbon monolith electrode (cell C) was carried out. The specific power and specific energy were 178 W/kg and 427 Wh/kg respectively (Figure 3(c)). While the specific capacitance (C_{sp}) calculated from CV and GCD data were 180 F/g and 154 F/g respectively.

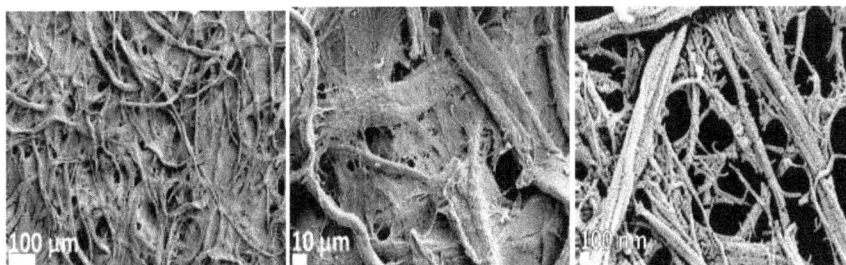

Figure 1. FESEM micrograph of filter paper nanoporous separator with different magnification "Reprinted from [25], with the permission of AIP Publishing".

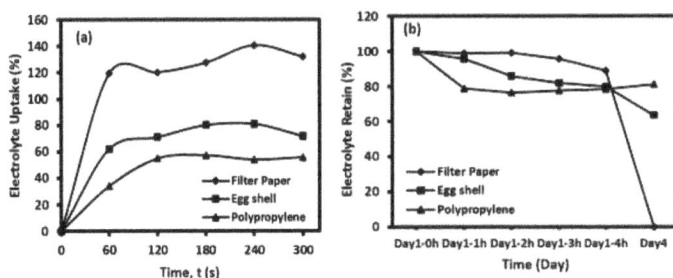

Figure 2. (a) Electrolyte uptake (%) (b) Electrolyte retention (%) of different separator in 1 M H_2SO_4 electrolyte at room temperature "Reprinted from [25]), with the permission of AIP Publishing".

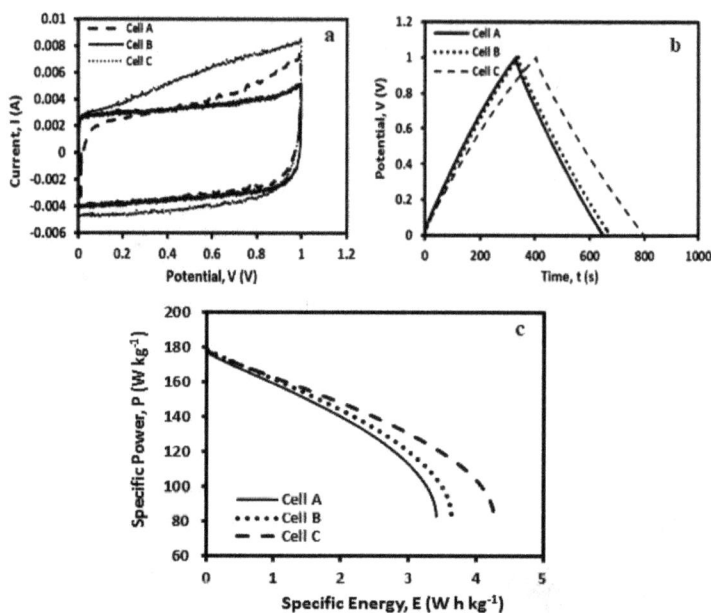

Figure 3. (a) Voltammogram data obtained at 1 mV s⁻¹, (b) GCD curves, (c) Ragone plots for different separator-based SCs, with Cell C is filter paper based SCs "Reprinted from [25], with the permission of AIP Publishing".

6.   Glass wool as a separator

Noorden et al. [16, 17] reported the use of glass wool as a noncorrosive separator for SCs application with high concentration aqueous electrolytes. The SCs were filled and investigated with aqueous sulfuric acid (H_2SO_4) and an activated carbon electrode. As evident from Figure 4, glass wool has inferior fibrous structure with higher porosity value (84 ± 2 %) compared to cellulose (36 ± 8 %). The separator materials reaction in 18 mol/L H_2SO_4 for 60 days was studied and the results are shown in Figure 5. Glass wool as a noncorrosive separator was able to preserve its physical properties even after 60 days compared to the cellulose separator that was immediately dissolved in the high concentration of H_2SO_4 suggesting that the glass wool has high reliability and capability in withstanding high concentration electrolytes. The CV behaviour of both the glass wool and cellulose-based SCs, showed quasi-rectangular and symmetrical CV curves. The glass wool separator based SCs exhibited greater specific capacitance $C_{s,cv}$ and $C_{s,g}$ of 126 F/g and 131 F/g respectively and low equivalent series resistance of 13Ω.

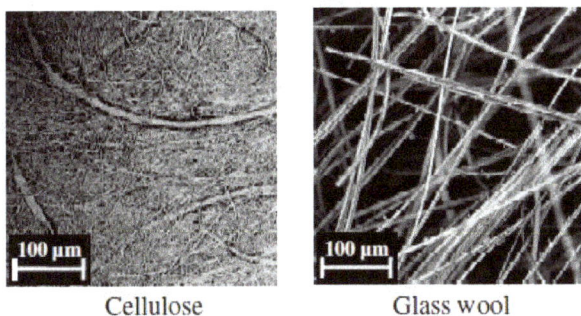

Cellulose   Glass wool

Figure 4 SEM micrographs of separator material Reprinted with permission from [16], Copyright [2013]. The Electrochemical Society.

After
9 months

Cellulose   Glass wool   Cellulose   Glass wool

Figure 5 High concentration acid test on separator material. Reprinted with permission from[16]. Copyright [2013]. The Electrochemical Society.

7. Polymer

The use of polymer separators in SCs has been very favorable because of their high chemical and mechanical resistance, lower price and ease of processing. However, poor wetting ability with various electrolytes due to strong hydrophobic character has limited the usage of polymer such as PE, PP, and PA as separators. Polypropylene (Celgard®) is the most used polymer as macroporous separator whereas, polyvinylidene difluoride (PVDF), polyvinylidene fluoride-co-hexafluoropropylene (PVDF-HFP), and polymethyl methacrylate (PMMA) have also been in synthesis of gel polymer electrolytes because of having a high melting point, mechanical integrity, dielectric properties and good affinity towards the electrolyte. PVDF tends to lose its mechanical strength when it is swollen in solvents and plasticized. While in PVDF copolymer, the effect is more profound, because it is low in crystallinity [9].

7.1 Poly (vinylidene fluoride)-based separator

Tonurist et al. [26] reported the self-made separator materials using electrospinning at different electric field strengths and polymer solutions in N,N-dimethylacetamide or N,N-dimethylformamide and acetone mixture (8:2 mass ratio)) feed rates. The high-frequency series and low-frequency parallel resistance, characteristic time constant and power density values influenced by the separator structure, thickness, total porosity, specific surface area, and chemical composition.

Karabelli et al. [27] prepared macroporous polymer separator based on poly(vinylidene fluoride)(PVDF) by a phase inversion process using acetone as solvent. The PVDF separator exhibited highly porous structure (80%) and good mechanical properties. However, among the fluorinated polymers, the homopolymer PVDF was the most suitable grade to be employed in a supercapacitor. By decreasing the separator porosity, the mechanical strength of the separator and liquid electrolyte could be improved.

7.2 Poly (aryl ether sulfone) separator

Huo et al. [12] described quaternary ammonium functionalized poly(aryl ether sulfone) as a separator for supercapacitor. The separator was obtained via polymerization, bromination, and derivatization of methylated poly(aryl ether sulfone)(PAES) with quaternary-ammonium groups. The separator demonstrated excellent thermal stability and mechanical properties. Nevertheless, the separator still exhibits good tensile strength in the range of 38.87-43.50 MPa, elongation at break of 9.96-13.36% and tensile modulus of 1.46-1.69 GPa. The Coulombic efficiency of the supercapacitor was 95% over 5000 cycles. Increasing the quaternization degree of PAES, has increased the discharge-specific capacitance, energy density and power density of the SCs.

7.3 Poly (vinyl alcohol) (PVA)-based separator

The production of thin film separator based on PVA is simple and cost-efficient because the material is cheap, widely available, biodegradable and environmentally friendly [28]. Bon et al. proposed the flexible poly (vinyl alcohol)(PVA)-ceramic composite (PVA-CC) separator for SCs. The PVA aqueous solution was mixed with ceramic particles such as aluminum oxide (Al_2O_3), silicon dioxide (SiO2) and titanium dioxide (TiO_2).

7.4   Sulfonated poly(ether ether ketone)

Sivaraman et al. [29] described sulfonated poly (ether ether ketone) SPEEK membrane as both a separator and electrolyte. The supercapacitor was fabricated by hot pressed electrode and SPEEK membrane at 180 °C for 30 s to minimize the ohmic resistance and ensuring good contact between the two. The thickness of the membrane was 50 µm and the concentration of SPEEK was 20wt% of the fabricated cell. A good capacitor performance was shown, which demonstrated that the SPEEK membrane can be used as both separator and electrolyte. The cyclic voltammogram of the separator showed close to rectangular shape and the capacitance was 0.6 F, which corresponds to 27 F per g of the active material. The impedance value of about 1.67 Ω has been attributed to low ohmic resistance between the electrode and the SPEEK membrane.

7.5   Conductive polymer

Organic polymers like polyaniline, polypyrrole, and derivatives of polythiophene are conductive polymers (CPs) which conduct electricity through a conjugated bond system along polymer chain [30] and hence making them excellent materials for energy storage.

Lv et al. [31] reported polyaniline (PANI) as a conductive polymer deposited on bacterial cellulose (BC) substrate as a separator for integral electrode-separator. BC has three-dimensional multilayer network structure and excellent mechanical strength which has been chosen as the flexible supporting substrate for depositing PANI, which is usually fragile and has poor stability, the effect by its polymeric backbone structure. However, PANI on its own possesses high specific capacitance, environmental stability, low-cost and easily controlled structures [31]. The integral electrode-separator showed a well bending and stretching performance with a device thickness of 15µm. The SCs showed a cycle life of 28.3 F/cm^3 volumetric capacitance, as well as ability to retain 100% at 2500 charge/discharge cycles of 0.1 A/g current density. Furthermore, it has shown 2.48 Ω solution resistance (R_s) at 1 mol PVA/H_2SO_4 electrolyte. This flexible SCs has potential as wearable devices.

7.6   Polymer modification

Polymer modification has been performed to improve the electrolyte limitation within the SCs using radiation-induced techniques (cross-linking or grafting) and by chemical (sulfonation) [9]. The cross-linking methods reinforce the separator by improving its mechanical properties at higher temperatures and decreasing the swelling of the polymer. The cross-linking methods involved either chemical treatment or irradiation.

The most used polymer for synthesizing gel polymer electrolytes have been PVDF, PVDF-HFP, and PMMA. PVDF and its copolymer were selected as separator/polymer electrolytes substrates because of their high melting point, mechanical integrity, dielectric properties and good affinity towards the electrolytes solvent. However, PVDF can be plasticized when it is swollen in solvents, thus losing its mechanical strength.

Radiation-induced crosslinking has been used to improve the mechanical properties of cellulosic material by forming hydrogels. Thus, the liquid electrolyte is absorbed within the cellulose-based separators network.

According to Stepniak et al. [32], the grafting of acrylic acid (AAc) on polypropylene (PP) was achieved by using the plasma-induced grafting technique. The AAc was polymerized after the treatment of plasma activated PP under UV irradiation. When the degree of grafting increased, the specific capacitance also increased. However, the surface area decreases upon excessive grafting and the double layer formation caused a decrease in the capacitance. Regardless of the contact angle measurement, the PP turned from hydrophobic into the hydrophilic structure. This helped the PP as a separator to absorb more electrolytes and decrease in the electrolytic area resistance. The resistance decreased from 120,000 to 40-50 $\Omega m/cm^2$. The CV profiles of the grafted sample showing a nearly rectangular shape suggested a good capacitive supercapacitor. From the voltammogram, the specific capacitance value for the AC3 and AC4 electrodes were calculated to be 114 and 103 F/g respectively.

Sivaraman et al. [33] have reported sulfonated fluorinated ethylene propylene copolymer membrane (FEP-g-AA-SO$_3$H) as a separator for SCs. The FEP membrane was exposed to γ-irradiation (50 kGy doses). The grafted film was then washed and regenerated, followed by sulfonation. The FEP-g-AA-SO$_3$H membrane showed very good mechanical properties, that the tensile strength was 7.5 MPa, coupled with 125% elongation, strong and tough enough to separate anode and cathode in the SCs. The cyclic voltammogram showed almost a rectangular shape and capacitance of 98 F/g at 10 mV/s scan rate. The capacitance decay was approximately 0.013% per charge/discharge cycle. The impedance spectra for the FEP-g-AA-SO$_3$H separator-based SCs consist of a small semicircle at high frequency due to the contact impedance generated from the electrical connection between electrode materials as well as the charge transfer at the contact interface between the electrode and the electrolyte solution. Furthermore, the supercapacitor showed a low solution resistance of 1.35 Ω due to the good contact between the electrode and separator interface, and the low thickness of the separator membrane. A transition to a linear nature at low frequency gives the charge transfer resistance of 2.25Ω.

Supercapacitor Technology: Materials, Processes and Architectures   Materials Research Forum LLC
Materials Research Foundations **61** (2019) 95-120   https://doi.org/10.21741/9781644900499-5

Xie et al. [23] studied the modification of poly(vinylidene fluoride) (PVDF) porous membranes with polyvinyl alcohol (PVA) and glutaraldehyde (GA) via hydrogen bonds and aldol cross-linking reaction respectively to increase their surface hydrophilicity and fabricated the PVDF membrane separator. A hydrophilic surface was produced when PVA molecules attached on the PVDF membrane via strong hydrogen bonds (i.e F····H—O) of hydroxyl groups (—OH) and highly electronegative fluorine atoms. However, the cross-linking of PVA with GA resulted in increase of surface hydrophilicity leading to improvement of the thermal and mechanical strength of the PVDF. The SEM images show pore structure of PVDF membrane without any significant change before and after hydrophilic modification. Furthermore, the membranes show a typical asymmetrical structure that consists of a skin layer with honey-comb-like nano-sized pores and a sub-layer consists of finger-like large pores. Sufficient electrolyte supply as well as ion distribution channels were achieved when different types of pores were interconnected. The membrane porosities were 72.4% and 64.4 % for pristine membrane and modified membrane respectively. After modification, porosity decrease appeared due to the occupation of the membrane pores space by PVA and GA cross-linked molecules. The contact angles of water drops on the surface of PVDF and modified PVDF (mPVDF) membranes were 73.5° and 61.0° respectively. Due to poor hydrophilicity, the pristine PVDF membrane was not able to accommodate sufficient electrolyte. The CV curves of the pristine PVDF separator based SCs exhibited lower specific capacitances than the PVDF separator. During the charge/discharge process, the electrolyte ions were unable to be transported freely through the hydrophobic pores from one side to the other sides of the electrode. The ions can only approach the outer surface of the electrode. The pristine PVDF-based SCs showed deviation in CV curves from the ideal rectangular shape because the interior pores of pristine PVDF hardly contribute to the capacitive behavior. The GCD curves for the pristine-based PVDF SCs, demonstrated high internal resistance from their significant IR drops. For the mPVDF based SCs, no obvious IR drops were observed as the hydrophilic PVDF membranes were able to provide efficient electrolyte supply and facilitate the ion distribution in the separator. The assembled mPVDF-based SCs showed the highest specific capacitance, excellent rate performance and cycling stability attributed to the cooperative effect of the hydrophilic porous separator and electrodes.

Commercially, polyolefin separators have good mechanical and thermal properties. They hinder thermal runway triggered by electrical short-circuits or rapid overcharging. Nevertheless, polyolefin separators cannot effectively absorb the organic electrolyte solvents like ethylene carbonate and propylene carbonate and aqueous electrolytes like KOH and H_2SO_4 solutions, because of hydrophobic nature low surface energy. In addition, the solvent leakage from the interfaces between electrodes or from the opposite sides of current collectors often causes the deterioration of cycle performance. High-performance polyolefin separators can be achieved by modifying their surface hydrophilic by sulfonation to improve compatibility with various electrolyte solutions. Moreover, inorganic (ceramic) nanoparticles could influence the improvements of the separator ionic conductivity, the mechanical and chemical stability, safety and the interfacial activity of cells [34].

However, commercial separators such as polyolefin-based porous films and cellulose paper are not suitable for high-temperature supercapacitors (HTSCs) because they tend to shrink at high temperature and consequently can lead to electrical short circuits.

Stepnaik et al. [35] have reported a thin microporous PP separator (25μm) partly filled with a modifier (acetone oligomers) and finally with polymer hydrogel (PAAM-KOH-H$_2$O) electrolytes. The modifier changed the hydrophobicity of PP into hydrophilic character making it wettable with aqueous or polymer hydrogel electrolytes. The SCs exhibit an excellent higher capacitance and high-rate dischargeability.

7.7   Composite Gel/Membrane

Poly (vinyl alcohol) (PVA) hydrogel has been used previously as gel polymer electrolyte in SCs but it has shown poor mechanical strength and low ionic conductivity [36]. However, the PVA-based gel polymer electrolytes can be adopted with cellulose to provide high mechanical strength.

Liang et al. [37] have reported the modification of sulfonated poly(ether ether ketone) (PEEK) membrane through adoption of an excellent film-forming property of poly(vinyl alcohol) (PVA). The modification was done via fabrication of chemically crosslinked SPEEK/PVA and SPEEK/PVA semi-interpenetrating polymer networks (sIPIN). The composite membranes have demonstrated moderate water uptake, proton conductivity, and elongation at break that improved the solid/liquid interfaces. The high specific capacitance of 134 F/g at current density 1 A/g and long cyclic stability was achieved.

Ji et al. [36] developed a novel flexible cellulose/PVA composite gel via solution freeze-thawing method and applied as a separator for quasi-solid-state SCs. The composite gels, CP 5.0 with 5% cellulose and 95% PVA showed a network structure and improved mechanical properties. As evident from Figure 6, s CP 5.0 composite gel has a high ionic conductivity around 1.85×10^{-2} S/cm. Figure 7(a) illustrates the SEM image of CP-0 with the circular morphology of the gel surface distribution. The pores diameters higher than 1 μm have been suitable for electrolyte diffusion but not for mechanical properties enhancement. CP 5.0 sample reveals homogeneous porous structure as clear from Figure

7(b). It is due to good miscibility between cellulose and PVA. The pores sizes were in the range of a few hundred nanometers to micrometers. The CP 5.0 composite gel gained a balanced performance because of the cellulose reinforcement effect. The CP-5.0 composite gel stress-strain was improved and it has good toughness (Figure 8). This shows that its mechanical strength increased compared to PVA hydrogel (CP 0), which is important for the safety of the separator. The charge-discharge of CP-5.0 composite based quasi-solid-state SCs exhibited by voltage window in the range of 0-1.8 V, can be attributed to the reversible ion adsorption/desorption process. The specific capacitance was 125.1 F/g at 1 A/g, that was 91% over the value of 0.25 A/g, because of high stability and high rate capability of SCs. Nearly rectangular-shape cyclic voltammetry (CV) profiles were recorded which showed the charge/discharge process as high reversible and kinetically straightforward. The CP -5.0 composite gel based quasi-solid-state SCs showed the smallest semi-circle indicating SCs lowest charge transfer resistance. Moreover, bulk resistance and charge transfer resistance values were lowest. Long term cycling stability of the CP-5.0 composite gel based quasi-solid-state SCs showed an increased specific capacitance compared to CP-0, which enhanced by ~ 11.2%. Furthermore, the capacitance retention after 1500 cycles (after equilibration within first 400 cycles), at a current density of 1 A/g was as high as 98.52% (capacitance retention of 88%) suggesting a high cycling stability of CP-5.0 composite gel based quasi-solid-state SCs after improving the PVA gel with cellulose. After charging, a series of CP-5.0 composite gel based quasi-solid-state SCs could light up a LED as shown in Figure 9.

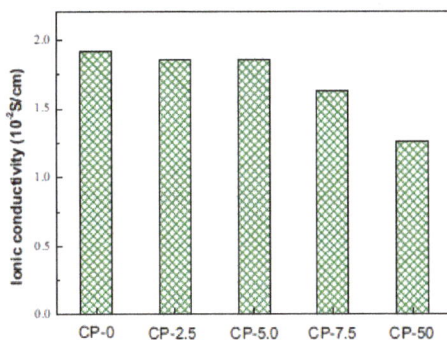

Figure 6 The ionic conductivity test of different separator after immersed in 1.5 M Li$_2$SO$_4$ solution Reprinted/adapted with permission from Springer Nature: [Cellulose][36], [COPYRIGHT](2018)

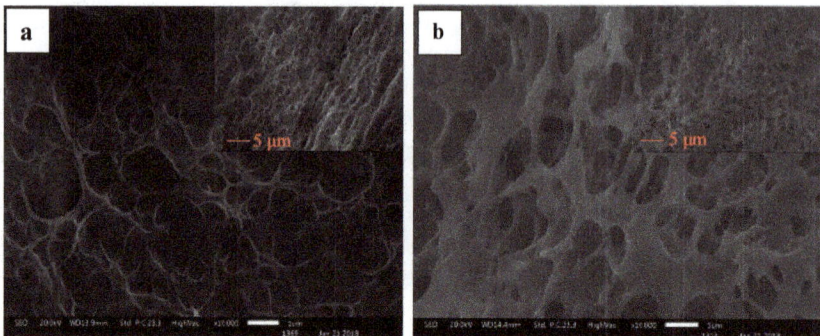

*Figure 7 SEM micrographs of the cellulose/PVA composite gels of (a) CP-0 (b) CP-5.0.
The insets are the low magnification images Reprinted/adapted with permission from
Springer Nature: [Cellulose][36], [COPYRIGHT](2018)*

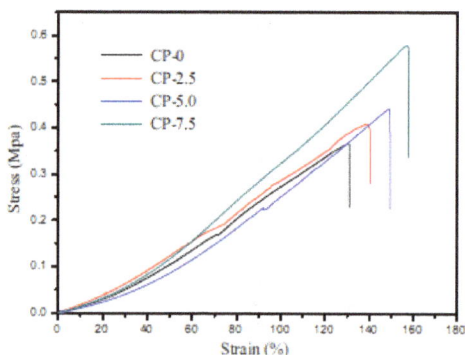

*Figure 8 Stress-strain behaviour of cellulose/PVA composite gels. Reprinted/adapted
with permission from Springer Nature: [Cellulose][36], [COPYRIGHT](2018)*

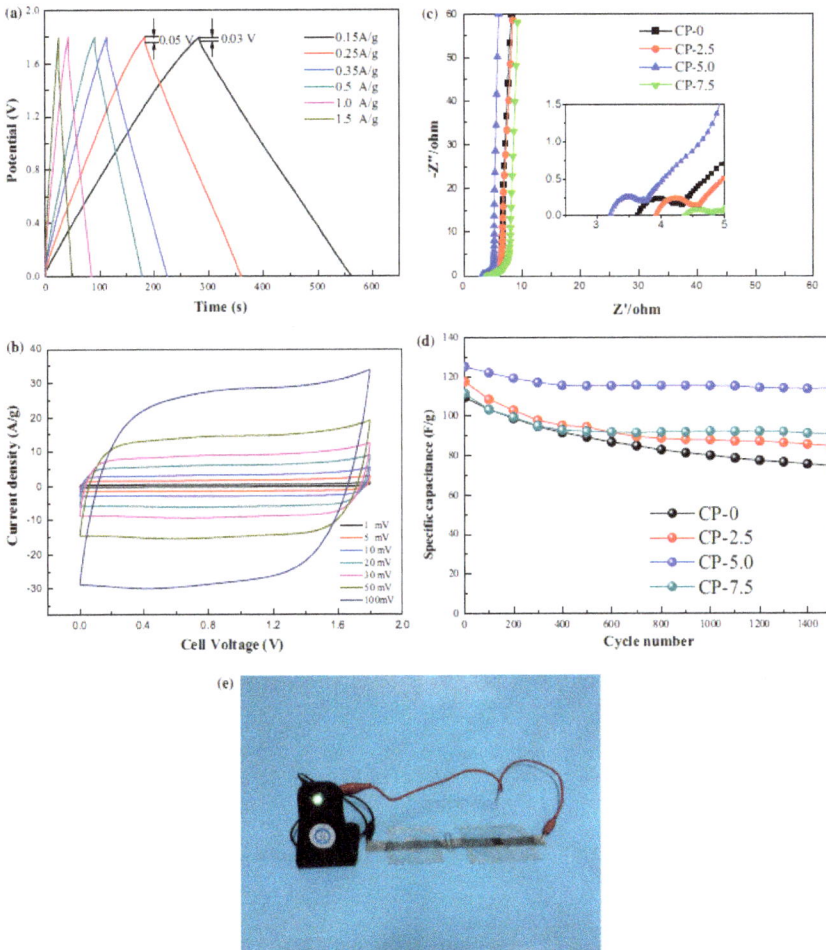

Figure 9. (a) GCD curves of the CP-5.0 composite gel based SCs, (b) CV curves of the CP-5.0 composite gel based SCs,(c) EIS of the different composite gel based SCs, and close up of high frequency region, (d) Specific capacitance with different composite gel based SCs at current density 1.0 A g⁻¹, voltage range 0-1.8 V, (e) LED powered by CP-5.0 composite gel based SCs. (Reprinted/adapted with permission from Springer Nature: [Cellulose][36], [COPYRIGHT](2018).

Supercapacitor Technology: Materials, Processes and Architectures   Materials Research Forum LLC
Materials Research Foundations **61** (2019) 95-120   https://doi.org/10.21741/9781644900499-5

7.8   Biomass and biowaste materials as a separator

Yao et al.[18] described the biomass waste, soybean leaf (SL) as a separator. SEM images showed that the morphology of the SL was highly porous along with well-ordered tube arrays. Furthermore, a dense upper surface made of few grinning-mouth-like open macropores (stomata) as well as many folded platelets as evident in Figures 10(a) and (b). The porous structures make the electrolyte transportation and storage easy. The thickness, porosity, tensile strength, contact angle (CA) and degree of water uptake (D_w) of the SL separator were compared with PP and cellulose separators (Table 1). The electrolyte CA showed that the SL separator has good wettability to the electrolyte compared to PP separator. The SL separator can quickly be wetted because the electrolyte droplets can reach onto the surface of the SL separator easily. Other than that, the permeability of the SL separator increases because of its porous structure. The ionic transportation accelerated due to excellent D_w properties of the SL separator that favours to hold enough electrolyte. The nearly rectangular CV shape of the SL separator based SCs at 100 mV/s, compared to the PP separator, showed obvious distortion. The inset in CV curves shows a LEDs powered by two SL separator based SCs (Figure 10(c)). The GCD measurements of SL separator based SCs show a smaller IR drop of 0.13 V compared to the PP separator (0.28 V)(Figure 10 (d)). The inner resistance did not exhaust its output capacity due to good wettability of SL separator. The SCs showed a high capacitance (358 F/g) at 1 A/g), superior cycle stability of 91% over 8000 cycles, and a superior rate capability in a symmetric BJPC-1.5 based electrode in 6M KOH (Figure 10(e)). Moreover, ordered channels in SL separator provided spaces for the electrolytes storage.

Table 1. Physical characterisation of different separator material [18].

Separator	Thickness (μm)	Porosity (%)	Tensile strength (MPa)	CA (°)	D_w (%)
SL	129	85	16	11	52
Cellulose	100	75	24	29	41
PP	25	50	13	74	4

Supercapacitor Technology: Materials, Processes and Architectures   Materials Research Forum LLC
Materials Research Foundations **61** (2019) 95-120   https://doi.org/10.21741/9781644900499-5

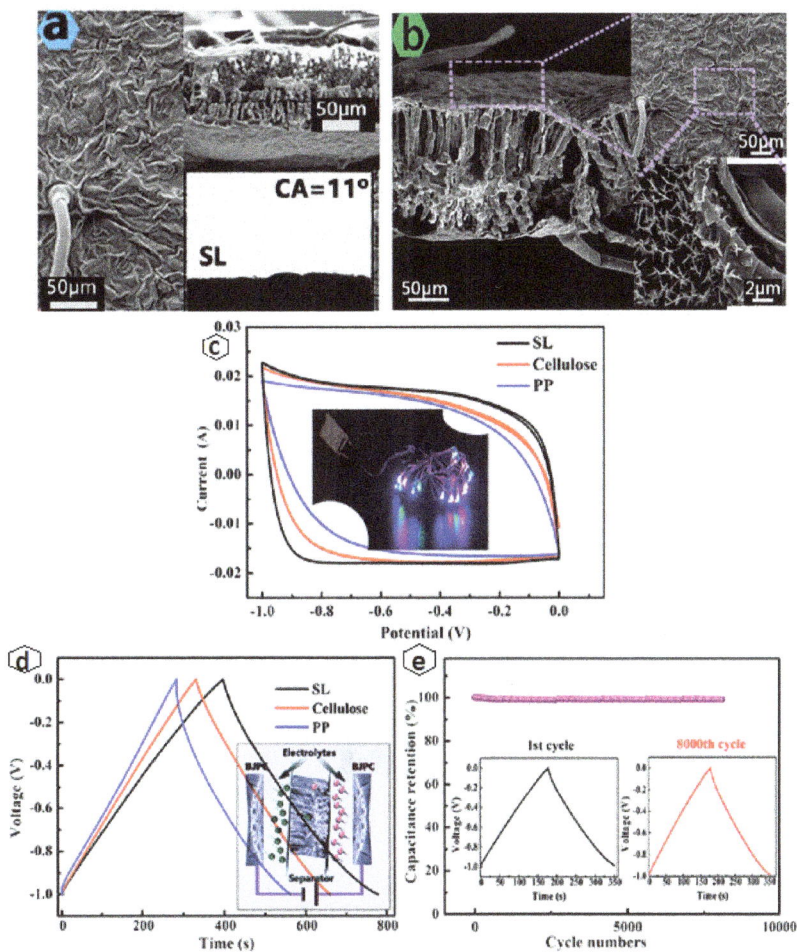

Figure 10. (a) SEM micrographs of the soybean leaf separator(b) SEM micrographs of the soybean leaf, (c) CV curves at 100 mV s^{-1} of different separator based SCs and BJPC 1.5 as electrodes (inset digital photograph of LEDs powered by the SCs) (d) GCD curves different separator based SCs and BJPC 1.5 as electrodes at 0.5 A g^{-1} (inset: illustration for SCs with a simplified electric circuit with a SL separator and BJPC 1.5 electrode, (e) Cyclic stability at 1 A g^{-1} for 8000 cycles of SL based separator SCs and BJPC 1.5 as electrodes "Reprinted (adapted) with permission from ([18]).Copyright (2018) American Chemical Society".

Yang et al. [38], reported biowaste eggshell membrane (ESM) as a separator to assemble aqueous electrolyte based asymmetric supercapacitor. The individual ESM fibre showed a random oriented structure with diameter in the range of 1-4 μm (Figure 11). The biopolymeric fibrous macropores network of 1-10 μm and random orientation of the fibre have provided a favourable way to ESM for mass transportation of electrolytes with low ion diffusion resistance. Figures 12(a-d) presents the comparison of ESM and filter separators based asymmetric SCs. Charge/discharge curve of the ESM separator based asymmetric supercapacitors indicates a slightly higher capacitance and longer discharge time compared to the filter separator. The higher energy density of 14 W h/kg was achieved by ESM separator based SCs compared to the modest energy density of 9.3 W h/kg delivered by the filter based SCs at a power density of 150 W/kg. Moreover, the ESM based SCs has a superior energy density of 14 Wh/kg with a power density of 150 W/kg and cycling stability (79% capacitance retention after 1000 cycles). Furthermore, EIS carried out showed that ESM separator based supercapacitor has a bigger semicircle, due to its semipermeable property. From Figure 12(e), the equivalent series resistant (ESR) for the ESM separator (1.83 Ω) compared to filter based separator (1.99 Ω) shows that the ESM separator displays an impressive ion conductivity and cyclability.

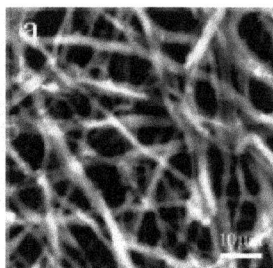

Figure 11. (a) SEM micrographs of natural ESM. "Reprinted (adapted) with permission from ([38]). Copyright (2018) American Chemical Society".

According to Dahlan et al. [39], the separator derived from duck shell membrane coated with TiO$_2$ nanoparticles showed improved electrolyte uptake due to smaller pore size. Specific capacitance achieved with current 0.007 mA and voltage 0.02V was 2.3 F/g.

Similarly, Taer at al. [40] reported the chicken eggshell membranes (CESM) separator for SCs.

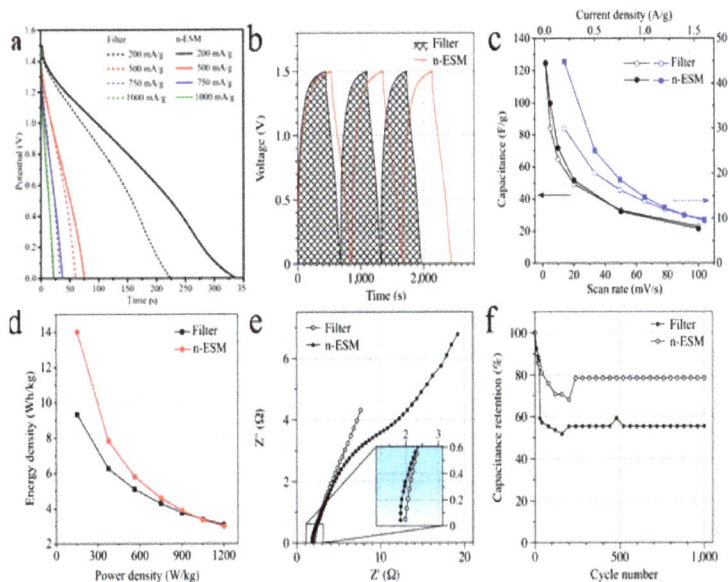

Figure 12. Comparative of (a) Discharge curves of the asymmetric SCs based on filter and ESM separators at current densities of 200, 500, 750 and 1000 mA g^{-1} (b) GCD curves of different separator-based asymmetric SCs at 200 mA g^{-1} (d) Energy and power density of different separator-based asymmetric SCs (e) Nyquist plots of different separator-based asymmetric SCs in the frequency range 100 kHz to 1 Hz, (f) Charge/discharge cycling performance of different separator-based asymmetric SCs at 500 mA $^{-1}$ "Reprinted (adapted) with permission from ([38]). Copyright (2018) American Chemical Society".

Yu et al. [19] described the natural and hierarchically ordered macroporous eggshell membrane (ESM) separator for SCs. The ESM possess a macroporous network with interwoven and coalescing shell membrane fibre with diameter ranging from 0.5 to 1 μm. The macroporous favoured for diffusing ions with a low resistance to the SCs. The CV curve showed a rectangular shape, showing that the charge was accumulated at the electrolyte/electrode interface through an electric double layer, to retain a relative shape without obvious distortion when scan rates increased. This revealed a high performance of supercapacitor due to the low contact resistance among electrode, current collector, and the separator. ESM possesses promising potential as a separator in the current commercial supercapacitor (<100 °C) as well as a high-temperature capacitor (>100 °C) because of its favourable porous structures, high decomposition temperature (over 200

°C), low swelling degree, and good mechanical strength. In addition, from the electrochemical performance, by using ESM as a separator, the supercapacitor shows a low resistance, quick charge-discharge ability (τ is 4.76 s) and outstanding cycling stability (92% retention after 10,000 cycles).

8.   Graphene oxide (GO) separator

Flexible graphene oxide papers (GOPs) with a maximal size of pores parallel to the electrodes have been designed and used as separator [41]. The GOPs were obtained after precipitation of water suspension of graphene oxide. Baskakov et al. [42] emphasized on usage of a graphene oxide separator for a metal-free SCs. It exhibited that under similar conditions, the capacity of the graphene oxide separator based SCs exceeded the Nafion separator based SCs. Chen at al. [43] reported the fabrication of graphene-based tandem integrated SCs. A novel direct electrode deposition on the separator (DEES) method was used. The supercapacitor exhibited high volumetric energy density of 52.5 W h /L at a power density of 6037 W /L and excellent gravimetric energy density of 26.1 W h /L at a power density of 3002 W /kg with outstanding electrochemical stability over 10,000 cycles.

Conclusions and perspectives

Despite the rapid development of supercapacitor, numerous challenges exist to be tackled and considered in developing an ideal separator. The porous and hydrophilicity separator ease the wettability of electrolytes, smooth ion transportation and conduction without undergoing any chemical or mechanical changes. Hence, to overcome the limitation discussed previously, possible surface modification is essential to modify the chemical, physical and mechanical properties of the commercial separator. As such, the influence of the separator characteristic and material have to be carefully optimized as they can give a significant contribution to the overall series resistance, thus limiting or increasing the power capability of the supercapacitor.

Acknowledgments

The authors would like to acknowledge the financial support provided by University Malaya grant: Equitable Society Research Cluster (ESRC) Research Grant GC002A-15SBS, Postgraduate Research Grant PG057-2015B, University Malaya Centre for Commercialization and Innovation (UMCIC) Prototype Grant RU005F-2016 and Impact-Oriented Interdisciplinary Research Grant (No. IIRG018A-2019). The authors also thank the Ministry of Higher Education for MyBrain 15 Scholarship and National Taipei University of Technology (Taipei Tech), Taiwan for the Student Exchange Program

References

[1] C.M. Costa, M.M. Silva, S. Lanceros-Mendez, Battery separators based on vinylidene fluoride (VDF) polymers and copolymers for lithium ion battery applications, RSC Adv. 3 (2013) 11404-11417. https://doi.org/10.1039/c3ra40732b

[2] L. Zhang, Z. Liu, G. Cui, L. Chen, Biomass-derived materials for electrochemical energy storages, Prog. Polym. Sci. 43 (2015) 136-164. https://doi.org/10.1016/j.progpolymsci.2014.09.003

[3] B. Ding, J. Yu, Electrospun Nanofibers for Energy and Environmental Applications, Springer Berlin Heidelberg 2014. https://doi.org/10.1007/978-3-642-54160-5

[4] A. Balakrishnan, K.R.V. Subramanian, Nanostructured Ceramic Oxides for Supercapacitor Applications, Taylor & Francis 2014. https://doi.org/10.1201/b16522

[5] M.F. Ahmer, A.M. Asiri, S. Zaidi, Electrochemical Capacitors: Theory, Materials and Applications, Materials Research Forum LLC 2018.

[6] A. González, E. Goikolea, J.A. Barrena, R. Mysyk, Review on supercapacitors: Technologies and materials, Renew. Sust. Energy Rev. 58 (2016) 1189-1206. https://doi.org/10.1016/j.rser.2015.12.249

[7] B. Szubzda, A. Szmaja, M. Ozimek, S. Mazurkiewicz, Polymer membranes as separators for supercapacitors, Appl. Phys. A Mater. Sci. Process. 117 (2014) 1801-1809. https://doi.org/10.1007/s00339-014-8674-y

[8] N.S.M. Nor, M. Deraman, R. Omar, E. Taer, Awitdrus, R. Farma, N.H. Basri, B.N.M. Dolah, Nanoporous separators for supercapacitor using activated carbon monolith electrode from oil palm empty fruit bunches C3-AIP Conference Proceedings, 1586 (2014) 68-73.

[9] M.M. Nasef, S.A. Gürsel, D. Karabelli, O. Güven, Radiation-grafted materials for energy conversion and energy storage applications, Prog. Polym. Sci. 63 (2016) 1-41. https://doi.org/10.1016/j.progpolymsci.2016.05.002

[10] K.I. Ozoemena, S. Chen, Nanomaterials in Advanced Batteries and Supercapacitors, Springer International Publishing 2016. https://doi.org/10.1007/978-3-319-26082-2

[11] M. Lu, F. Beguin, E. Frackowiak, Supercapacitors: Materials, Systems, and Applications, Wiley 2013. https://doi.org/10.1002/9783527646661

[12] P.F. Huo, S.L. Zhang, X.R. Zhang, Z. Geng, J.S. Luan, G.B. Wang, Quaternary ammonium functionalized poly(aryl ether sulfone)s as separators for supercapacitors based on activated carbon electrodes, J. Membrane Sci. 475 (2015) 562-570. https://doi.org/10.1016/j.memsci.2014.10.047

[13] A. Muzaffar, M.B. Ahamed, K. Deshmukh, J. Thirumalai, A review on recent advances in hybrid supercapacitors: Design, fabrication and applications, Renew. Sust. Energy Rev. 101 (2019) 123-145. https://doi.org/10.1016/j.rser.2018.10.026

[14] A. Laforgue, L. Robitaille, Electrochemical testing of ultraporous membranes as separators in mild aqueous supercapacitors, J. Electrochem. Soc. 159 (2012) A929-A936. https://doi.org/10.1149/2.020207jes

[15] A. Jabbarnia, W.S. Khan, A. Ghazinezami, R. Asmatulu, Investigating the thermal, mechanical, and electrochemical properties of PVdF/PVP nanofibrous membranes for supercapacitor applications, J.Appl. Polym. Sci. 133 (2016) 43707. https://doi.org/10.1002/app.43707

[16] Z.B. Ahmad Noorden, S. Sugawara, S. Matsumoto, Glass wool material as alternative separator for higher rating electric double layer capacitor, ECS Transactions 53 (2013) 43-51. https://doi.org/10.1149/05331.0043ecst

[17] Z.A. Noorden, S. Sugawara, S. Matsumoto, Noncorrosive separator materials for electric double layer capacitor, IEEJ T. Electr. Electr. 9 (2014) 235-240. https://doi.org/10.1002/tee.21961

[18] Q. Yao, H. Wang, C. Wang, C. Jin, Q. Sun, One step construction of nitrogen–carbon derived from bradyrhizobium japonicum for supercapacitor applications with a soybean leaf as a separator, ACS Sust. Chem. Eng. 6 (2018) 4695-4704. https://doi.org/10.1021/acssuschemeng.7b03777

[19] H.J. Yu, Q.Q. Tang, J.H. Wu, Y.Z. Lin, L.Q. Fan, M.L. Huang, J.M. Lin, Y. Li, F.D. Yu, Using eggshell membrane as a separator in supercapacitor, J. Power Sources 206 (2012) 463-468. https://doi.org/10.1016/j.jpowsour.2012.01.116

[20] K. Liivand, T. Thomberg, A. Jänes, E. Lust, Separator materials influence on supercapacitors performance in viscous electrolytes, ECS Trans. 64 (2015) 41-49. https://doi.org/10.1149/06420.0041ecst

[21] B. Qin, Y. Han, Y. Ren, D. Sui, Y. Zhou, M. Zhang, Z. Sun, Y. Ma, Y. Chen, A Ceramic-based separator for high-temperature supercapacitors, Energy Technol. 6 (2018) 306-311. https://doi.org/10.1002/ente.201700438

[22] X. Wang, D. Kong, Y. Zhang, B. Wang, X. Li, T. Qiu, Q. Song, J. Ning, Y. Song, L. Zhi, All-biomaterial supercapacitor derived from bacterial cellulose, Nanoscale 8 (2016) 9146-9150. https://doi.org/10.1039/C6NR01485B

[23] Q. Xie, X. Huang, Y. Zhang, S. Wu, P. Zhao, High performance aqueous symmetric supercapacitors based on advanced carbon electrodes and hydrophilic poly(vinylidene fluoride) porous separator, Appl. Surf. Sci. 443 (2018) 412-420. https://doi.org/10.1016/j.apsusc.2018.02.274

[24] S. Tuukkanen, S. Lehtimäki, F. Jahangir, A.-P. Eskelinen, D. Lupo, S. Franssila, Printable and disposable supercapacitor from nanocellulose and carbon nanotubes, Electronics System-Integration Technology Conference (ESTC), 2014, IEEE, 2014, pp. 1-6. https://doi.org/10.1109/ESTC.2014.6962740

[25] N.S.M. Nor, M. Deraman, R. Omar, E. Taer, Awitdrus, R. Farma, N.H. Basri, B.N.M. Dolah, Nanoporous Separators for Supercapacitor Using Activated Carbon Monolith Electrode from Oil Palm Empty Fruit Bunches, in: H. Setyawan, Widiyastuti, S. Machmudah (Eds.) 5th Nanoscience and Nanotechnology Symposium2014, pp. 68-73.

[26] K. Tõnurist, T. Thomberg, A. Jänes, T. Romann, V. Sammelselg, E. Lust, Influence of separator properties on electrochemical performance of electrical double-layer capacitors, J. Electroanal. Chem. 689 (2013) 8-20. https://doi.org/10.1016/j.jelechem.2012.11.024

[27] D. Karabelli, J.C. Lepretre, F. Alloin, J.Y. Sanchez, Poly(vinylidene fluoride)-based macroporous separators for supercapacitors, Electrochim. Acta 57 (2011) 98-103. https://doi.org/10.1016/j.electacta.2011.03.033

[28] C.Y. Bon, L. Mohammed, S. Kim, M. Manasi, P. Isheunesu, K.S. Lee, J.M. Ko, Flexible poly(vinyl alcohol)-ceramic composite separators for supercapacitor applications, J. Ind. Eng. Chem. 68 (2018) 173-179. https://doi.org/10.1016/j.jiec.2018.07.043

[29] P. Sivaraman, V. Hande, V. Mishra, C.S. Rao, A. Samui, All-solid supercapacitor based on polyaniline and sulfonated poly (ether ether ketone), J. Power Sources 124 (2003) 351-354. https://doi.org/10.1016/S0378-7753(03)00606-2

[30] I. Shown, A. Ganguly, L.-C. Chen, K.-H. Chen, Conducting polymer-based flexible supercapacitor, Energy Sci. Eng. 3 (2015) 2-26. https://doi.org/10.1002/ese3.50

[31] X. Lv, G. Li, D. Li, F. Huang, W. Liu, Q. Wei, A new method to prepare no-binder, integral electrodes-separator, asymmetric all-solid-state flexible supercapacitor derived from bacterial cellulose, J. Phys. Chem. Solids 110 (2017) 202-210. https://doi.org/10.1016/j.jpcs.2017.06.017

[32] I. Stepniak, A. Ciszewski, Grafting effect on the wetting and electrochemical performance of carbon cloth electrode and polypropylene separator in electric double layer capacitor, J. Power Sources 195 (2010) 5130-5137. https://doi.org/10.1016/j.jpowsour.2010.02.032

[33] P. Sivaraman, S. Rath, V. Hande, A. Thakur, M. Patri, A. Samui, All-solid-supercapacitor based on polyaniline and sulfonated polymers, Synth. Met.156 (2006) 1057-1064. https://doi.org/10.1016/j.synthmet.2006.06.017

[34] K.M. Kim, M. Latifatu, Y.-G. Lee, J.M. Ko, J.H. Kim, W.I. Cho, Effect of ceramic filler-containing polymer hydrogel electrolytes coated on the polyolefin separator on the electrochemical properties of activated carbon supercapacitor, J. Electroceramics 32 (2014) 146-153. https://doi.org/10.1007/s10832-013-9860-6

[35] I. Stepniak, A. Ciszewski, Electric double layer capacitors with polymer hydrogel electrolyte based on poly(acrylamide) and modified electrode and separator materials, Electrochim. Acta 54 (2009) 7396-7400.
https://doi.org/10.1016/j.electacta.2009.07.072

[36] Y. Ji, N. Liang, J. Xu, D. Zuo, D. Chen, H. Zhang, Cellulose and poly(vinyl alcohol) composite gels as separators for quasi-solid-state electric double layer capacitors, Cellulose 26 (2018) 1055-1065. https://doi.org/10.1007/s10570-018-2123-6

[37] N. Liang, Y. Ji, D. Zuo, H. Zhang, J. Xu, Improved performance of carbon-based supercapacitors with sulfonated poly(ether ether ketone)/poly(vinyl alcohol) composite membranes as separators, Polym. Int. 68 (2019) 120-124.
https://doi.org/10.1002/pi.5704

[38] P. Yang, J. Xie, C. Zhong, Biowaste-derived three-dimensional porous network carbon and bioseparator for high-performance asymmetric supercapacitor, ACS Appl. Energy Mater. 1 (2018) 616-622. https://doi.org/10.1021/acsaem.7b00150

[39] D. Dahlan, N. Sartika, Astuti, E.L. Namigo, E. Taer, Effect of TiO$_2$ on duck eggshell membrane as separators in supercapacitor applications, Materials Science Forum 827 (2015) 151-155. https://doi.org/10.4028/www.scientific.net/MSF.827.151

[40] E. Taer, Sugianto, M.A. Sumantre, R. Taslim, Iwantono, D. Dahlan, M. Deraman, Eggs shell membrane as natural separator for supercapacitor applications C3- Adv. Mater. Res. 896 (2014) 66-69.
https://doi.org/10.4028/www.scientific.net/AMR.896.66

[41] Y.M. Shulga, S.A. Baskakov, Y.V. Baskakova, Y.M. Volfkovich, N.Y. Shulga, E.A. Skryleva, Y.N. Parkhomenko, K.G. Belay, G.L. Gutsev, A.Y. Rychagov, V.E. Sosenkin, I.D. Kovalev, Supercapacitors with graphene oxide separators and reduced graphite oxide electrodes, J. Power Sources 279 (2015) 722-730.
https://doi.org/10.1016/j.jpowsour.2015.01.032

[42] S.A. Baskakov, Y.V. Baskakova, N.V. Lyskov, N.N. Dremova, A.V. Irzhak, Y. Kumar, A. Michtchenok, Y.M. Shulga, Fabrication of current collector using a composite of polylactic acid and carbon nano-material for metal-free supercapacitors with graphene oxide separators and microwave exfoliated graphite oxide electrodes, Electrochim. Acta 260 (2018) 557-563. https://doi.org/10.1016/j.electacta.2017.12.102

[43] W. Chen, C. Xia, H.N. Alshareef, Graphene based integrated tandem supercapacitors fabricated directly on separators, Nano Energy 15 (2015) 1-8.
https://doi.org/10.1016/j.nanoen.2015.03.040

Supercapacitor Technology: Materials, Processes and Architectures   Materials Research Forum LLC
Materials Research Foundations **61** (2019) 121-140   https://doi.org/10.21741/9781644900499-6

Chapter 6

Starch-Based Electrolytes: An Eco-Friendly and Economical Option for Flexible Supercapacitor

Neelam Srivastava

Department of Physics (MMV), Banaras Hindu University, Varanasi, India

*neelamsrivastava_bhu@yahoo.co.in, neel@bhu.ac.in

Abstract

Environment challenges have attracted the attention of scientists to utilize renewable materials in order to adopt a green-process. The electrochemists have also switched to utilizing renewable polymers for device fabrication. For commercial devices, the quality of the fabrication materials is determined by various factors such as i) cost, ii) eco-friendliness and iii) ease of design, etc. Glutaraldehyde crosslinked starch-based electrolytes show impressive results with all these criteria along with having excellent electrochemical properties such as wide electrochemical stability window (up to 4.5V) low equivalent series resistance (ESR<2Ω for 0.8mm thickness and 1x1cm^2 area) and fast ion transport (ion relaxation time of the order of μsec). These electrolytes are easy to prepare and they cost around 15 INR for 0.8mm thickness and 1x1cm^2 area. Hence, starch-based electrolytes are very promising for flexible supercapacitor fabrication.

Keywords

Starch, Economical, Eco-Friendly, Electrolytes, ESR, ESW

Contents

1.   Introduction

With the ever-accelerating technology of flexible devices, its presence has been registered in every sphere of life such as flexible health monitoring systems, flexible sensors, flexible display devices, e-textile and human interactivity, etc. [1,2]. With the increasing interest in portable and wearable electronic equipment, the research and development in flexible supercapacitors (FSCs) [3,4,5,6] and batteries are also receiving scientific attention. The two main components of energy devices (namely supercapacitors and batteries) are i) electrode and ii) electrolytes and they have different roles in different devices. Capacity, power density ($V^2/4R$ where V is electrochemical stability window and R is equivalent series resistance) and energy density ($CV^2/2$ where C is capacity) are the important merit parameters for supercapacitors along with their stability, charging/discharging cycle and Columbic efficiency, etc. Although these three key parameters are jointly achieved by the combination of electrode and electrolytes but dominantly the capacity is decided by the electrode material and ESR (equivalent series resistance) is decided by the electrolyte whereas on the electrochemical stability window (ESW=V) effect of electrode/electrolyte interface can be seen. To fabricate an efficient supercapacitor at commercial level, an electrolyte is required which has high electrochemical stability, wide ESW, high ionic concentration, low ESR, minimal toxicity, adequate viscosity and low volatility. Electrolyte must be economically viable and available in abundance. Thus electrolyte is very important for supercapacitors [7] but these supercapacitors specific electrolytes are not explored as required whereas electrode related research has received great attention and has seen tremendous advancement for rigid as well as flexible devices [8-17]. The main difference between batteries and supercapacitors is that though the energy density of a supercapacitor is much lower (more than 10 times) than batteries but they have a higher power density [18] and it can be achieved when the ESR is low and that's why the liquid electrolytes have dominated in R&D of supercapacitors to investigate new electrodes [19]. Unfortunately along with designing issues such as leakage, corrosion, packing, etc. liquid electrolytes suffer from low ESW value. Therefore, scientists have been looking for a better electrolytes system which not only provides designing ease but also have better electrochemical parameters. Polymer electrolytes, who possess physique of solid electrolytes but being amorphous in nature have conductivity comparable to liquid electrolytes, came as a promising alternative and have been well studied and commercially used worldwide for electrochemical devices. Tremendous work carried out in this field has led to the subdivision of the field in different categories i) gel polymer electrolytes, ii) composite polymer electrolytes, iii) solvent-swollen polymer electrolytes, and iv) solvent-free polymer electrolytes, etc. All have their advantages and disadvantages but the last one i.e. solvent-free (or polymer-salt complex) polymer electrolytes have the benefit of being easily modified by simply varying of their composition [20]. Success of polymer-salt depends upon the selection of suitable polymer matrix which not only holds and keeps the added salt dissociated but also assists the ion transport through its segmental motion. The added salt should have small dissociation energy, small cation size and good affinity for the selected polymer. The present discussion is regarding the use of renewable polymers as host for electrolyte preparation.

Polymer host matrix has to i) provide a solid matrix and simultaneously remain flexible enough for easy designing, ii) accept the added salt in matrix, iii) dissociate the salt and facilitate the ion transport and iv) accommodate the volume changes during the charging/discharging of devices, etc. Hence many polymers have been explored since the 1970s to improve the quality and achieve as many advantageous parameters as possible. Many synthetic polymers such as PEO, PVA, PVAc, PVP, PMMA and PAN, etc. [21-28] along with PC, EC, PEG, etc. as plasticizers have been studied with variety of salts in different compositions. Being synthetic, these polymers hosts end up being chemical garbage after the use of the device. The environment awareness has attracted the scientific community to utilize natural/renewable polymers alternates. Hence natural polymers like chitosan, gum acacia, agar-agar, cellulose, gelatin and starch, etc. [29-36] have been examined but the desired success has yet to be achieved for using them at a commercial level.

Our group got impressed by starch when we decided to work with renewable polymers because it is economical, easy to modify, abundant in nature, available in variety but literature survey didn't gave the same impression and it was observed that starch has not received as much attention as it deserves to be, although many researchers have commented on its potentiality [34, 37]. Scientists have worked upon starch-based electrolytes following different routes of preparation and composition. The present chapter has been divided into two parts i) starch-based electrolytes where blending/mixing process with different polymers has been adopted and ii) starch-based electrolytes crosslinked with glutaraldehyde.

2. Starch-based mixed polymer electrolytes

Recent literature witnesses the efforts of researcher to modify starch for electrochemical applications where electrolytes are synthesized using pure starches and/or chemically/ physically modified starches. Starch modification is carried out using techniques like blending, plasticization (a physical method) and functional group modifications (chemical method) etc. Pure starch-based polymer-salt complexes are very brittle and low conducting along with having small shelf life. Hence the starch-based electrolytes are always prepared either with an extra polymer which may be biopolymer and/or synthetic one. Since the starch-based materials are prone to be attacked by microorganism and hence their physique and electrical properties may change with time. This seems to be the reason why pure starch-based electrolytes, are rarely seen in literature. Same can be associated with selection of solvent like acetic acid because water-based electrolytes are easily affected by microorganisms. The conductivity of polymer-salt complexes achieves higher value at temperature higher than the glass transition temperature (T_g) because only above this temperature polymer segments are free for vibrational movements and hence segmental motion assisted ion transport improves the conductivity. Hence chemical and physical methods have been used to increase the conductivity by decreasing the Tg. In chemical reaction, starch's functional group is modified whereas in physical modification the Tg is lowered by adding the plasticizers. Plasticizers themselves also support the ion transport and hence poly(ethylene glycol) and glycerol are added for enhancing the conductivity. The films of starch-based electrolytes have been obtained using different techniques including reactive extrusion, solution casting of gelatinized starch [35, 38].

Starches from different botanical origins have different composition especially in terms of % amylose and amylopectin and hence they behave differently on addition of salt. Variety of starches such as Potato, corn, wheat, arrowroot, rice, etc. have commonly been studied for electrolyte synthesis, out of which corn starch has received greatest attention probably because of its highest (~76%) production. Table 1 summarizes the amylose and amylopectin contents of few common starches. Lithium salts, have been the obvious choice because of their good electrochemical parameters. Starch has very favorable atmosphere for hydrogen bonding hence it retains good amount of water in its matrix which supports the ionic conductivity by introducing proton conduction in the matrix and/or by dissociating the added salt hence ammonium salts (NH_4Br, NH_4Cl, NH_4NO_3 etc.) have also been used by many researchers with the hope of increasing the proton conductivity. Unfortunately pure starch-based electrolytes have never been reported to have good conductivity which is generally much $< 10^{-5}$S/cm and hence glycerol, and ionic liquids etc. have been added as plasticizers to it and the maximum achieved conductivity has been reported to be $\sim10^{-4}$ S/cm. This not only requires extra chemicals but also makes the synthesis process complicated and hence these materials have not been considered good. Table 2 summarizes some recent studies related to starch-based electrolytes. The supercapacitors for commercial applications need ESR of the order of few mΩ that means these electrolytes have limited scope for commercial fabrication of supercapacitors. Although laboratory scale supercapacitors having high capacity

(~F/gram) and power density >1000W/kg are being fabricated and tested with suitable electrode materials but the ESR value is either missing in these reports or it has high value which is expected from not too high conductivity values. Hence these kinds of electrolyte may be good for laboratory scale studies and/or understanding the science of materials/devices but they are not suitable for commercial applications.

Table 1. Percentage of Amylose and Amylopectin in some common starches

Type of Starch	% Amylopectin	%Amylose
Arrowroot	85	15
Corn	73	27
Potato	78	22
Rice	83	17
Tapioca	82	18
Wheat	76	24

Table 2. Recent studies on starch based electrolytes

Biopolymer electrolyte System	Conductivity [S/cm]	Reference No
Chitosan + starch (60:40)Glycerol as plasticizer (0%) LiClO4 +acetic acid and water	3.7×10^{-4}	40
Corn starch + LiPF6	3.21×10^{-4}	41
Corn starch+ LiOAc+ acetic acid	1.04×10^{-3}	42
Corn starch+ chitosan+ NH4Cl+acetic acid	5.11×10^{-4}	43
Corn starch+ fumed silica+ LiClO4+ distilled water	1.24	44
Starch (not mentioned which type)+ chitosan+ NH4Br+ acetic acid	9.72×10^{-5}	45
Potato starch+ methyl cellulose + glycerol+ NH4NO3+ acetic acid	1.26×10^{-3}	46
Corn starch+ inorganic filler BaTiO3+ LiClO4+ distilled water	1.28×10^{-2} (at 75°C)	47
Corn+LiClO$_4$	1.28×10^{-4}	48
Corn+LiI	1.83×10^{-4}	49
Corn+LiClO$_4$+SiO$_2$	1.23×10^{-4}	50
Corn+LiT$_f$SI	4.56×10^{-3} (at 50°C)	51

Potato+LiCf$_3$SO$_3$+glycerol	1.32×10^{-3}	52
Potato+LiClO$_4$+glycerol	4.25×10^{-4}	53
Potato+ magnesium acetate+ glycerol +BmImCl	1.12×10^{-5}	54
Corn+BmImPF$_6$	1.47×10^{-4}	55
Rice+LiI	4.68×10^{-5}	56
Agar + acetic acid	1.1×10^{-4}	38
Corn starch+chitosan+NH$_4$I	3.04×10^{-4}	39

3.   Starch-based electrolyte using glutaraldehyde as crosslinker

Starch being direct food for microorganism seems to be the main reason for it not being as popular and successful as it deserves. In the last decade extensive studies have been carried out in our laboratory and preparation protocol has been optimized to get flexible free-standing electrolyte films having high conductivity ($>10^{-2}$S/cm). The following discussion reviews starch-based polymer-salt complex electrolytes which have been synthesized by crosslinking with glutaraldehyde (GA).

3.1   Synthesis protocol of glutaraldehyde crosslinked starch-based electrolyte

Polymer-salt complexes have been prepared using solution cast technique where host polymer and salt are dissolved separately in a common solvent and then these are mixed and stirred for hours (>6 hours) to get a homogenous solution. These solutions are then left in Petri-dishes to dry. The same procedure is followed for starch-based electrolyte synthesis also. Since the target was to develop economical electrolyte, hence at initial stage the distilled water was selected as solvent, unfortunately the film was very brittle and fungal growth could be seen within a week [57]. Fig.1 depicts photograph of one such film having fungal growth on it. Hence other common solvents (methanol and acetone) were tried. Methanol and acetone chemically modify the starch [59] and hence the fungal growths were not seen in these systems but the material remains brittle and for electrical measurements pellets have to be prepared [58]. To address these problems glutaraldehyde (GA) was used as a crosslinker which resulted in formation of fungal free, flexible and free-standing films. As discussed in one of our papers [57] the GA engages same chemical moieties which are supposed to assist the transportation of ions. Hence amount of GA has two-fold effects on the material i) it improves the morphology and makes it fungus free and ii) it hinders the ion transport by engaging the hanging moieties. Therefore the optimization of its amount was essential and 2 mL of GA was found to be good enough for 1 gram of starch to have flexible free-standing long lasting electrolyte films with high conductivity. The starch with high amylose content needs a little higher amount of GA to get better electrochemical properties. The optimized synthesis process is schematically represented in Fig. 2a. Fig. 2b shows the photographs of stretching and bending behavior of synthesized films. Using the optimized procedure, many starch-based electrolyte films have been successfully prepared [57-67]. Table 3 summarizes the starches and salts combinations which have been studied by our group. Electrolytes having salt: starch ratio from 0.1:1 to 3:1 have been synthesized and characterized. It is a well-known fact that salt breaks the starch molecules and polymers with smaller molecules are more amorphous hence as the salt content increases the flexibility of starch-based electrolyte film increases and after a certain limit it is physically not robust enough for electrical characterization in solid form. Physicochemical and functional changes (including change of gelatinization temperature and breaking of starch molecules etc.) brought by addition of salt [68-71], make this system very interesting and important as it gives us flexibility to design the desired morphology.

Table 3. Starch-Salt combinations explored by our group

Starches	Salts
Potato starch	NaI, NaSCN, NaClO$_4$, NH$_4$I
Arrowroot starch	NaI, NaSCN, NaClO$_4$, NH$_4$I
Corn starch	NaI, NaSCN, NaClO$_4$, NH$_4$I, MgClO$_4$
Rice starch	NaI, MgClO$_4$ MgCl$_2$
Wheat starch	NaI

Fig.1 Photograph of starch film after fungal growth

The synthesized electrolytes can be divided into two groups i) salt-in-polymer electrolytes (SIPE) where the amount of salt added is less than the amount of starch and ii) polymer-in-salt electrolytes (PISE) where the amount of salt is greater than the polymer. For flexible supercapacitors, starch-based PISEs are very promising electrolytes as these have good flexibility and can be easily twisted, bent and stretched. The developed synthesis protocol has many advantages as it does not require any sophisticated instrument and the electrolyte in different shapes and sizes can be easily prepared. The approximate cost of the 1x1 cm^2 electrolyte of 0.7 mm thickness is about Rs.15. They are very fascinating from a designing point-of-view as well as they can be molded in different shape and size by pouring the solution in specific containers. They can easily be cut into different shapes and sizes. This protocol also facilitates different procedures (such as dip coating, pouring the electrolyte on electrode, spin coating, etc.) to get good electrode/electrolyte contacts.

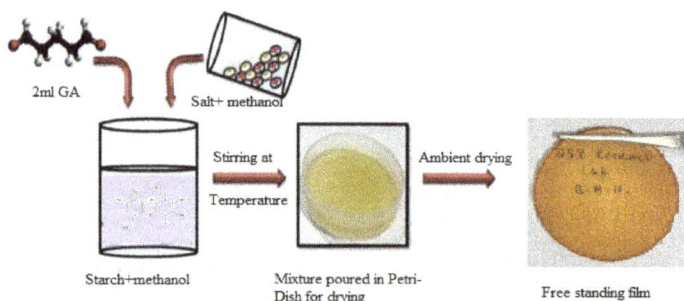

Fig.2a Schematic presentation of GA crosslinked starch based electolyte systems

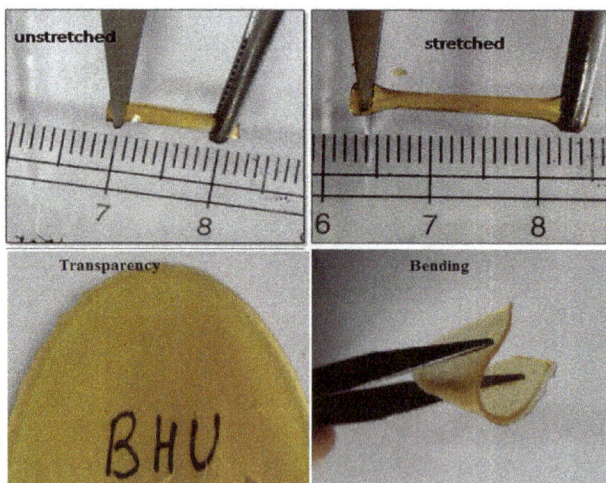

Fig.2b. Photographs showing the tranperancy and flexibility of GA crosslinked electrolytes

3.2   Electrochemical behavior of GA crosslinked starch electrolytes

The electrochemical behavior of GA crosslinked starch electrolytes is also very promising. Conductivity of salt-in-polymer electrolytes (SIPE) has been $\leq 10^{-4}$S/cm whereas polymer-in-salt electrolytes (PISE) are more conducting and their conductivity reached to $>10^{-2}$S/cm. Table 4 lists a comparative conductivity data for different salts with potato starch in SIPE range. Despite the fact that the electrolytes are quite stable and free-standing films, because of poor conductivity, these cannot be used in any device without further modification with fillers/plasticizers/ionic liquid, etc. The value of resistance has great relevance in case of supercapacitors because it directly controls the power density hence even a very small difference in conductivity when studied in terms of the resistance seems to be considerable big for supercapacitor applications. In the present series of materials quite small value of resistance ($<2\Omega$) has been achieved for the studied dimension i.e. 1x1 cm^2 area with 0.07 cm thickness. As discussed before these electrolytes can be easily molded in different shapes and sizes hence it is obvious that using different techniques such as dip coating etc., resistance can be further lowered to a great extent resulting in tremendous increase in power density. Table 5 sums different synthesized materials, their conductivity and ESR values.

Table 4.Conductivity of potato starch based electrolyte systems having different sodium salts in SIPE concentration range

Salt/starch ratio	σ (S/cm) (NaI)	σ (S/cm) (NaSCN)	σ (S/cm) (NaClO$_4$)
0.5:1	4.79×10^{-6}	3.75×10^{-6}	7.10×10^{-6}
0.1:1	3.59×10^{-6}	3.67×10^{-6}	4.59×10^{-6}
0.15:1	4.43×10^{-6}	2.51×10^{-6}	4.44×10^{-6}
0.2:1	3.57×10^{-6}	3.82×10^{-6}	4.72×10^{-6}
0.25:1	1.58×10^{-5}	4.34×10^{-6}	3.07×10^{-6}
0.30:1	2.99×10^{-6}	1.83×10^{-5}	4.07×10^{-6}
0.35:1	4.94×10^{-6}	4.63×10^{-6}	3.23×10^{-6}
0.40:1	5.17×10^{-5}	5.58×10^{-6}	3.93×10^{-6}
0.45:1	1.25×10^{-5}	4.71×10^{-6}	7.72×10^{-6}
0.50:1	6.72×10^{-6}	1.10×10^{-4}	3.42×10^{-6}

Table 5. *Electrochemical parameters of studied starch-salt combination having different salt concentration*

Table 5a. (Corn starch + NaClO₄)					
Salt/starch ratio	ESR [Ω]	ESW [V]	Conductivity [S/cm]	τ [μs]	f_r [Hz]
0.6:1	38.82	3.67	1.94×10^{-3}	268	471600
0.7:1	16.09	3.47	4.63×10^{-3}	167	294100
0.8:1	11.62	2.89	6.22×10^{-3}	73	261300
0.9:1	6.24	2.18	1.47×10^{-2}	65	114400
1:1	6.34	2.55	1.00×10^{-2}	70	206400

Table 5b. (Arrowroot starch +NaClO₄)					
Salt/starch ratio	ESR [Ω]	ESW [V]	Conductivity [S/cm]	τ [μs]	f_r [Hz]
0.6:1	38.21	2.93	1.03×10^{-3}	321	107800
0.7:1	21.31	2.63	2.11×10^{-3}	132	530700
0.8:1	21.26	2.47	3.26×10^{-3}	117	471600
0.9:1	6.73	2.46	9.03×10^{-3}	68	206400
1:1	7.14	2.57	5.60×10^{-3}	75	261300

Table 5c. (Corn starch+ Mg(ClO₄)₂)					
Salt/starch ratio	ESR [Ω]	ESW [V]	Conductivity [S/cm]	τ [μs]	f_r [Hz]
0.6:1	101.79	3.88	5.59×10^{-4}	167	1070000
0.7:1	37.79	3.73	1.11×10^{-3}	132	957600
0.8:1	19.02	3.47	3.24×10^{-3}	65	419100
0.9:1	6.03	3.35	6.02×10^{-3}	54	206400
1:1	10.35	3.42	5.57×10^{-3}	58	261300

Table 5c. (Corn starch+ Mg(ClO₄)₂)					
Salt/starch ratio	ESR [Ω]	ESW [V]	Conductivity [S/cm]	τ [μs]	f_r [Hz]
0.6:1	101.79	3.88	5.59×10^{-4}	167	1070000
0.7:1	37.79	3.73	1.11×10^{-3}	132	957600
0.8:1	19.02	3.47	3.24×10^{-3}	65	419100
0.9:1	6.03	3.35	6.02×10^{-3}	54	206400
1:1	10.35	3.42	5.57×10^{-3}	58	261300

Table 5f. (Arrowroot starch+ NaI)					
Salt/starch ratio	ESR [Ω]	ESW [V]	Conductivity [S/cm]	τ [μs]	f_r [Hz]
0.6:1	31.17	2.44	2.11×10^{-3}	301	471600
0.7:1	18.45	2.35	7.30×10^{-3}	188	330900
0.8:1	4.91	2.08	1.08×10^{-3}	104	144800

| 0.9:1 | 3.02 | 2.03 | 1.89×10^{-2} | 17 | 114400 |
| 1:1 | 3.92 | 1.87 | 1.72×10^{-2} | 14 | 183400 |

Table 5g. (Wheat starch + NaI)

Salt/starch ratio	ESR [Ω]	ESW [V]	Conductivity [S/cm]	τ [μs]	f_r [Hz]
0.8:1	38	2.6	1.55×10^{-3}	--	252090
1:1	3.5	2.4	1.63×10^{-2}	15	160454
2:1	1.5	2.2	3.57×10^{-2}	13	106972
3:1	1.1	2.4	3.61×10^{-2}	13	92017

Small electrochemical stability window (ESW) of liquid electrolytes is their major drawback which affects the energy density as well as the power density. Although the use of ionic liquids results in quite a wider ESW but they are not economical and hence have limitation for commercial use. The GA crosslinked starch electrolytes have advantage as they have wider ESW (>2V which reached up to 4.5V in many systems). Here it is obvious to see that with increased conductivity ESW may decrease and vice versa. A novice to the field of supercapacitor should go through the literature with caution as many times researchers have reported a wide ESW value without commenting on the ESR value of the system. Such systems are useless because the effective power density will be reduce to a great extent. For example, in present series itself the magnesium salt containing electrolytes have wider ESW (as summarized in table 5) but the ESR values are high. In case of GA crosslinked starch-salt electrolytes, ESW and ESR strongly depend upon the type and amount of salt used hence electrochemical parameters can be modified by selecting proper salt and its concentration in particular starch. The ESW and ESR values are comparable (in some cases better) to other electrolytes as reported in literature. Table 6 illustrates an approximate comparison of different electrolyte systems.

Charging or discharging speed of devices depends upon the ion relaxation time which has two parts i) time taken by the ions to move in the electrolyte and reach up to the electrodes and ii) the time taken by the ions in going inside the pores of the electrodes. Of course the second one dominantly depends upon the nature and structure of the electrode but the first one is solely decided by the electrolyte. This parameter is estimated from the plot of imaginary capacitance vs. frequency curves [65, 67]. The estimated values for Al/electrolyte/Al cell are summarized in Table 5. Since the Al electrode is non-porous flat blocking electrode hence the estimated relaxation time is basically associated with movement of ions in electrolyte. This value is of the order of micro seconds and it varies with the type and amount of salt used. It is quite quick, indicating that starch is a favorable matrix for fast ion transportation.

Table 6:A Comparison of electrochemical parameters of GA crosslinked starch electrolytes with other common electrolytes

Parameters	Ionic liquid	Hydrogel electrolyte	Liquid electrolytes	Other solid polymer electrolyte	GA crosslinked starch electrolytes
ESW	3V-4V	1.23V	1V-1.3V	2.2V-2.7V	2.5V-4.5V
ESR(ohm)	3-8	10-15	3-10	20-40	<2
Quality of film	Gel type	Gel type	Liquid type	Free standing films	Flexible, twistable, bendable, stretchable free standing films
Relaxation time	5s	1.4s	0.5s	NA	14μs-70μs
σ(S/cm)	10^{-4}-10^{-2}	10^{-3}	10^{-4}-10^{-3}	10^{-3}	10^{-3}-10^{-2}
f_r	1.2Hz	1Hz	1Hz	NA	10kHz

After finalizing the synthesis protocol and understanding the changes in electrochemical parameters with respect to different salts and the salt concentration, its use in electrochemical devices has been tested. The supercapacitors have been fabricated with two different electrode systems namely teflon coated carbon cloths and anodized aluminum. The fabricated devices [67] have electrochemical double layer capacitance type behavior and capacitance ~50 mF/cm^2 has been achieved at 5V/sec scan rate, with good cycling stability. The device can be recycled easily at 10000V/sec scan rate, indicating that ions movement is very fast and easy leading to fast charging/discharging devices. The time constant for devices remained in micro seconds to few msec confirming the fast charging/ discharging ability of fabricated devices. The Columbic efficiency of the devices was nearly 97% and the parallel capacitance to series capacitance ratio (C_p/C_s) remains closure to 1 for a wide range of frequency indicating good polarizability of the devices. In addition, starch chains provide sufficient intramolecular and intermolecular hydrogen bonding, which pave the way for self-healing performance for changing volume at interface during charging/discharging [5].

Summary

The advancement in flexible electronic technology demands cost-effective and flexible energy storage systems. Hence research and development in the area of flexible electrolytes have a increased tremendously. Generally, flexible electrolytes are developed

Supercapacitor Technology: Materials, Processes and Architectures   Materials Research Forum LLC
Materials Research Foundations **61** (2019) 121-140   https://doi.org/10.21741/9781644900499-6

by soaking a suitable solid matrix in liquid electrolytes but they have many inherent limitations and hence the need of flexible solid electrolytes are felt. In this regard, the GA crosslinked starch-based electrolytes seem to be a promising alternative. The ESW of these electrolytes are comparable to other known electrolytes and the ESR value and relaxation time are quite low. These electrolytes are easy to synthesize in different shapes and sizes. The technical parameters of these electrolytes can be easily manipulated, as per requirement, by using different chemical constitutes and by varying their concentration, and this ease adds up to their advantages. Starch is a cost-effective and renewable polymer hence these electrolytes have an edge over other electrolytes.

Acknowledgement

Some of the work summarized here were carried out through two projects i) "Synthesis & Electrical Characterization of Starch-based Electrolyte Systems" project sanction no 42-814/2013 (SR) dated 22.03.2013 University Grant Commission (New Delhi) and ii) "Study of Potato starch and Magnesium salt based biodegradable Polymer Electrolyte systems" project sanction no CST/D-6173 dated 16/02/2017Council of Science and Technology, U.P. Author is thankful to both the funding agency for financial support. Author is also thankful to Ms Madhavi Yadav (PhD student) for helping in manuscript preparation.

References

[1] A. Nathan, A. Ahnood, M. T. Cole, S. Lee, Y. Suzuki, P. Hiralal, F. Bonaccorso, T. Hasan, L. Garcia-Gancedo, A. Dyadyusha, S. Haque, P. Andrew, S. Hofmann, J. Moultrie, D. Chu, A. J. Flewitt, A. C. Ferrari, M. J. Kelly, J. Robertson, G. A. J. Amaratunga, W. I. Milne, Flexible electronics: The next ubiquitous platform, Proceedings of the IEEE 100 (2012) 1486-1517. https://doi.org/10.1109/JPROC.2012.2190168.

[2] X. Wang, K. Jiang, G. Shen, Flexible fiber energy storage and integrated devices: recent progress and perspectives, Mater. Today 18 (2015) 265-272. http://dx.doi.org/10.1016/j.mattod.2015.01.002.

[3] C. Wang, G. G. Wallace, Flexible electrodes and electrolytes for energy storage, Electrochim. Acta 175 (2015) 87-9. http://dx.doi.org/10.1016/j.electacta.2015.04.067.

[4] L. Dong, C. Xu, Y. Li, Z. Huang, F. Kang, Q. Yang, X. Zhao, Flexible electrodes and supercapacitors for wearable energy storage: A review by category, J. Mater. Chem. A 4 (2016) 4659-4685. http://dx.doi.org/10.1039/C5TA10582J.

[5] Q. Xue, J. Sun, Y. Huang, M. Zhu, Z. Pei, H. Li, Y.Wang, N. Li, H. Zhang, C. Zhi, Recent progress on flexible and wearable supercapacitors, Small 1701827 (2017) 1-11. https://doi.org/10.1002/smll.201701827.

[6] S. Kiruthika, C. Sow, G. U. Kulkarni, Transparent and flexible supercapacitors with networked electrodes, Small 1701906 (2017) 1-9. https://doi.org/10.1002/smll.201701906.

[7] C. Zhong, Y. Deng, W. Hu, J. Qiao, L. Zhangd, J. Zhang, A review of electrolyte materials and compositions for electrochemical supercapacitors, Chem. Soc. Rev. 44 (2015) 7484-7539. http://dx.doi.org/10.1039/C5CS00303B.

[8] X. Li, J.Shao, S. Kim, C. Yao, J. Wang, Y.Miao, Q. Zheng, P. Sun, R. Zhang, P. V. Braun, High energy flexible supercapacitors formed via bottom-up infilling of gel electrolytes into thick porous electrodes, Nat. Commun. 2578 (2018) 1-8. http://dx.doi.org/10.1038/s41467-018-04937-8.

[9] L. V. Thekkekara, X. Chen, M. Gu, Two-photon-induced stretchable graphene supercapacitors, Sci. Rep. 11722 (2018) 1-9. http://dx.doi.org/10.1038/s41598-018-30194-2.

[10] L. Basirico, G. Lanzara, Moving towards high-power, high-frequency and low-resistance CNT supercapacitors by tuning the CNT length, axial deformation and contact resistance, Nanotechnology, 23 (2012) 1-13. http://dx.doi.org/10.1088/0957-4484/23/30/305401.

[11] G. Wang, L. Zhang, J. Zhang, A review of electrode materials for electrochemical supercapacitors, Chem. Soc. Rev. 41 (2012) 797–828. http://dx.doi.org/10.1039/C1CS15060J.

[12] T. Purkait, G. Singh, D. Kumar, M. Singh, R. S. Dey, High-performance flexible supercapacitors based on electrochemically tailored three-dimensional reduced graphene oxide networks, Sci. Rep. 640 (2018) 1-13. http://dx.doi.org/10.1038/s41598-017-18593-3.

[13] S. Y. Liew, W. Thielemans, S. Freunberger, S. Spirk, Polysaccharides in Supercapacitors, in: Polysaccharide Based Supercapacitors, Springer Briefs in Molecular Science, Springer, Cham, 2017, 15-53. https://doi.org/10.1007/978-3-319-50754-5_2.

[14] P. K. Jha, S. K. Singh, V. Kumar, S. Rana, S. Kurungot, N. Ballav, High-level supercapacitive performance of chemically reduced graphene oxide, Chem 3 (2017) 846- 860. http://dx.doi.org/10.1016/j.chempr.2017.0.

[15] B. K. Kim, S. Sy, A. Yu, J. Zhang, Electrochemical Supercapacitors for Energy Storage and Conversion, in: Handbook of Clean Energy Systems, John Wiley & Sons, Ltd.2015, pp.1-25. https://doi.org/10.1002/9781118991978.hces112.

[16] R. Zhang, J. Ding, C. Liu, E.Yan, Highly stretchable supercapacitors enabled by interwoven CNTs partially embedded in pdms, ACS Appl. Energy Mater. 5 (2018) 2048–2055. http://dx.doi.org/10.1021/acsaem.8b00156.

[17] G.P. Pandey, A.C. Rastogi, C. R. Westgate, All-solid-state supercapacitors with poly(3,4-ethylenedioxythiophene)- coated carbon fiber paper electrodes and ionic liquid gel polymer electrolyte J. Power Sources 245 (2014) 857-865. https://doi.org/10.1016/ j.jpowsour.2013.07.017.

[18] F. Rafik, H. Gualous, R. Gallay, A. Crausaz, A. Berthon, Frequency, thermal and voltage supercapacitor characterization and modeling, J. Power Sources 165 (2007) 928–934. https://doi.org/10.1016/j.jpowsour.2006.12.021.

[19] J. Xue, Y. Zhao, H.Cheng, C. Hu, Y. Hu, Y. Meng, H. Shao, Z.Zhang, L.Qu, An all-cotton-derived, arbitrarily foldable, high-rate, electrochemical supercapacitor, Phys. Chem. Chem. Phys., 2013, 15, 8042; https://doi.org/10.1039/C3CP51571K.

[20] M. Armand, Polymer electrolytes, Ann. Rev. Mater. Sci. 16 (1986) 245-261. https://doi.org/10.1146/annurev.ms.16.080186.001333.

[21] C.W. Liew, K.H.Arifin, J.Kawamura, Y.Iwai, S.Ramesh, A.K.Arof, Effect of halide anions in ionic liquid added poly(vinyl alcohol) based ion conductors for electrical double layer capacitor, J. Non-Cryst. Solids 458 (2017) 97-106. https://doi.org/10.1016/j.jnoncrysol.2016.12.022.

[22] C.C. Yang, S. Chiu, S.J. Kuo, S.Ch., Preparation of poly(vinyl alcohol)/montmrillonile /poly(styrene sulfonic acid) composite membrane for hydrogen–oxygen polymer electrolyte fuel cell. Current Applied Physics 11(2011) S229-S237.

[23] F. Deng, X. Wang, D. He, J. Hu, C. Gong, Y. S.Ye, X. Xie, Z. Xue, Microporous polymer electrolyte based on PVdF/PEO star polymer blends for lithium ion batteries. J. Membrane Sci. 491 (2015) 82-89. https://doi.org/10.1016/ j.memsci.2015.05.021.

[24] S. Rajendran, O. Mahendran, R. Kannan, Characterisation of $[(1-x)$ PMMA–xPVdF] polymer blend electrolyte with Li^+ ion, Fuel 81 (2002) 1077-1081. https://doi.org/10.1016/S0016-2361(01)00178-8.

[25] S. B. Aziz, T.J. Woo, M.F.Z. Kadir, H.M. Ahmed, A conceptual review on polymer electrolytes and ion transport models, J. Sci. Adv. Mater. Devices 3 (2018) 1-17. https://doi.org/10.1016/j.jsamd.2018.01.002.

[26] R. C. Agrawal, G.P. Pandey, Solid polymer electrolytes: Materials designing and all-solid-state battery applications: An overview, J. Phys. D: Appl. Phys. 41 (2008) 1-18. https://doi.org/10.1088/0022-3727/41/22/223001.

[27] A.B. Samui, P. Sivaraman Solid polymer electrolytes for supercapacitors, in: C. Sequeira, D. Santos (Eds.), Polymer Electrolytes: Fundamentals and Applications, Woodhead Publishing Series in Electronic and Optical Materials, 2010, pp. 431-470.

[28] J. Qiao, J. Zhang, Electrolytes for electrochemical supercapacitors, editors: C. Zhang, Y. Deng, W. Hu, D. Sun, X. Han, CRC press, 2016, ISBN 13:978-1-4987-4757-8 e-book

[29] D. F.Vieira, C. O.Avellaneda, A. Pawlicka, Conductivity study of a gelatin-based polymer electrolyte, Electrochim. Acta 53 (2007) 1404-1408. https://doi.org/10.1016/j.electacta.2007.04.034.

[30] Y. Wu, F. Geng, P. R.Chang, J.Yu, X. Ma, Effect of agar on the microstructure and performance of potato starch film. Carbohyd. Polym. 76 (2009) 299-304. https://doi.org/10.1016/j.carbpol.2008.10.031

[31] T. Basu, M. M. Goswami, T. R. Middya, S. Tarafdar, Morphology and ion-conductivity of gelatin–$LiClO_4$ films: Fractional diffusion analysis, J. Phys. Chem. B 116 (2012) 11362–11369. https://doi.org/10.1021/jp306205h.

[32] X. Kang, J. Wang, Z. Tang, H. Wu,Y. Lin, Direct electrochemistry and electrocatalysis of horseradish peroxidase immobilized in hybrid organic–inorganic film of chitosan/sol–gel/carbon nanotubes, Talanta 78 (2009) 120-125. https://doi.org/10.1016/j.talanta.2008.10.063.

[33] R. I. Mattos, C. E. Tambelli, J. P. Donoso, R. G. F. Costa & A. Pawlicka, NMR and Conductivity Study of Starch Based Polymer Gel Electrolytes, Mol. Cryst. Liq. Cryst. 447 (2006) 55-64. https://doi.org/10.1080/15421400500385308.

[34] V.L. Finkenstadt, Natural polysaccharides as electroactive polymers, Appl. Microbiol. Biotechnol. 67 (2005) 735-745. https://doi.org/10.1007/s00253-005-1931-4.

[35] A.S.A. Khiar, A.K. Arof, Conductivity studies of starch-based polymer electrolytes, Ionics 16 (2010)123-129. https://doi.org/10.1007/s11581-009-0356-y.

[36] R.I.Mattos, C.E.Tambelli, J.P.Donoso, A.Pawlicka, NMR study of starch based polymer gel electrolytes: Humidity effects, Electrochim. Acta 53 (2007)1461-1465. https://doi.org/10.1016/j.electacta.2007.05.061

[37] V.L. Finkenstadt, J.L. Willett, Electroactive materials composed of starch, J. Polym. Environ. 12 (2004) 43-46. https://doi.org/10.1023/B:JOOE.0000010049.33284.08.

[38] E. Raphael, C. O. Avellaneda, B. Manzoli, A. K. Pawlika, Agar-based films for application as polymer electrolyte, Electrochim. Acta 55 (2010) 1455-1459. https://doi.org/10.1016/j.electacta.2009.06.010.

[39] Y.M. Yusof, M.F. Shukur, H.A. Illias, M.F.Z. Kadir, Conductivity and chemical properties of corn starch-chitosan blend biopolymer electrolyte incorporated with ammonium iodide, Phys. Scr. 89 (2014) 035701. http://dx.doi.org/10.1088/0031-8949/89/03/035701.

[40] Y.N. Sudhakar, M. Selvakumar, Lithium perchlorate doped plasticized chitosan and starch blend as biodegradable polymer electrolyte for supercapacitors, Electrochim. Acta 78 (2012) 398–405. https://doi.org/10.1016/j.electacta.2012.06.032.

[41] C.W. Liew, S. Ramesh, Comparing triflate and hexafluorophosphate anions of ionic liquids in polymer electrolytes for supercapacitor applications, Materials 7 (2014) 4019-4033. https://doi.org/10.3390/ma7054019.

[42] M.F. Shukur, R. Ithnin, M.F.Z. Kadir, Electrical characterization of corn starch-LiOAc electrolytes and application in electrochemical double layer capacitor, Electrochim. Acta 136 (2014) 204-216. https://doi.org/10.1016/j.electacta.2014.05.075.

[43] M.F. Shukur, M.F.Z. Kadir, Hydrogen ion conducting starch-chitosan blend based electrolyte for application in electrochemical devices, Electrochim.Acta 158 (2015) 152-165. https://doi.org/10.1016/j.electacta.2015.01.167.

[44] K. H. Teoh, Chin-Shen Lim, Chiam-Wen Liew, S. Ramesh, S. Ramesh, Electric double-layer capacitors with corn starch-based biopolymer electrolytes incorporating silica as filler, Ionics 21 (2015) 2061–2068. https://doi.org/10.1007/s11581-014-1359-x.

[45] M.F. Shukur, R. Ithnin, M.F.Z. Kadir, Protonic transport analysis of starch-chitosan blend based electrolytes and application in electrochemical device, Mol. Cryst. Liq. Cryst. 603 (2014), 12th International Conference on Polymers and Advanced Materials (ICFPAM 2013). https://doi.org/10.1080/15421406.2014.966259.

[46] M.H. Hamsan, M.F. Shukur, M.F.Z. Kadir, NH_4NO_3 as charge carrier contributor in glycerolized potato starch-methyl cellulose blend-based polymer electrolyte and the application in electrochemical double-layer capacitor, Ionics 23 (2017) 3429-3453. https://doi.org/10.1007/s11581-017-2155-1.

[47] K.H. Teoh, C.S. Lim, C.W. Liew, S. Ramesh, Preparation and performance analysis of barium titanate incorporated in corn starch-based polymer electrolytes for electric double layer capacitor application, J. Appl. Polym. Sci. 133 (2016) 1-18. https://doi.org/10.1002/app.43275.

[48] K.H. Teoh, C.S. Lim, S. Ramesh, Lithium ion conduction in corn starch based solid polymer electrolytes. Measurement 48 (2014) 87-95. https://doi.org/10.1016/j.measurement.2013.10.040.

[49] M.F. Shukur, F.M. Ibrahim, N.A. Majid, R. Ithnin, M.F.Z. Kadir, Electrical analysis of amorphous corn starch-based polymer electrolyte membrane doped with LiI, Phys. Scr. 88 (2013) 1-9. http://stacks.iop.org/PhysScr/88/025601.

[50] K.H. Teoh, C.S. Lim, C.W. Liew, S. Ramesh, S. Ramesh, Elecrical double-layer capacitors with corn starch-based biopolymer electrolytes incorporating silica as filler, Ionics 21 (2015) 2061-2068. https://doi.org/10.1007/s11581-014-1359-x.

[51] S. Ramesh, R. Shanti, E. Morris, Exerted influence of deep eutectic solvent concentration in the room temperature ionic conductivity and thermal behaviour of corn starch based polymer electrolytes, J. Mol. Liq. 166 (2012) 40-43. https://doi.org/10.1016/j.molliq.2011.11.010.

[52] N.N.A. Amran, N.S.A. Manan, M.F.Z. Kadir, The effect of $LiCF_3SO_3$ on the complexation with potato starch-chitosan blend polymer electrolytes, Ionics 22 (2016)1647-1658. https://doi.org/10.1007/s11581-016-1684-3.

[53] Y.M. Yusof, M.F.Z. Kadir, Electrochemical characterizations and the effect of glycerol in biopolymer electrolytes based on methylcellulose-potato starch blend. Mol. Cryst. Liq. Cryst. 627 (2016) 220-233. https://doi.org/10.1080/15421406.2015.1137115.

[54] M.F. Shukur, R. Ithnin, M.F.Z. Kadir, Ionic conductivity and dielectric properties of potato starch-magnesium acetate biopolymer electrolytes: the effect of glycerol and 1-butyl-3-methylimidazolium chloride, Ionics 22 (2016)1113-1123. https://doi.org/ 10.1007/s11581-015-1627-4

[55] S. Ramesh, C.W. Liew, A. K. Arof, Ion conducting corn starch biopolymer electrolytes doped with ionic liquid 1-butyl-3-methylimidazolium

hexafluorophosphate, J. Non-Cryst. Solids 357 (2011)3654-3660.
https://doi.org/10.1016/j.jnoncrysol.2011.06.030.

[56] M.H. Khanmirzaei, S. Ramesh, Ionic transport and FTIR properties of lithium iodide doped biodegradable rice starch based polymer electrolytes, Int. J. Electrochem. Sci. 8 (2013) 9977-9991.

[57] T. Tiwari, K. Pandey, N. Srivastava, P.C. Srivastava, Effect of glutaraldehyde on electrical properties of arrowroot starch + NaI electrolyte system, J. Appl. Polym. Sci. 121(2011) 1-7. https://doi.org/10.1002/app.33559.

[58] T. Tiwari, M. Kumar, N. Srivastava, P.C. Srivastava, Electrical transport study of potato starch-based electrolyte system II, Mater. Sci. Eng. B 182 (2014) 6-13. https://doi.org/10.1016/j.mseb.2013.11.010.

[59] T. Tiwari, N. Srivastava, P.C. Srivastava, Electrical transport study of potato starch based electrolyte system, Ionics 17 (2011) 353-360. https://doi.org/ 10.1007/s11581-010-0516-0.

[60] M. Kumar, T.Tiwari, N. Srivastava , Electrical transport behaviour of bio-polymer electrolyte system: Potato starch+ammonium iodide, Carbohydr. Polym. 88 (2012) 54-60. https://doi.org/10.1016/j.carbpol.2011.11.059.

[61] T. Tiwari, M. Kumar, N. Srivastava, Study of arrowroot starch based polymer electrolytes and its application in MFC, (2019). http://doi.wiley.com/10.1002/star.201800313

[62] M. Yadav, G. Nautiyal, A. Verma, M. Kumar,T. Tiwari, N. Srivastava, Electrochemical characterization of $NaClO_4$–mixed rice starch as a cost-effective and environment-friendly electrolyte. (Accepted in Ionics) https://doi.org/10.1007/s11581-018-2794-x.

[63] J.K. Chauhan, M. Kumar, M. Yadav, T. Tiwari, N. Srivastava, Effect of $NaClO_4$ concentration on electrolytic behaviour of corn starch film for supercapacitor application, Ionics 23 (2017) 2943–2949. https://doi.org/10.1007/s11581-017-2136-4.

[64] T. Tiwari, J. K. Chauhan, M. Yadav, M. Kumar, N. Srivastava, Arrowroot+ NaI: A low cost, fast ion conducting eco-friendly polymer electrolyte system, Ionics 23 (2017) 2809–2815. https://doi.org/10.1007/s11581-017-2028-7.

[65] M. Yadav, M. Kumar, T.Tiwari, N. Srivastava, Wheat Starch + NaI: A high conducting environment friendly electrolyte system for energy devices, Ionics 23 (2017) 2871-2880. https://doi.org/10.1007/s11581-016-1930-8.

[66] T. Tiwari, N. Srivastava, P.C. Srivastava, Ion dynamics study of potato starch + sodium salts electrolyte system, Int. J. Electrochem. 2013 (2013) 1-8. http://dx.doi.org/10.1155/2013/670914.

[67] M. Yadav, M. Kumar, N. Srivastava, Supercapacitive performance analysis of low cost and environment friendly potato starch based electrolyte system with anodized aluminium and teflon coated carbon cloth as electrode, Electrochim. Acta 283 (2018) 1551-1559. https://doi.org/10.1016/j.electacta.2018.07.060.

[68] A. Railanmaa, S. Lehtimäki, D. Lupo, Comparison of starch and gelatin hydrogels for non-toxic supercapacitor electrolytes, Appl. Phys. A 459 (2017) 1-8. https://doi.org/10.1007/s00339-017-1068-1.

[69] A.C. Eliasson, Starch: Physicochemical and Functional Aspects, in: A.C. Eliasson (Eds.) Carbohydrates in Food, Marcel Decker Inc., New York 1996, pp. 431-503.

[70] Y. Chen, C. Wang, T. Chang, L. Shi, H. Yang, M. Cui, Effect of salts on textural, color, and rheological properties of potato starch gels, Starch/Stärke 65 (2013) 1-8. https://doi.org/10.1002/star.201300041.

[71] V.K.Villwock, J. N. BeMiller, Effects of salts on the reaction of normal corn starch with propylene oxide, Starch/Stärke 57 (2005) 281–290. https://doi.org/10.1002/star.200400384.

Supercapacitor Technology: Materials, Processes and Architectures   Materials Research Forum LLC
Materials Research Foundations **61** (2019) 141-170   https://doi.org/10.21741/9781644900499-7

Chapter 7

Pseudocapacitors

P.M. Anjana[1] and R.B. Rakhi[1,2,*]

[1]Chemical Sciences and Technology Division, CSIR- National Institute of Interdisciplinary Science and Technology (CSIR-NIIST), Thiruvananthapuram, Kerala, India, 695019

[2]Department of Physics, University of Kerala, Thiruvananthapuram, Kerala, India, 695019

*rakhiraghavanbaby@niist.res.in, rbrakhi@keralauniversity.ac.in

Abstract

Supercapacitors are expected to be the integral parts of power devices in a broad range of future applications. Supercapacitors must deliver high energy and power densities, safe mode of operation, and long cycle life, for their widespread practical applications. A meaningful way to achieve these parameters is to develop advanced electrode materials, which can provide high specific capacitance values. Pseudocapacitive materials are emerging as favorable candidates as electrode materials for supercapacitors, as they are capable of promoting the capacitance of the device by the surface Faradic redox reactions. The present chapter starts with a brief overview of electrochemical capacitors followed by the description of different energy storage mechanisms contributing to pseudocapacitive charge storage. This chapter also covers the details of different types of pseudocapacitive electrode materials. Making hybrid composites of carbon and pseudocapacitive material is a promising approach to overcome the poor cycle life of pseudocapacitors.

Keywords

Pseudocapacitors, Energy Density, Cycle Life, Electrode Materials, Hybrid Composites

Contents

1.   Introduction

Supercapacitors or ultracapacitors have attracted the energy storage industry over the last decade due to their high power density and outstanding cyclic stability [1, 2]. Supercapacitors are capable of providing higher energy densities than dielectric capacitors and greater power densities than batteries [3]. Supercapacitor possesses many desirable qualities that make them an exciting energy storage option. These devices are having excellent cycle life as the charge storage mechanism is completely reversible. Supercapacitors are devices that can store and release electric charge much more quickly [4, 5]. A Supercapacitor consists of two electrodes, separated by an insulating separator soaked in an electrolyte. Depending on the charge storage mechanism, capacitors can be classified into two types: EDLC (Electrical Double Layer Capacitors) and Faradaic pseudocapacitors. In EDLCs, the electrostatic charge storage occurs across the electrode-electrolyte interface [6]. The electrode materials in EDLC are mainly carbon-based. The second type is pseudocapacitors, in which surface Faradaic reactions on electrodes also contribute to the total capacitance along with the electrostatic capacitance. Transition metal oxides, conducting polymers, and transition metal chalcogenides are widely used as the electrode materials in pseudocapacitors. [5, 7-9]. The pseudocapacitive electrodes accumulate charges via both electric double layer and Faradaic electrochemical process, which allow pseudocapacitors to achieve higher capacitance than EDLCs [10-13].

There are mainly three different types of Faradaic mechanisms which contribute pseudocapacitive characteristics; (1) underpotential deposition, (2) redox pseudocapacitance and (3) intercalation pseudocapacitance. Fig. 1 illustrates these processes. In electrodeposition process, "Nernst" potential or the equilibrium potential refers to the potential at which a metal will deposit onto itself. In underpotential deposition, the electrodeposition of a metal occurs at a potential less negative than the Nernst potential and the metal can be easily deposited on the surface of another material that onto itself.

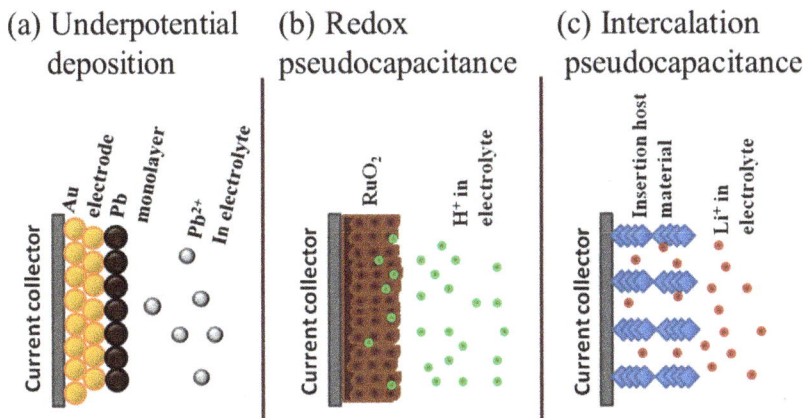

Fig.1 Different types of Faradaic mechanisms contributing to pseudocapacitance

Fig. 1 (a) demonstrates the electrosorption of Pb^{2+} ions on Au electrode, which can further be represented using the following equation [14];

$$2e + xPb^{2+} + Au \leftrightarrow Au. xPb$$

The concentration of anions in the electrolyte determines resulting surface structure and reaction rate.

The redox pseudocapacitance mechanism is purely Faradaic, the surface redox reactions on the electrode lead to the charge transfer between electrodes and electrolyte. Fig. 1(b) shows the pseudocapacitive charge storage taking place at ruthenium oxide electrode by the absorption and release of protons from the electrolyte [15]. In Intercalation pseudocapacitance, electrolyte ions intercalate into the space between the layers of a pseudocapacitive electrode material. The process is accompanied by a surface Faradaic charge-transfer, retaining the crystallographic phase of the electrode. The morphology and crystal structure of the electrode materials play a crucial role in all these three mechanisms.

2.   Redox pseudocapacitive electrode materials

Transition metal oxides, hydroxides, nitrides, and sulfides, and conducting polymers are the most popular electrode materials for pseudocapacitors. Except for conducting polymers, all other materials have poor electrical conductivity and low electrolyte accessible surface area. To overcome this, high surface area pseudocapacitive materials need to be developed by controlling the surface morphology and microstructure. The conductivity can be improved by either the direct growth of these materials on conducting substrates or making composite electrode materials by combining them with conducting materials.

2.1   Transition metal oxides

Transition metal oxides (TMO) are the commonly used electrode materials for pseudocapacitors [16, 17]. Ruthenium oxide (RuO_2) is the most promising TMO electrode material due to its high theoretical capacitance, excellent conductivity, good electrochemical reversibility, high rate capability, and long cycle life [18]. The electrochemical charge storage properties of RuO_2 electrodes in aqueous H_2SO_4 electrolyte were first reported by Trasatti and Buzzanca, in 1971 [19]. Intercalation of protons into the amorphous structure of RuO_2 was responsible for the superior electrochemical properties of RuO_2. Due to the lower resistivity, hydrous RuO_2 exhibited higher specific capacitance (C_{sp}) than conducting polymer or nanocarbons [20, 21]. Hu *et al.* synthesized hydrous ruthenium oxide (denoted as $RuO_x.nH_2O$) and mixed it with activated carbon. The as-prepared hydrous ruthenium oxide exhibited a C_{sp} of 1340 F g^{-1} at a scan rate of 25 mVs^{-1}, which is close to its theoretical value of 1400 F g^{-1} [22]. The widespread application of RuO_2 as an electrode material for pseudocapacitors is limited due to its high cost, lack of availability of low cost production technologies, and toxicity of Ru. Hence, as an alternative, MnO_2, SnO_2, Co_3O_4, NiO, V_2O_5, Fe_3O_4, etc. were investigated to reduce the cost factor [23-25].

Natural abundance, eco-friendliness, high theoretical C_{sp} (1232 F g^{-1}), and low cost have made manganese oxide, MnO_2 as the most promising alternative transition metal oxide pseudocapacitive material. Lee and Goodenough first reported the electrochemical properties of amorphous MnO_2. nH_2O in aqueous KCl electrolyte [26]. Their findings suggested that MnO_2 electrodes were capable of having C_{sp} values in the range of 200 to 250 F g^{-1} [27]. C_{sp} of MnO_2 can be dramatically improved by modifying the surface morphology to obtain maximum electrolyte accessible surface area. Ultrathin films of MnO_2 exhibited a C_{sp} value above 1000 F g^{-1}[28]. MnO_2 can act as an efficient electrode material both in basic and neutral aqueous electrolytes through the mechanism of surface

Supercapacitor Technology: Materials, Processes and Architectures   Materials Research Forum LLC
Materials Research Foundations **61** (2019) 141-170   https://doi.org/10.21741/9781644900499-7

adsorption of electrolytic cation along with proton incorporation, which can be
represented by the following equation [29]:

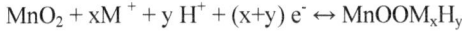

$$MnO_2 + xM^+ + y\,H^+ + (x+y)\,e^- \leftrightarrow MnOOM_xH_y$$

The equation indicates that the charge storage mechanism of MnO_2 is similar to that of
hydrous RuO_2. MnO_2 crystalizes in a variety of crystal structures (α, β, γ, ε) [30, 31]. The
C_{sp} of MnO_2 is highly influenced by the crystal structure and surface area (Fig. 2). The
poor electrical conductivity and cycle life hamper the practical applications of MnO_2. To
overcome these disadvantages, hybrid composites were prepared by incorporating MnO_2
in conducting carbon materials. Li *et al.* used MnO_2/carbon nanotubes (CNTs)
composites for supercapacitor electrodes and the experimental results indicated that the
composite electrodes exhibited excellent cycle stability and high C_{sp} (Fig. 3) [32].

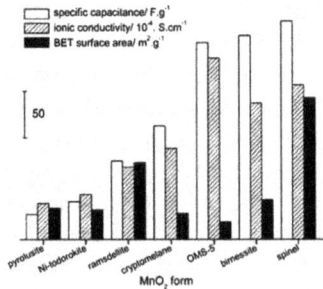

*Fig. 2 Comparison of the C_{sp}, ionic conductivity and BET surface area of different MnO_2
structures. [Adapted and reprinted with permission from [31]copyright 2009, American
Chemical Society].*

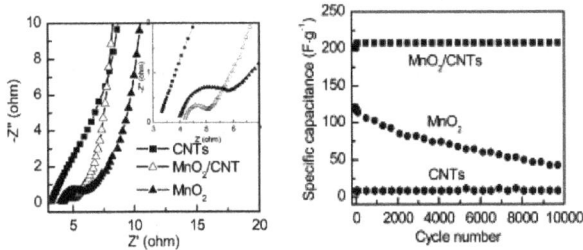

*Fig. 3 (a) Nyquist plots of the MnO_2/CNTs, MnO_2, and CNTs and (b) Cycle lifetime of the
MnO_2/CNTs, MnO_2, and pure CNTs at 1 Ag^{-1} [Adapted and reprinted with permission
[32] from copyright 2014, American Chemical Society].*

TMOs with the spinel structure, having general formula AB_2O_4 are also well accepted as supercapacitor electrode materials. The popular ones in this category are Co_3O_4, Mn_3O_4, Fe_3O_4 and ternary transition metal oxides like $NiCo_2O_4$, $NiMn_2O_4$, $ZnCo_2O_4$, and $ZnMn_2O_4$, etc. These materials are reported to have high theoretical capacitance (>3000 F/g) and energy densities [33-36]. The surface Faradaic redox reaction taking place at Co_3O_4 can be represented as following:

$$Co_3O_4 + H_2O + OH^- \leftrightarrow 3CoOOH + e^-$$

Liu et al. reported a high C_{sp} of 480 F g^{-1} for $Co_3O_4@MnO_2$ core/shell nanowire arrays [37]. Rakhi et al. reported the influence on the substrate morphology in tuning the electrochemical performance of the mesoporous cobalt oxide [38] (Fig. 4). Huang et al. fabricated supercapacitors based on $NiCo_2O_4$ nanowires coated with cobalt and nickel double hydroxide (DH) nanosheets. The composite electrode material exhibited high power (41.25 kW/kg) and energy (33 Wh/kg) densities [39]. Shen et al. reported a high C_{sp} value of 1010 F g^{-1} at 20 A g^{-1} for supercapacitor electrodes based on $NiCo_2O_4$ nanowire arrays [35]. Liang Kuo et al. synthesized $MnFe_2O_4$–carbon black (CB) composite powders by co-precipitation method for supercapacitor applications. The composite electrode exhibited an operating potential window of 1.0 V with superior cycling stability [36]. Xu et al. constructed flexible asymmetric supercapacitor (ASC) using flower - like Bi_2O_3 as the negative electrode and MnO_2 grown on carbon nanofiber (CNF) paper as the positive electrode. The device exhibited a potential window of 1.8 V and delivered a high energy density of 43.4 μWh cm^{-2} (11.3 W h kg^{-1}) [40]. A C_{sp} of 614 F g^{-1} was reported for a supercapacitor based on urchin-like Co_3O_4 micro spherical hierarchical superstructures [41]. Phase changes occurred during the charge/discharge process contribute to the improvement in the capacitive performance at the expense of the lifetime of the spinel pseudocapacitive materials.

Apart from the above mentioned TMOs, TiO_2, ZnO, NiO, Fe_3O_4, Fe_2O_3, CuO, V_2O_5, VO_2, etc. have also been widely used as supercapacitor electrodes. Li et al. fabricated hydrogenated TiO_2 nanotube arrays for supercapacitor application [42]. Rakhi et al. reported maximum energy density of 46Wh kg^{-1} for a symmetric capacitor based on VO_2 nanosheet electrodes [43]. Anjana et al. reported the fabrication of Fe_3O_4 and Zn-doped Fe_3O_4 (Zn/Fe_3O_4) nanoparticles by co-precipitation method for supercapacitor electrode material [44].

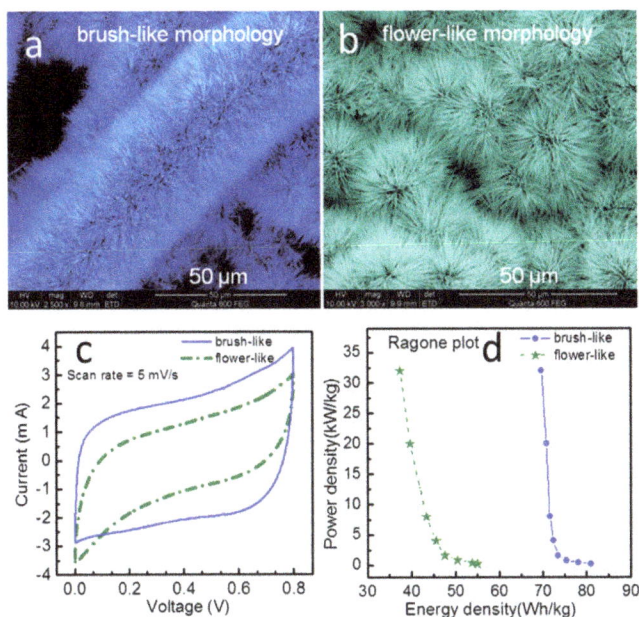

Fig. 4 SEM images of Co₃O₄ nanowires with (a) brush-like morphology and (b) flower-like morphology. Comparison of (c) Cyclic voltammograms and (d) Ragone plots of Co₃O₄ nanowires self arranged into two different morphologies. [Adapted and reprinted with permission from [38]copyright 2012, American Chemical Society].

2.2   Transition metal sulfides

Transition metal sulfides or metal chalcogenides have received considerable attention as electrode materials for energy storage devices due to their long cycle life, flexibility, large reactive sites, good conductivity and short path lengths for electron transport. Among the transition metal chalcogenides, the most popular ones are cobalt sulfides (CoS_x) and nickel sulfides (NiS_x) [45-47]. The supercapacitor property of cobalt sulfide was studied first time by Tao *et al.* in 2007 and as per their report, the material showed a high C_{sp} of 369 F g^{-1} [48].

Chou *et al.* synthesized flaky nickel sulfide (Ni_3S_2) nanostructure for supercapacitor electrode material by a simple potentiodynamic deposition approach and demonstrated a high C_{sp} of 717 F g^{-1} for the obtained material [49]. Poor cycling performance is the major disadvantage associated with metal sulfides. The stability can be improved by directly growing active materials on current collectors or by making nanocarbon/metal sulfide composites. Shang *et al.* synthesized bacteria-reduced graphene oxide (rGO)-nickel sulfide nanoparticles composite having a remarkably high C_{sp} of 1424 F g^{-1}.

Ternary metal sulfides offer high electronic conductivity and high theoretical capacity as compared to mono metal sulfides. Alshareef *et al.* reported a high C_{sp} of 1418 F g^{-1} for ternary $CoNi_2S_4$ nanosheets [50]. An asymmetric supercapacitor constructed using the ternary $CoNi_2S_4$ as the cathode and porous graphene as anode has shown an impressive performance (Figs. 5 (a) and (b)). However, the exact reaction mechanism taking place in ternary metal sulfides is still unknown and need to be explored [51, 52].

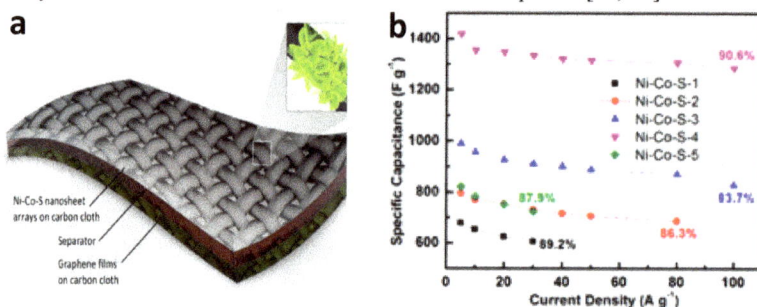

Fig. 5 (a) Schematic of an asymmetric supercapacitors fabricated using Ni-Co-S nanosheet arrays on carbon cloth as cathode and porous graphene film anode, (b) C_{sp} values for different Ni-Co-S cathodes as a function of current density. [Adapted and reprinted with permission from [50] copyright 2014, American Chemical Society].

Transition metal dichalcogenides (represented as MX_2, where M is the transition metal and X is S or Se or Te) have layered structures in which chalcogen atoms are sandwiched between hexagonally arranged layers of transition metal atoms. These materials exhibit high conductivity like graphene [53]. Materials such as MoS_2, VS_2, and WS_2, have been widely studied as active materials in pseudocapacitor electrodes [53-55]. Feng *et al.* reported an areal capacitance of 4760 $\mu F/cm^2$ for VS_2 electrode [51]. Zhou *et al.* reported a C_{sp} value of 122 F g^{-1} at 1 A g^{-1} for hydrothermally synthesized flower like MoS_2 nanospheres [56]. The incorporation of conducting polymers can further enhance the electrochemical performance of MoS_2. Ma *et al.* reported a high C_{sp} of 553.7 F g^{-1} for PPy/MoS_2 nanocomposite prepared by hydrothermal method [52].

2.3　　Metal nitrides nanomaterials

Metal nitrides have received high research interest as supercapacitor electrode materials due to their excellent intrinsic electrical conductivity ($\approx 10^6$ Ω^{-1} m^{-1}), catalytic activity, corrosion resistance to chemical attack, excellent electrochemical stability, and faster charge-discharge kinetics. MoN, Mo_2N, TiN, VN, NbN, Nb_4N_5, RuN, CrN, CoN, LaN, Ni_3N, Fe_2N, and W_2N have been popular candidates among various metals nitrides [57, 58]. Choi *et al.* synthesized nanocrystalline vanadium nitride electrode for supercapacitor, and the electrode exhibited a high C_{sp} value of about 1340 F g^{-1} [24]. The size, surface area, and morphology of nanoparticles have great influence on charge storage properties of VN. Zhang *et al.* reported an excellent capacitance of 430.7 F/g for VTiN /carbon nanofibers electrodes [59]. Molybdenum nitride has high conductivity and low leakage current in the alkaline medium. Durai *et al.* investigated the electrochemical properties of un-doped and Cu doped Mo_3N_2 thin films using 1 M KOH electrolyte [60]. The significant drawbacks of nitride-based pseudocapacitive electrode materials are high cost, severe agglomeration during charge-discharge cycling and low yield. Besides, more facile and controllable synthesis methods need to be developed.

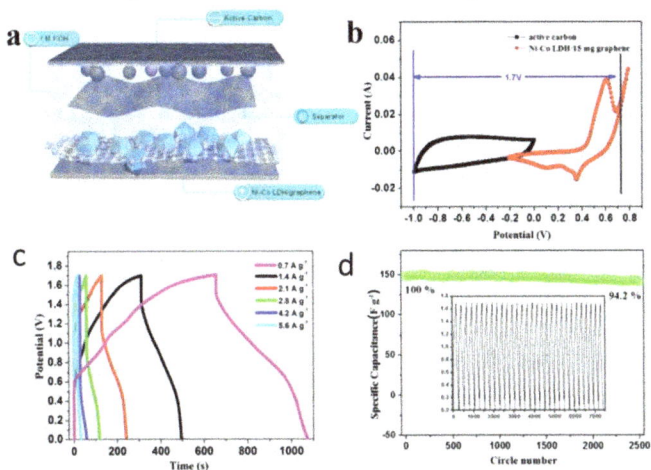

Fig. 6 (a) Schematic of an asymmetric supercapacitor with NiCo-LDH/graphene composite cathode and activated carbon anode. (b) comparison of CV curves of cathode and anode at a constant scan rate of 2 mV s^{-1}. (c) CD curves of the fabricated asymmetric supercapacitor at various current densities. (d) Stability curve of the assembled asymmetric supercapacitor at a constant current density of 1.4 A g^{-1}. [Adapted and reprinted with permission from [69] copyright 2017, American Chemical Society].

2.4 Layered metal hydroxides

The hydroxides of divalent transition metal oxides M^{2+} $(OH)_2$ possess typical 2D nanostructured lamellar structures [61, 62]. Layered metal hydroxides have attained great demand as pseudocapacitive electrode material due to their eco-friendly nature, high redox activity, and effective use of homogeneously dispersed transition metal atoms. Also, they can be easily exfoliated into monolayer nanosheets and make them stand out among metal hydroxide based electrodes [63, 64]. Various methods such as, electrodeposition, *in situ* growth, co-precipitation, etc. are available for the preparation of layered double hydroxides. Zhao *et al.* reported a high C_{sp} of 1079 F g^{-1} for CoMn - layered double hydroxide (LDH) grown on flexible carbon cloth [65]. This work demonstrated the importance of hierarchical configuration based on layered double hydroxide which increased the number of active sites and allowed a fast charge transfer. Anandan *et al.* fabricated MnNi layered double hydroxide nanoparticle $(MnNi(OH)_2)$ by a sonochemical irradiation method. The electrode prepared using $MnNi(OH)_2$ nanoparticles showed a C_{sp} of 160 F g^{-1} [66]. Jagadale *et al.* reported the fabrication of CoMn-Layered double hydroxide (CoMnLDH) electrode by a cost-effective electrodeposition method and evaluated the pseudocapacitive performance of the electrode. The as-prepared CoMnLDH electrode could deliver a maximum C_{sp} of 1062.6 F g^{-1} at 0.7 A g^{-1} with excellent cyclic stability of 96.3% over 5000 continuous charge/ discharge cycles [67]. Apart from Ni, Co and Mn researches have also reported the effect of Fe and Ti-based layered double hydroxide electrodes. For example, Ma *et al.* fabricated Co-Fe layered double hydroxides through the co-precipitation method, and pseudocapacitive studies were carried out. The electrochemical data demonstrated that Co-Fe LDH electrode can deliver a C_{sp} of 728 F g^{-1} at 1 A g^{-1} and is strongly dependent on Co/Fe ratios [68]. Combining layered double hydroxide with carbon nanotubes (CNT's), graphene, and carbon nano fibers (CNF's) can enhance electrical conductivity and surface area of the active material. Bai *et al.* constructed an asymmetric supercapacitor with Ni–Co LDH hollow nanocages incorporated with graphene composite as the cathode and activated carbon (AC) based anode. The device could deliver a maximum energy density of 68.0 Wh kg^{-1} at a power density of 594.9 W kg^{-1} [69]. (Fig. 6). Zhao *et al.* stated a maximum C_{sp} of 2960 F g^{-1} (at 1.5 A g^{-1}) for NiMn - LDH/CNT nanocomposite electrode [70]. Yang et al. successfully demonstrated a C_{sp} of 1035 F g^{-1} (at 1 A g^{-1}) for NiCoAl-LDH/ MWCNT composite electrode [71]. The addition of graphene can further enhance the capacitive properties of layered double hydroxides. Wu *et al.* synthesized porous Co-Al layered double hydroxide/graphene (GSP-LDH) film with a dual support system. The GSP - LDH electrode exhibited remarkably high C_{sp} (1043 F g^{-1} at 1 A g^{-1}) [72]. For Co-Ni double hydroxides/graphene binary composites synthesized by co-precipitation process, Cheng *et al.* reported an excellent C_{sp} of 2360 F g^{-1} (close to the theoretical capacitance value) and a remarkably improved rate capability of 86% at 20 A g^{-1} (comparable to that of carbon materials) [73].

The cycling performance of LDH based supercapacitor still needs improvement to make them more practical oriented. The capacitive properties of the LDHs can be enhanced by tuning metal ions in the host solution, and the conductivity can be improved by combining LDH's by conductive materials (conducting polymers, carbon materials, etc.).

2.5   Conducting polymers

Conducting polymer-based supercapacitors were first realized in the mid-1990s [5]. Due to the high conductivity, as a supercapacitor electrode material, the conducting polymers offer low equivalent series resistance (ESR) [74]. Conducting polymers provide conductivity through a conjugated bond system along the polymer backbone. Polyaniline (PANI), polypyrrole (PPY), poly(3,4-ethylenedioxythiophene) (PEDOT), and derivatives of polythiophene have been most commonly used conducting polymers for energy storage applications [75]. Polyaniline can deliver a charge density of 140 mAh g^{-1}, which is lower than that obtained from other expensive metal oxides but much higher than carbon-based devices that exhibit less than 15 mAh g^{-1} [76]. The pseudo capacitive property of these materials can be substantially increased by synthesizing composites of conducting polymers with other materials such as carbon (CNTs, activated carbon, graphene, etc.), inorganic oxides, hydroxides, and other metal compounds [77-80]. Kumar et al. demonstrated a C_{sp} of 250 F g^{-1} for highly conductive PANI grafted rGO (Fig. 7) [81]. Lu *et al.* stated a C_{sp} of 211 F g^{-1} for flexible PPy/CNT composite films with excellent cycling stability [82]. Palsania *et al.* synthesized nanocomposites of polyaniline (PANI), with equal weight % of graphene (G) and MoS_2 (PANI-G-MoS_2), prepared via *in-situ* oxidative polymerization method. The ternary PANI-G-MoS_2 composites displayed a remarkably high C_{sp} of 142.30 F g^{-1} with improved cycling stability in symmetric two-electrode configuration [83]. Reddy *et al.* fabricated hybrid films of poly (3,4-ethylenedioxypyrrole) (PEDOP) combined with Co_3O_4 nanorods (PEDOP@Co_3O_4) and PEDOP@Fe_3O_4 nanostructures. The symmetric supercapacitor constructed using PEDOP@Fe_3O_4 based cell delivered a C_{sp} of 673 F g^{-1} [84]. The composite formation enhanced conductivity, cycle life, mechanical stability, C_{sp} and processability of polymer-based supercapacitor electrodes.

Conducting polymers have many advantages as the electrode materials for supercapacitors. These are easily processable, highly flexible, and can be converted into thin films. Lower cycle life due to the mechanical stress experienced by the conducting polymers during reduction-oxidation reactions is the major drawback associated with these materials as supercapacitor electrodes [74, 75, 85]. The main solutions for enhancing cycle life include combining of conducting polymer with nanocarbon materials [86].

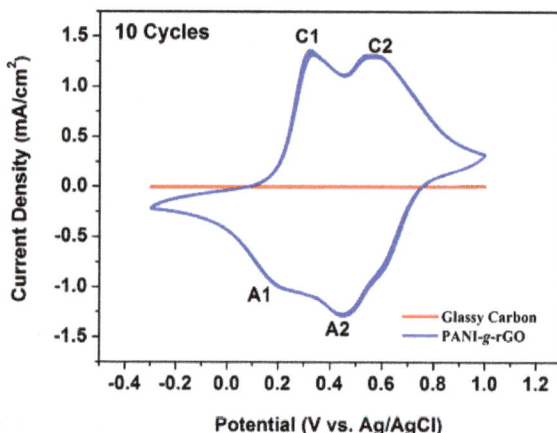

Fig. 7 Cyclic voltammograms (CV) of PANI-g-rGO recorded in 1 M H_2SO_4 solution as the electrolyte at a sweep rate of 100 mV s^{-1}. [Adapted and reprinted with permission from [81] copyright 2012, American Chemical Society].

3.   Intercalation Pseudocapacitive electrode materials

Intercalation type pseudocapacitive electrode materials have layered structures and the electrolyte ions intercalate into space between the layers. Transition metal carbides, some metal oxides like rutile RuO_2, and lithium/sodium metal oxide based composites, etc. are the popular electrode materials in this category. These materials can be used in hybrid capacitors due to their high rate capabilities. The following section explains the applicability of these materials as supercapacitor electrodes.

3.1   Metal oxides

Titanium oxides (TiO_2) have been widely used as intercalation type pseudocapacitive materials. TiO_2 is the most popular insertion-type anode for lithium-ion capacitors. Wu *et al.* reported TiO_2 nanotube electrodes subjected to the electrochemical hydrogenation doping (TiO_2-H) for supercapacitor applications [87]. The TiO_2-H electrode displayed a very high C_{sp} of 20.08 mF cm^{-2} at 0.05 mA cm^{-2} with outstanding rate capability. The major problem faced by TiO_2 based electrodes is their low electrical conductivity. This can be improved by adding conductive additives to different polymorphs of TiO_2. Wang *et al.* reported Li-ion capacitors made up of TiO_2 nanobelt arrays anode and graphene hydrogel cathode [88]. The device exhibited a high energy density of 82 Wh kg^{-1}. Kim *et al.* fabricated a hybrid supercapacitor based on TiO_2/rGO composite anode and activated carbon cathode [89]. The hybrid supercapacitor delivered a high energy density of 42 Wh kg^{-1}. The enhanced energy density of the hybrid supercapacitor can be attributed to the combination of Faradaic intercalation and non‐Faradaic surface reaction. Hydrogen titanates ($H_2Ti_3O_7$, $H_2Ti_6O_{13}$) with large interlayer spacing are another kind of Ti-based metal oxides used as electrodes for hybrid electrochemical capacitors [90, 91]. Wu *et al.* conducted a detailed study on the energy storage properties of $H_2Ti_3O_7$ nanowires [92]. Li *et al.* reported intercalation pseudocapacitive properties of layered hydrogen titanate nanowires via an alkaline-hydrothermal process followed by acid treatment [93].

Niobium pentoxide (Nb_2O_5) is another intercalation pseudocapacitive metal oxide known since the 1980s [94]. Luo *et al.* reported the superior pseudocapacitive response of niobium pentoxide/carbon (Nb_2O_5/C) [95]. Orthorhombic $T-Nb_2O_5$ is another promising intercalation pseudocapacitive anode [96, 97]. Lou *et al.* fabricated three-dimensionally ordered macroporous (3DOM) orthorhombic niobium oxide ($T-Nb_2O_5$) for lithium-ion batteries [98]. The 3DOM $T-Nb_2O_5$ demonstrated an excellent capacity value with remarkable cycle life. Ouendi *et al.* fabricated $T-Nb_2O_5$, and the electrochemical analysis indicated the presence of a lithium ion intercalation redox mechanism with excellent capacitive property [99]. Poor electronic conductivity of $T-Nb_2O_5$ nanocrystals can be modified by combining $T-Nb_2O_5$ with graphene. Kong *et al.* reported excellent gravimetric and volumetric capacitance values (620.5 F g^{-1} and 961.8 F cm^{-3}) for free-standing $T-Nb_2O_5$/graphene composite paper [100] .

Vanadium pentoxide (V_2O_5) is another interesting candidate for pseudocapacitive electrodes because of the multiple oxidation states of V with high theoretical capacity (325 mAh/g) and the ability to form layered compounds. Pseudocapacitive behavior of V_2O_5 is depended upon the reaction between active sites and electrolyte. Yang *et al.* described the synthesis of hollow V_2O_5 spheres which exhibited a high capacitance of 479 F g^{-1} at 5 mV s^{-1} [101]. However, the poor electronic conductivity V_2O_5 limited its practical application as a supercapacitor electrode. Li *et al.* fabricated V_2O_5/ rGO nanocomposites for supercapacitors using neutral aqueous electrolytes [102]. The V_2O_5/rGO nanocomposites exhibited a C_{sp} of 537 F g^{-1} at 1 A g^{-1}. Wang *et al.* reported the fabrication of supercapacitor based on V_2O_5/carbon nanotubes/super activated carbon (V_2O_5/CNTs-SAC), which displayed a C_{sp} of 357.5 F g^{-1} at 10 A g^{-1}.

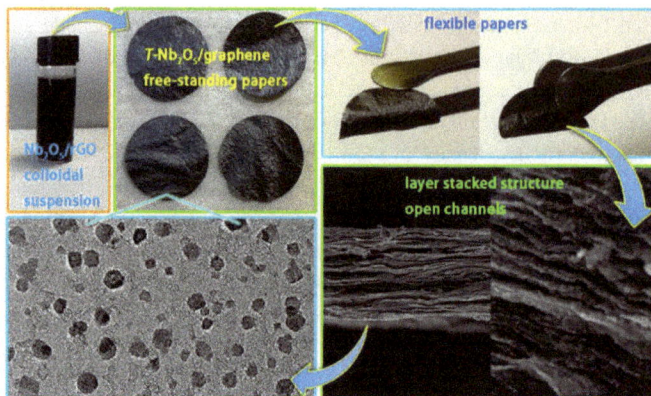

Fig. 8 Digital photographs of Nb_2O_5/rGO colloidal suspensions in EG and T-Nb_2O_5/graphene composite papers. TEM and SEM images reveal layer stalked structure [Adapted and reprinted with permission from [100] copyright 2015, American Chemical Society].

3.2   Lithium/Sodium metal oxide-based electrode

Capacitors fabricated using lithium titanate ($Li_4Ti_5O_{12}$) were introduced by Amatucci *et al.* in 2001 [103]. The low electronic conductivity was the major problem for nanostructured $Li_4Ti_5O_{12}$ which can be modified by the surface deposition of conductive layers and doping with suitable ions/atoms. Kim *et al.* synthesized graphene-wrapped $Li_4Ti_5O_{12}$ electrode for hybrid supercapacitor [104]. The device delivered an energy density of 50 Wh kg^{-1}. Naoi et al. reported the synthesis of nano-crystalline $Li_4Ti_5O_{12}$ grafted onto carbon nano-fiber anchors (nc-$Li_4Ti_5O_{12}$/CNF) nanocomposite [105]. The hybrid capacitor constructed using nc-Li4Ti5O12/CNF composite as the negative electrode and activated carbon as the positive electrode was capable of providing a high energy density of 40 Wh L^{-1} and high power density of 7.5 kW L^{-1}. Mladenov and co-workers fabricated lithium titanate ($Li_4Ti_5O_{12}$) via thermal co-decomposition of oxalates for supercapacitor application [106]. Asymmetric hybrid supercapacitor fabricated using activated graphitized carbon as the negative electrode and activated carbon-$Li_4Ti_5O_{12}$ oxide composite as the positive electrode delivered greater C_{sp} of 150 F g^{-1} with good cyclic stability. Fu *et al.* reported the synthesis of $Li_4Ti_5O_{12}$-rGO (LTO-RGO) composite for a hybrid supercapacitor using LTO as the negative electrode and activated carbon (AC) as the positive electrode, and the result indicated LTO-RGO composite as the promising anode material for hybrid supercapacitor [107]. Lee *et al.* demonstrated a hybrid supercapacitor electrode based on granule $Li_4Ti_5O_{12}$ (LTO) as the anode and activated carbon as the cathode (Fig.9) [108]. The electrochemical analysis revealed good capacitive performance.

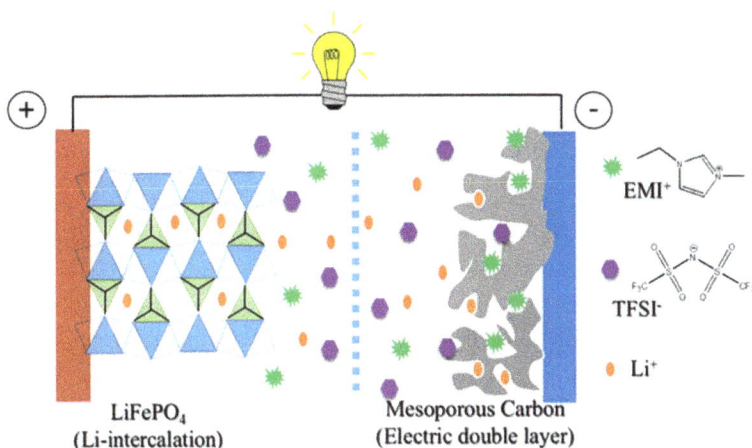

Fig. 9 Schematic of a hybrid supercapacitor based on imidazolium ionic liquid [Adapted and reprinted with permission from [114] copyright 2018, American Chemical Society].

Sodium titanate ($Na_2Ti_3O_7$) has emerged as a promising electrode material for hybrid electrochemical capacitors (HEC), especially for sodium-ion capacitors. Radhiyah and co-workers fabricated a flower-shaped hydrated layered sodium titanate material, $Na_2Ti_2O_4(OH)_2$, through a facile hydrothermal method [109]. The $Na_2Ti_2O_4(OH)_2$ electrode exhibited an excellent C_{sp} (300 F g^{-1}) in an aqueous electrolyte. Similar to other intercalation type metal oxides $Na_2Ti_3O_7$ also showed poor conductivity which restricted its application as energy storage devices. CNTs or rGO, or AC can be used in combination with $Na_2Ti_3O_7$ to overcome this difficulty. Wang *et al.* synthesized carbon modified $Na_2Ti_3O_7 \cdot 2H_2O$ nanobelts (Na-TNBs) with layered structure by a simple hydrothermal method [110]. The flexible electrode fabricated using C/Na-TNBs delivered a C_{sp} of 459.71 F g^{-1}. Also, the electrode displayed a capacitance retention of 74.3% after 6000 charge-discharge cycles. Qiu *et al.* fabricated two-dimensional $Na_2Ti_3O_7$ nanosheets on activated carbon fiber ($Na_2Ti_3O_7$/activated carbon fiber) as an electrode for energy storage application [111]. Sodium ion hybrid capacitor assembled using $Na_2Ti_3O_7$/activated carbon fiber nanocomposite anode and activated carbon fiber cathode was able to deliver a maximum C_{sp} of 76.8 F g^{-1} at 1 A g^{-1} in a wide operating voltage of 3.0 V [111].

Two non-aqueous hybrid electrode materials- $LiFePO_4$ and $Na_3V_2(PO_4)_3$-have attained great demand in energy storage application. These materials have a structure that supports fast ion diffusion and wide potential window for operation. Jian *et al.* constructed a supercapacitor electrode based on $Na_3V_2(PO_4)_3$ material uniformly coated with a layer of 6 nm carbon ($Na_3V_2(PO_4)_3$/C) [112]. Yu *et al.* developed a cathode for energy storage application using the carbon-coated $Na_3V_2(PO_4)_3$ grown over graphite foam [113]. The fabricated cathode exhibited an excellent sodium-ion storage performance and long cycle life. Lavall *et al.* developed hybrid supercapacitor with mesoporous carbon as the negative electrode and $LiFePO_4$ as the positive electrode, using an ionic liquid electrolyte [114]. The supercapacitor cell delivered an energy density of 43.3 W h kg^{-1} at 0.010 A g^{-1}, while maintaining an excellent cycling performance (Fig. 9).

3.3   Transition metal carbides

Transition metal carbides (MXenes) belong to another group of active materials for supercapacitor application due to their high in-plane electrical conductivity, superior reactivity, and strong mechanical properties. The application of MXenes in electrochemical energy storage was first reported by the Gogotsi group [115]. MXenes are capable of giving maximum volumetric capacitances up to 900 F cm^{-3}. These are prepared by selective removal of the A (Al, Si, In, Sn, etc.) element from the ternary transition metal carbides (MAX phases)[116-119]. Rakhi *et al.* reported the effectiveness of the post-etch annealing ambient on the structure and electrochemical properties of the Ti_2CT_x MXenes (Fig. 10) [120]. The study demonstrated that annealing in N_2/H_2 atmosphere improved the capacitive performance of the MXene sample (C_{sp} value 51 F g^{-1} at 1A g^{-1} in symmetric two electrode configuration). Recently Fu *et al.* developed a graphene-encapsulated MXene $Ti_2CTx@$ polyaniline composite (GMP) material for supercapacitor electrodes [121]. The obtained GMP electrode exhibited a gravimetric capacitance of 635 F g^{-1} at 1 A g^{-1} with excellent cycling stability of 97.54% after 10000 cycles.

High in-plane electrical conductivity and the 2D-layered structure of MXenes have made them strong candidates for electrochemical energy storage application. Additionally, transition metal carbides have lower anode potentials and high rate capability than the corresponding transition metal oxides. Except for all these advantages, high production cost involved in the large-scale scale synthesis limits their practical applications. Hence, low-cost synthesis methods are needed to be developed for the preparation of high-quality metal carbides.

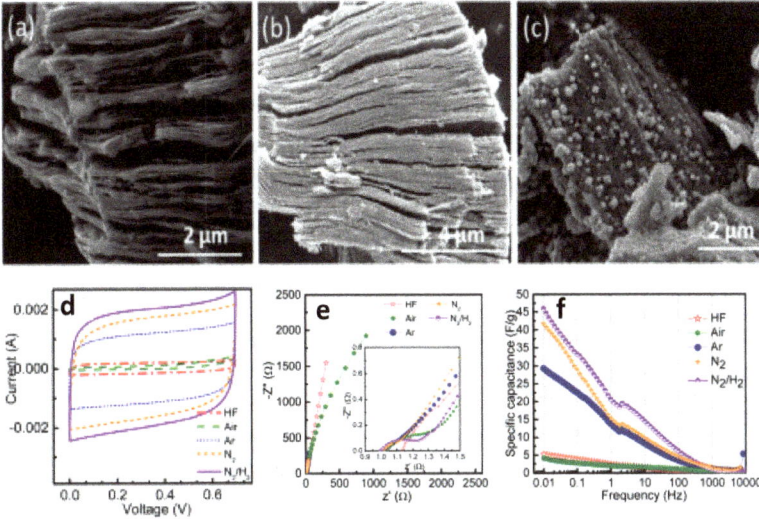

Fig. 10 (a) SEM images of (a) HF treated MAX phase and MXene samples annealed in (b) N_2/H_2 atmosphere and (c) air. (d) Comparison of cyclic voltammograms, (e) Nyquist plots, and (f) Variation of C_{sp} as a function of frequency of MXene samples based supercapacitors. [Adapted and reprinted with permission from[120] copyright 2009, American Chemical Society].

Conclusions

Development of nanostructured electrode materials with exceptional electronic conductivity is crucial in the design of high-performance supercapacitors. Pseudocapacitive materials with their Faradaic charge storage mechanism can provide high specific capacitance as supercapacitor electrodes. In this chapter, we have presented the information about different pseudocapacitive materials, pointing out their advantages and disadvantages in supercapacitors. Based on the type of the Faradaic charge storage mechanism responsible for the pseudocapacitive charge storage, the pseudocapacitive materials can be broadly grouped under either redox or intercalation pseudocapacitive electrode materials. Low conductivity and poor cyclic stability are the major disadvantages of pseudocapacitive electrode materials. Pseudocapacitive materials are usually combined with highly conductive materials, for improving the conductivity, ion transfer rate, and cycle life. For real applications, along with electrode material, electrolyte and separator used in the supercapacitor cell fabrication also have got high significance in determining the capacitive property.

Acknowledgement

R.B. Rakhi is thankful to SERB, DST for the Ramanujan Fellowship, CSIR-NIIST and UGC, Govt. of India .

Rrferences

[1] M.S. Halper, J.C. Ellenbogen, Supercapacitors : A Brief Overview, The MITRE Corporation, McLean, Virginia, USA, 1-34, 2006.

[2] A. Burke, Ultracapacitors: why, how, and where is the technology, J. Power Sources 91 (2000) 37-50. https://doi.org/10.1016/S0378-7753(00)00485-7

[3] M. Winter, R.J. Brodd, What are batteries, fuel cells, and supercapacitors?, Chem. Rev. 104 (2004) 4245-4270. https://doi.org/10.1021/cr020730k

[4] P.J. Hall, E.J. Bain, Energy-storage technologies and electricity generation, Energy Policy 36 (2008) 4352-4355. https://doi.org/10.1016/j.enpol.2008.09.037

[5] B.E.Convey, Electrochemical super capacitors: Scientific fundamentals and Technological Applications, 2 ed., Kluwer academic/Plenum Publishers, New York, 1999.

[6] I. Danaee, M. Jafarian, F. Forouzandeh, F. Gobal, M.G. Mahjani, Electrochemical impedance studies of methanol oxidation on GC/Ni and GC/NiCu electrode, Int. J. Hydrogen Energy 34 (2009) 859-869. https://doi.org/10.1016/j.ijhydene.2008.10.067

[7] M.-S. Wu, P.-C. Julia Chiang, Fabrication of nanostructured manganese oxide electrodes for electrochemical capacitors, Electrochem. solid-state Lett. 7 (2004) A123-A126. https://doi.org/10.1149/1.1695533

[8] W. Sugimoto, H. Iwata, Y. Murakami, Y. Takasu, Electrochemical capacitor behavior of layered ruthenic acid hydrate, J. Electrochem. Soc. 151 (2004) A1181-A1187. https://doi.org/10.1149/1.1765681

[9] X. Dong, W. Shen, J. Gu, L. Xiong, Y. Zhu, H. Li, J. Shi, MnO_2-embedded-in-mesoporous-carbon-wall structure for use as electrochemical capacitors, J. Phys. Chem. B 110 (2006) 6015-6019. https://doi.org/10.1021/jp056754n

[10] B.E. Conway, V. Birss, J. Wojtowicz, The role and utilization of pseudocapacitance for energy storage by supercapacitors, J Power Sources 66 (1997) 1-14. https://doi.org/10.1016/S0378-7753(96)02474-3

[11] I.-H. Kim, K.-B. Kim, Ruthenium oxide thin film electrodes for supercapacitors, electrochem, Solid-State Lett. 4 (2001) A62-A64. https://doi.org/10.1149/1.1359956

[12] M. Mastragostino, C. Arbizzani, F. Soavi, Polymer-based supercapacitors, J. Power Sources 97-98 (2001) 812-815. https://doi.org/10.1016/S0378-7753(01)00613-9

[13] K.S. Ryu, K.M. Kim, N.-G. Park, Y.J. Park, S.H. Chang, Symmetric redox supercapacitor with conducting polyaniline electrodes, J. Power Sources 103 (2002) 305-309. https://doi.org/10.1016/S0378-7753(01)00862-X

[14] V. Augustyn, P. Simon, B. Dunn, Pseudocapacitive oxide materials for high-rate electrochemical energy storage, Energy & Environ. Sci. 7 (2014) 1597-1614. https://doi.org/10.1039/c3ee44164d

[15] S. Trasatti, G. Buzzanca, Ruthenium dioxide: A new interesting electrode material. Solid state structure and electrochemical behaviour, J. Electroanal. Chem. Interfacial Electrochem. 29 (1971) A1-A5. https://doi.org/10.1016/S0022-0728(71)80111-0

[16] M.-S. Wu, Y.-A. Huang, C.-H. Yang, J.-J. Jow, Electrodeposition of nanoporous nickel oxide film for electrochemical capacitors, Int. J. Hydrogen Energy 32 (2007) 4153-4159. https://doi.org/10.1016/j.ijhydene.2007.06.001

[17] B. Wahdame, D. Candusso, X. François, F. Harel, J.-M. Kauffmann, G. Coquery, Design of experiment techniques for fuel cell characterisation and development, Int. J. Hydrogen Energy 34 (2009) 967-980. https://doi.org/10.1016/j.ijhydene.2008.10.066

[18] I.-H. Kim, K.-B. Kim, Electrochemical characterization of hydrous ruthenium oxide thin-film electrodes for electrochemical capacitor applications, J. Electrochem. Soc. 153 (2006) A383-A389. https://doi.org/10.1149/1.2147406

[19] L.M. Doubova, S. Daolio, A. De Battisti, Examination of RuO_2 single-crystal surfaces: charge storage mechanism in H_2SO_4 aqueous solution, J. Electroanal. Chem. 532 (2002) 25-33. https://doi.org/10.1016/S0022-0728(02)00981-6

[20] J.P. Zheng, T.R. Jow, A new charge storage mechanism for electrochemical capacitors, J. Electrochem. Soc. 142 (1995) L6-L8. https://doi.org/10.1149/1.2043984

[21] J.P. Zheng, P.J. Cygan, T.R. Jow, Hydrous ruthenium oxide as an electrode material for electrochemical capacitors, J. Electrochem. Soc. 142 (1995) 2699-2703. https://doi.org/10.1149/1.2050077

[22] C.-C. Hu, W.-C. Chen, K.-H. Chang, How to achieve maximum utilization of hydrous ruthenium oxide for supercapacitors, J. Electrochem. Soc.151(2004) A281-A290. https://doi.org/10.1149/1.1639020

[23] D.J. Ham, R. Ganesan, J.S. Lee, Tungsten carbide microsphere as an electrode for cathodic hydrogen evolution from water, Int. J. Hydrogen Energy 33 (2008) 6865-6872. https://doi.org/10.1016/j.ijhydene.2008.05.045

[24] D. Choi, G.E. Blomgren, P.N. Kumta, Fast and reversible surface redox reaction in nanocrystalline vanadium nitride supercapacitors, Adv. Mater.18 (2006) 1178-1182. https://doi.org/10.1002/adma.200502471

[25] E.B. Castro, S.G. Real, L.F. Pinheiro Dick, Electrochemical characterization of porous nickel–cobalt oxide electrodes, Int. J. Hydrogen Energy 29 (2004) 255-261. https://doi.org/10.1016/S0360-3199(03)00133-2

[26] H.Y. Lee, J.B. Goodenough, Supercapacitor behavior with KCl electrolyte, J. Solid State Chem.144 (1999) 220-223. https://doi.org/10.1006/jssc.1998.8128

[27] X. Wang, X. Wang, W. Huang, P.J. Sebastian, S. Gamboa, Sol–gel template synthesis of highly ordered MnO_2 nanowire arrays, J. Power Sources 140 (2005) 211-215. https://doi.org/10.1016/j.jpowsour.2004.07.033

[28] M. Toupin, T. Brousse, D. Bélanger, Charge storage mechanism of MnO_2 electrode used in aqueous electrochemical capacitor, Chem. Mater. 6 (2004) 3184-3190. https://doi.org/10.1021/cm049649j

[29] P. Simon, Y. Gogotsi, Materials for electrochemical capacitors, Nat. Mater. 7 (2008) 845-854. https://doi.org/10.1038/nmat2297

[30] W.F. Wei, X.W. Cui, W.X. Chen, D.G. Ivey, Manganese oxide-based materials as electrochemical supercapacitor electrodes, Chem. Soc. Rev. 40 (2011) 1697-1721. https://doi.org/10.1039/C0CS00127A

[31] O. Ghodbane, J.-L. Pascal, F. Favier, Microstructural effects on charge-storage properties in MnO_2-based electrochemical supercapacitors, ACS Appl. Mater. Interfaces 1 (2009) 1130-1139. https://doi.org/10.1021/am900094e

[32] L. Li, Z.A. Hu, N. An, Y.Y. Yang, Z.M. Li, H.Y. Wu, Facile Synthesis of MnO_2/CNTs composite for supercapacitor electrodes with long cycle stability, J. Phys. Chem. C 118 (2014) 22865-22872. https://doi.org/10.1021/jp505744p

[33] F. Zhang, C. Yuan, X. Lu, L. Zhang, Q. Che, X. Zhang, Facile growth of mesoporous Co_3O_4 nanowire arrays on Ni foam for high performance electrochemical

capacitors, J. Power Sources 203 (2012) 250-256.
https://doi.org/10.1016/j.jpowsour.2011.12.001

[34] C. Yuan, X. Zhang, L. Su, B. Gao, L. Shen, Facile synthesis and self-assembly of hierarchical porous NiO nano/micro spherical superstructures for high performance supercapacitors, J. Mater. Chem.19 (2009) 5772-5777.
https://doi.org/10.1039/b902221j

[35] L. Shen, Q. Che, H. Li, X. Zhang, Mesoporous $NiCo_2O_4$ nanowire arrays grown on carbon textiles as binder-free flexible electrodes for energy storage, Adv. Funct. Mater. 24 (2014) 2630-2637. https://doi.org/10.1002/adfm.201303138

[36] S.-L. Kuo, N.-L. Wu, Electrochemical characterization on $MnFe_2O_4$/carbon black composite aqueous supercapacitors, J. Power Sources 162 (2006) 1437-1443.
https://doi.org/10.1016/j.jpowsour.2006.07.056

[37] J. Liu, J. Jiang, C. Cheng, H. Li, J. Zhang, H. Gong, H.J. Fan, Co_3O_4 nanowire@MnO_2 ultrathin nanosheet core/shell arrays: A new class of high-performance pseudocapacitive materials, Adv. Mater. 23 (2011) 2076-2081.
https://doi.org/10.1002/adma.201100058

[38] R.B. Rakhi, W. Chen, D. Cha, H.N. Alshareef, Substrate dependent self-organization of mesoporous cobalt oxide nanowires with remarkable pseudocapacitance, Nano Lett. 12 (2012) 2559-2567. https://doi.org/10.1021/nl300779a

[39] L. Huang, D. Chen, Y. Ding, S. Feng, Z.L. Wang, M. Liu, Nickel–cobalt hydroxide nanosheets coated on $NiCo_2O_4$ nanowires grown on carbon fiber paper for high-performance pseudocapacitors, Nano Lett.13 (2013) 3135-3139.
https://doi.org/10.1021/nl401086t

[40] H. Xu, X. Hu, H. Yang, Y. Sun, C. Hu, Y. Huang, Flexible asymmetric micro-supercapacitors based on Bi_2O_3 and MnO_2 nanoflowers: Larger areal mass promises higher energy density, Advanced Energy Materials 5 (2015) 1401882.
https://doi.org/10.1002/aenm.201401882

[41] L. Hou, C. Yuan, L. Yang, L. Shen, F. Zhang, X. Zhang, Urchin-like Co_3O_4 microspherical hierarchical superstructures constructed by one-dimension nanowires toward electrochemical capacitors, RSC Adv. 1 (2011) 1521-1526.
https://doi.org/10.1039/c1ra00312g

[42] X. Lu, G. Wang, T. Zhai, M. Yu, J. Gan, Y. Tong, Y. Li, Hydrogenated TiO_2 nanotube arrays for supercapacitors, Nano Lett.12 (2012) 1690-1696.
https://doi.org/10.1021/nl300173j

[43] R.B. Rakhi, D.H. Nagaraju, P. Beaujuge, H.N. Alshareef, Supercapacitors based on two dimensional VO2 nanosheet electrodes in organic gel electrolyte, Electrochim. Acta 220 (2016) 601-608. https://doi.org/10.1016/j.electacta.2016.10.109

[44] P.M. Anjana, M.R. Bindhu, M. Umadevi, R.B. Rakhi, Antimicrobial, electrochemical and photo catalytic activities of Zn doped Fe_3O_4 nanoparticles, J. Mater. Sci. Mater. Electron. 29 (2018) 6040-6050. https://doi.org/10.1007/s10854-018-8578-2

[45] H. Wan, X. Ji, J. Jiang, J. Yu, L. Miao, L. Zhang, S. Bie, H. Chen, Y. Ruan, Hydrothermal synthesis of cobalt sulfide nanotubes: The size control and its application in supercapacitors, J. Power Sources 243 (2013) 396-402. https://doi.org/10.1016/j.jpowsour.2013.06.027

[46] X. Meng, H. Sun, J. Zhu, H. Bi, Q. Han, X. Liu, X. Wang, Graphene-based cobalt sulfide composite hydrogel with enhanced electrochemical properties for supercapacitors, New J. Chem. 40 (2016) 2843-2849. https://doi.org/10.1039/C5NJ03423J

[47] S. Peng, L. Li, H. Tan, R. Cai, W. Shi, C. Li, S.G. Mhaisalkar, M. Srinivasan, S. Ramakrishna, Q. Yan, MS_2 (M = Co and Ni) Hollow Spheres with Tunable Interiors for High-Performance Supercapacitors and Photovoltaics, Adv. Funct. Mater. 24 (2014) 2155-2162. https://doi.org/10.1002/adfm.201303273

[48] F. Tao, Y.-Q. Zhao, G.-Q. Zhang, H.-L. Li, Electrochemical characterization on cobalt sulfide for electrochemical supercapacitors, Electrochem. Commun. 9 (2007) 1282-1287. https://doi.org/10.1016/j.elecom.2006.11.022

[49] S.W. Chou, J.Y. Lin, Cathodic deposition of flaky nickel sulfide nanostructure as an electroactive material for high-performance supercapacitors, J. Electrochem. Soc. 160 (2013) D178-D182. https://doi.org/10.1149/2.078304jes

[50] W. Chen, C. Xia, H.N. Alshareef, One-step electrodeposited nickel cobalt sulfide nanosheet arrays for high-performance asymmetric supercapacitors, ACS Nano 8 (2014) 9531-9541. https://doi.org/10.1021/nn503814y

[51] J. Feng, X. Sun, C. Wu, L. Peng, C. Lin, S. Hu, J. Yang, Y. Xie, Metallic few-layered VS2 ultrathin nanosheets: high two-dimensional conductivity for in-plane supercapacitors, J. Am. Chem. Soc. 133 (2011) 17832-17838. https://doi.org/10.1021/ja207176c

[52] G. Ma, H. Peng, J. Mu, H. Huang, X. Zhou, Z. Lei, In situ intercalative polymerization of pyrrole in graphene analogue of MoS_2 as advanced electrode

material in supercapacitor, J. Power Sources 229 (2013) 72-78.
https://doi.org/10.1016/j.jpowsour.2012.11.088

[53] J.M. Soon, K.P. Loh, Electrochemical double-layer capacitance of MoS$_2$ nanowall films, Electrochem. Solid-State Lett.10 (2007) A250-A254.
https://doi.org/10.1149/1.2778851

[54] J. Feng, X. Sun, C.Z. Wu, L.L. Peng, C.W. Lin, S.L. Hu, J.L. Yang, Y. Xie, Metallic Few-layered VS$_2$ ultrathin nanosheets: high two-dimensional conductivity for in-plane supercapacitors, J. Am. Chem. Soc. 133 (2011) 17832-17838.
https://doi.org/10.1021/ja207176c

[55] M. Chhowalla, H.S. Shin, G. Eda, L.J. Li, K.P. Loh, H. Zhang, The chemistry of two-dimensional layered transition metal dichalcogenide nanosheets, Nat. Chem. 5 (2013) 263-275. https://doi.org/10.1038/nchem.1589

[56] X. Zhou, B. Xu, Z. Lin, D. Shu, L. Ma, Hydrothermal synthesis of flower-like MoS$_2$ nanospheres for electrochemical supercapacitors, J. Nanosci. Nanotech. 14 (2014) 7250. https://doi.org/10.1166/jnn.2014.8929

[57] C. Xia, Y. Xie, W. Wang, H. Du, Fabrication and electrochemical capacitance of polyaniline/titanium nitride core–shell nanowire arrays, Synth. Met. 192 (2014) 93-100. https://doi.org/10.1016/j.synthmet.2014.03.018

[58] S. Ghosh, S.M. Jeong, S.R. Polaki, A review on metal nitrides/oxynitrides as an emerging supercapacitor electrode beyond oxide, Korean J. Chem. Eng. 5 (2018) 1389-1408. https://doi.org/10.1007/s11814-018-0089-6

[59] Y. Xu, J. Wang, B. Ding, L. Shen, H. Dou, X. Zhang, General strategy to fabricate ternary metal nitride/carbon nanofibers for supercapacitors, ChemElectroChem 2 (2015) 2020-2026. https://doi.org/10.1002/celc.201500310

[60] G. Durai, P. Kuppusami, J. Theerthagiri, Microstructural and supercapacitive properties of reactive magnetron co-sputtered Mo$_3$N$_2$ electrodes: Effects of Cu doping, Mater. Lett. 220 (2018) 201-204. https://doi.org/10.1016/j.matlet.2018.02.120

[61] G. Abellan, J.A. Carrasco, E. Coronado, J. Romero, M. Varela, Alkoxide-intercalated CoFe-layered double hydroxides as precursors of colloidal nanosheet suspensions: structural, magnetic and electrochemical properties, J. Mater. Chem. C 2 (2014) 3723-3731. https://doi.org/10.1039/C3TC32578D

[62] Y. Wang, W.S. Yang, J.J. Yang, A Co-Al layered double hydroxides nanosheets thin-film electrode - Fabrication and electrochemical study, Electrochem. Solid-State Lett. 10 (2007) A233-A236. https://doi.org/10.1149/1.2768166

[63] M.E. Spahr, P. Novak, B. Schnyder, O. Haas, R. Nesper, Characterization of layered lithium nickel manganese oxides synthesized by a novel oxidative coprecipitation method and their electrochemical performance as lithium insertion electrode materials, J. Electrochem. Soc. 145 (1998) 1113-1121. https://doi.org/10.1149/1.1838425

[64] Y. Wang, W.S. Yang, S.C. Zhang, D.G. Evans, X. Duan, Synthesis and electrochemical characterization of Co-Al layered double hydroxides, J. Electrochem. Soc. 152 (2005) A2130-A2137. https://doi.org/10.1149/1.2041107

[65] J. Zhao, J. Chen, S. Xu, M. Shao, D. Yan, M. Wei, D.G. Evans, X. Duan, CoMn-layered double hydroxide nanowalls supported on carbon fibers for high-performance flexible energy storage devices, J. Mater. Chem. A 1 (2013) 8836-8843. https://doi.org/10.1039/c3ta11452j

[66] S. Anandan, C.-Y. Chen, J.J. Wu, Sonochemical synthesis and characterization of turbostratic MnNi(OH)$_2$ layered double hydroxide nanoparticles for supercapacitor applications, RSC Adv. 4 (2014) 55519-55523. https://doi.org/10.1039/C4RA10816G

[67] A.D. Jagadale, G. Guan, X. Li, X. Du, X. Ma, X. Hao, A. Abudula, Ultrathin nanoflakes of cobalt–manganese layered double hydroxide with high reversibility for asymmetric supercapacitor, Journal of Power Sources 306 (2016) 526-534. https://doi.org/10.1016/j.jpowsour.2015.12.097

[68] K. Ma, J.P. Cheng, J. Zhang, M. Li, F. Liu, X. Zhang, Dependence of Co/Fe ratios in Co-Fe layered double hydroxides on the structure and capacitive properties, Electrochim. Acta 198 (2016) 231-240. https://doi.org/10.1016/j.electacta.2016.03.082

[69] X. Bai, Q. Liu, Z. Lu, J. Liu, R. Chen, R. Li, D. Song, X. Jing, P. Liu, J. Wang, Rational design of sandwiched Ni–Co layered double hydroxides hollow nanocages/graphene derived from metal-organic framework for sustainable energy storage, ACS Sustain. Chem. Eng. 5 (2017) 9923-9934. https://doi.org/10.1021/acssuschemeng.7b01879

[70] J. Zhao, J. Chen, S. Xu, M. Shao, Q. Zhang, F. Wei, J. Ma, M. Wei, D.G. Evans, X. Duan, Hierarchical NiMn layered double hydroxide/carbon nanotubes architecture with superb energy density for flexible supercapacitors, Adv. Funct. Mater. 24 (2014) 2938-2946. https://doi.org/10.1002/adfm.201303638

[71] J. Yang, C. Yu, X. Fan, Z. Ling, J. Qiu, Y. Gogotsi, Facile fabrication of MWCNT-doped NiCoAl-layered double hydroxide nanosheets with enhanced electrochemical performances, J. Mater. Chem. A 1 (2013) 1963-1968. https://doi.org/10.1039/C2TA00832G

[72] X. Wu, L. Jiang, C. Long, T. Wei, Z. Fan, Dual Support system ensuring porous co–al hydroxide nanosheets with ultrahigh rate performance and high energy density for supercapacitors, Adv. Funct. Mater. 25 (2015) 1648-1655. https://doi.org/10.1002/adfm.201404142

[73] Y. Cheng, H. Zhang, C.V. Varanasi, J. Liu, Improving the performance of cobalt-nickel hydroxide-based self-supporting electrodes for supercapacitors using accumulative approaches, Energy Environ. Sci. 6 (2013) 3314-3321. https://doi.org/10.1039/c3ee41143e

[74] C. Arbizzani, M. Mastragostino, F. Soavi, New trends in electrochemical supercapacitors, J. Power Sources 100 (2001) 164-170. https://doi.org/10.1016/S0378-7753(01)00892-8

[75] T. Nohma, H. Kurokawa, M. Uehara, M. Takahashi, K. Nishio, T. Saito, Electrochemical characteristics of $LiNiO_2$ and $LiCoO_2$ as a positive material for lithium secondary batteries, J. Power Sources 54 (1995) 522-524. https://doi.org/10.1016/0378-7753(94)02140-X

[76] A. Rudge, I. Raistrick, S. Gottesfeld, J. Ferraris, A study of the electrochemical properties of conducting polymers for application in electrochemical capacitors, Electrochim. Acta 39 (1994) 273-287. https://doi.org/10.1016/0013-4686(94)80063-4

[77] C. Peng, S. Zhang, D. Jewell, G.Z. Chen, Carbon nanotube and conducting polymer composites for supercapacitors, Prog. Nat. Sci. 18 (2008) 777-788. https://doi.org/10.1016/j.pnsc.2008.03.002

[78] L.-M. Huang, T.-C. Wen, A. Gopalan, Electrochemical and spectroelectrochemical monitoring of supercapacitance and electrochromic properties of hydrous ruthenium oxide embedded poly(3,4-ethylenedioxythiophene)–poly(styrene sulfonic acid) composite, Electrochim. Acta 51 (2006) 3469-3476. https://doi.org/10.1016/j.electacta.2005.09.049

[79] M. Mallouki, F. Tran-Van, C. Sarrazin, P. Simon, B. Daffos, A. De, C. Chevrot, J.F. Fauvarque, Polypyrrole-Fe_2O_3 nanohybrid materials for electrochemical storage, J. Solid State Electrochem. 11 (2007) 398-406. https://doi.org/10.1007/s10008-006-0161-8

[80] P. Gómez-Romero, K. Cuentas-Gallegos, M. Lira-Cantú, N. Casañ-Pastor, Hybrid nanocomposite materials for energy storage and conversion applications, J. Mater. Sci. 40 (2005) 1423-1428. https://doi.org/10.1007/s10853-005-0578-y

[81] N.A. Kumar, H.-J. Choi, Y.R. Shin, D.W. Chang, L. Dai, J.-B. Baek, Polyaniline-grafted reduced graphene oxide for efficient electrochemical supercapacitors, ACS Nano 6 (2012) 1715-1723. https://doi.org/10.1021/nn204688c

[82] X. Lu, H. Dou, C. Yuan, S. Yang, L. Hao, F. Zhang, L. Shen, L. Zhang, X. Zhang, Polypyrrole/carbon nanotube nanocomposite enhanced the electrochemical capacitance of flexible graphene film for supercapacitors, J. Power Sources 197 (2012) 319-324. https://doi.org/10.1016/j.jpowsour.2011.08.112

[83] S. Palsaniya, H.B. Nemade, A.K. Dasmahapatra, Synthesis of polyaniline/graphene/MoS$_2$ nanocomposite for high performance supercapacitor electrode, Polymer 150 (2018) 150-158. https://doi.org/10.1016/j.polymer.2018.07.018

[84] B.N. Reddy, S. Deshagani, M. Deepa, P. Ghosal, Effective pseudocapacitive charge storage/release by hybrids of poly(3,4-ethylenedioxypyrrole) with Fe$_3$O$_4$ nanostructures or Co3O4 nanorods, Chem. Eng. J. 334 (2018) 1328-1340. https://doi.org/10.1016/j.cej.2017.11.068

[85] E. Frackowiak, V. Khomenko, K. Jurewicz, K. Lota, F. Béguin, Supercapacitors based on conducting polymers/nanotubes composites, J. Power Sources 153 (2006) 413-418. https://doi.org/10.1016/j.jpowsour.2005.05.030

[86] G.A. Snook, P. Kao, A.S. Best, Conducting-polymer-based supercapacitor devices and electrodes, J. Power Sources 196 (2011) 1-12. https://doi.org/10.1016/j.jpowsour.2010.06.084

[87] H. Wu, D. Li, X. Zhu, C. Yang, D. Liu, X. Chen, Y. Song, L. Lu, High-performance and renewable supercapacitors based on TiO$_2$ nanotube array electrodes treated by an electrochemical doping approach, Electrochim. Acta 116 (2014) 129-136. https://doi.org/10.1016/j.electacta.2013.10.092

[88] H. Wang, C. Guan, X. Wang, H.J. Fan, A high energy and power Li-Ion capacitor based on a TiO$_2$ nanobelt array anode and a graphene hydrogel cathode, Small 11 (2015) 1470-1477. https://doi.org/10.1002/smll.201402620

[89] H. Kim, M.-Y. Cho, M.-H. Kim, K.-Y. Park, H. Gwon, Y. Lee, K.C. Roh, K. Kang, A novel high-energy hybrid supercapacitor with an anatase TiO$_2$–reduced graphene oxide anode and an activated carbon cathode, Adv. Energy Mater. 3 (2013) 1500-1506. https://doi.org/10.1002/aenm.201300467

[90] Y. Wang, Z. Hong, M. Wei, Y. Xia, Layered H$_2$Ti$_6$O$_{13}$-nanowires: A new promising pseudocapacitive material in non-aqueous electrolyte, Adv. Funct. Mater. 22 (2012) 5185-5193. https://doi.org/10.1002/adfm.201200766

[91] M. Lübke, P. Marchand, D.J.L. Brett, P. Shearing, R. Gruar, Z. Liu, J.A. Darr, High power layered titanate nano-sheets as pseudocapacitive lithium-ion battery anodes, J. Power Sources 305 (2016) 115-121. https://doi.org/10.1016/j.jpowsour.2015.11.060

[92] F. Wu, Z. Wang, X. Li, H. Guo, Hydrogen titanate and TiO_2 nanowires as anode materials for lithium-ion batteries, J. Mater. Chem. 21 (2011) 12675-12681. https://doi.org/10.1039/c1jm11042j

[93] J. Li, Z. Tang, Z. Zhang, Layered hydrogen titanate nanowires with novel lithium intercalation properties, Chem. Mater. 17 (2005) 5848-5855. https://doi.org/10.1021/cm0516199

[94] B. Reichman, A.J. Bard, Electrochromism at niobium pentoxide electrodes in aqueous and acetonitrile solutions, J. Electrochem. Soc.127 (1980) 241-242. https://doi.org/10.1149/1.2129628

[95] G. Luo, H. Li, L. Gao, D. Zhang, T. Lin, Porous structured niobium pentoxide/carbon complex for lithium-ion intercalation pseudocapacitors, Mater. Sci. Eng. B 214 (2016) 74-80. https://doi.org/10.1016/j.mseb.2016.09.004

[96] K. Brezesinski, J. Wang, J. Haetge, C. Reitz, S.O. Steinmueller, S.H. Tolbert, B.M. Smarsly, B. Dunn, T. Brezesinski, Pseudocapacitive contributions to charge storage in highly ordered mesoporous group v transition metal oxides with iso-oriented layered nanocrystalline domains, J. Am. Chem. Soc. 132 (2010) 6982-6990. https://doi.org/10.1021/ja9106385

[97] J.W. Kim, V. Augustyn, B. Dunn, The effect of crystallinity on the rapid pseudocapacitive response of Nb_2O_5, Adv. Energy Mater. 2 (2012) 141-148. https://doi.org/10.1002/aenm.201100494

[98] S. Lou, X. Cheng, L. Wang, J. Gao, Q. Li, Y. Ma, Y. Gao, P. Zuo, C. Du, G. Yin, High-rate capability of three-dimensionally ordered macroporous $T-Nb_2O_5$ through Li^+ intercalation pseudocapacitance, J. Power Sources 361 (2017) 80-86. https://doi.org/10.1016/j.jpowsour.2017.06.023

[99] S. Ouendi, C. Arico, F. Blanchard, J.-L. Codron, X. Wallart, P.L. Taberna, P. Roussel, L. Clavier, P. Simon, C. Lethien, Synthesis of $T-Nb_2O_5$ thin-films deposited by atomic layer deposition for miniaturized electrochemical energy storage devices, Energy Storage Mater. (2018). https://doi.org/10.1016/j.ensm.2018.08.022

[100] L. Kong, C. Zhang, J. Wang, W. Qiao, L. Ling, D. Long, Free-standing T-Nb_2O_5/graphene composite papers with ultrahigh gravimetric/volumetric capacitance

for Li-Ion intercalation pseudocapacitor, ACS Nano 9 (2015) 11200-11208. https://doi.org/10.1021/acsnano.5b04737

[101] J. Yang, T. Lan, J. Liu, Y. Song, M. Wei, Supercapacitor electrode of hollow spherical V_2O_5 with a high pseudocapacitance in aqueous solution, Electrochim. Acta 105 (2013) 489-495. https://doi.org/10.1016/j.electacta.2013.05.023

[102] M. Li, G. Sun, P. Yin, C. Ruan, K. Ai, Controlling the formation of rodlike V_2O_5 nanocrystals on reduced graphene oxide for high-performance supercapacitors, ACS Appl. Mater. Interfaces 5 (2013) 11462-11470. https://doi.org/10.1021/am403739g

[103] G.G. Amatucci, F. Badway, A. Du Pasquier, T. Zheng, An asymmetric hybrid nonaqueous energy storage cell, J. Electrochem. Soc. 148 (2001) A930-A939. https://doi.org/10.1149/1.1383553

[104] H. Kim, K.-Y. Park, M.-Y. Cho, M.-H. Kim, J. Hong, S.-K. Jung, K. Chul Roh, K. Kang, High-performance hybrid supercapacitor based on graphene-wrapped $Li_4Ti_5O_{12}$ and Activated Carbon, ChemElectroChem 1 (2014) 125-130. https://doi.org/10.1002/celc.201300186

[105] K. Naoi, S. Ishimoto, Y. Isobe, S. Aoyagi, High-rate nano-crystalline $Li_4Ti_5O_{12}$ attached on carbon nano-fibers for hybrid supercapacitors, J. Power Sources 195 (2010) 6250-6254. https://doi.org/10.1016/j.jpowsour.2009.12.104

[106] M. Mladenov, K. Alexandrova, N.V. Petrov, B. Tsyntsarski, D. Kovacheva, N. Saliyski, R. Raicheff, Synthesis and electrochemical properties of activated carbons and $Li_4Ti_5O_{12}$ as electrode materials for supercapacitors, J. Solid State Electrochem. 17 (2013) 2101-2108. https://doi.org/10.1007/s10008-011-1424-6

[107] C.C. Fu, L.J. Zhang, J.H. Peng, H. Wang, H. Yan, Synthesis of $Li_4Ti_5O_{12}$-reduced graphene oxide composite and its application for hybrid supercapacitors, Ionics 22 (2016) 1829-1836. https://doi.org/10.1007/s11581-016-1726-x

[108] B.-G. Lee, S.-H. Lee, Application of hybrid supercapacitor using granule $Li_4Ti_5O_{12}$/activated carbon with variation of current density, J. Power Sources 343 (2017) 545-549. https://doi.org/10.1016/j.jpowsour.2017.01.094

[109] R.A. Aziz, I.I. Misnon, K.F. Chong, M.M. Yusoff, R. Jose, Layered sodium titanate nanostructures as a new electrode for high energy density supercapacitors, Electrochim. Acta 113 (2013) 141-148. https://doi.org/10.1016/j.electacta.2013.09.128

[110] C. Wang, Y. Xi, M. Wang, C. Zhang, X. Wang, Q. Yang, W. Li, C. Hu, D. Zhang, Carbon-modified $Na_2Ti_3O_7 \cdot 2H_2O$ nanobelts as redox active materials for high-

performance supercapacitor, Nano Energy 28 (2016) 115-123.
https://doi.org/10.1016/j.nanoen.2016.08.021

[111] X. Qiu, X. Zhang, L.-Z. Fan, In situ synthesis of a highly active $Na_2Ti_3O_7$ nanosheet on an activated carbon fiber as an anode for high-energy density supercapacitors, J. Mater. Chem. A 6 (2018) 16186-16195.
https://doi.org/10.1039/C8TA04982C

[112] Z. Jian, L. Zhao, H. Pan, Y.-S. Hu, H. Li, W. Chen, L. Chen, Carbon coated $Na_3V_2(PO_4)_3$ as novel electrode material for sodium ion batteries, Electrochem. Commun. 14 (2012) 86-89. https://doi.org/10.1016/j.elecom.2011.11.009

[113] X. Zhong, Z. Yang, Y. Jiang, W. Li, L. Gu, Y. Yu, Carbon-coated $Na_3V_2(PO_4)_3$ anchored on freestanding graphite foam for high-performance sodium-ion cathodes, ACS Appl. Mater. Interfaces 8 (2016) 32360-32365.
https://doi.org/10.1021/acsami.6b11873

[114] P.F.R. Ortega, G.A. dos Santos Junior, L.A. Montoro, G.G. Silva, C. Blanco, R. Santamaría, R.L. Lavall, $LiFePO_4$/mesoporous carbon hybrid supercapacitor based on LiTFSI/imidazolium ionic liquid electrolyte, J. Phys. Chem. C 122 (2018) 1456-1465.
https://doi.org/10.1021/acs.jpcc.7b09869

[115] M. Ghidiu, M.R. Lukatskaya, M.Q. Zhao, Y. Gogotsi, M.W. Barsoum, Conductive two-dimensional titanium carbide 'clay' with high volumetric capacitance, Nature 516 (2014) 78-171. https://doi.org/10.1038/nature13970

[116] M.R. Lukatskaya, O. Mashtalir, C.E. Ren, Y. Dall'Agnese, P. Rozier, P.L. Taberna, M. Naguib, P. Simon, M.W. Barsoum, Y. Gogotsi, Cation intercalation and high volumetric capacitance of two-dimensional titanium carbide, Science 341 (2013) 1502.
https://doi.org/10.1126/science.1241488

[117] A.H. Feng, Y. Yu, Y. Wang, F. Jiang, Y. Yu, L. Mi, L.X. Song, Two-dimensional MXene Ti3C2 produced by exfoliation of Ti_3AlC_2, Mater. Design 114 (2017) 161-166.
https://doi.org/10.1016/j.matdes.2016.10.053

[118] M. Naguib, M. Kurtoglu, V. Presser, J. Lu, J.J. Niu, M. Heon, L. Hultman, Y. Gogotsi, M.W. Barsoum, Two-dimensional nanocrystals produced by exfoliation of Ti_3AlC_2, Adv. Mater. 23 (2011) 4248-4253. https://doi.org/10.1002/adma.201102306

[119] Z. Ling, C.E. Ren, M.Q. Zhao, J. Yang, J.M. Giammarco, J.S. Qiu, M.W. Barsoum, Y. Gogotsi, Flexible and conductive MXene films and nanocomposites with high capacitance, PNAS111 (2014) 16676-16681. https://doi.org/10.1073/pnas.1414215111

[120] R.B. Rakhi, B. Ahmed, M.N. Hedhili, D.H. Anjum, H.N. Alshareef, Effect of postetch annealing gas composition on the structural and electrochemical properties of Ti_2CT_x MXene electrodes for supercapacitor applications, Chem. Mater. 27 (2015) 5314-5323. https://doi.org/10.1021/acs.chemmater.5b01623

[121] J. Fu, J. Yun, S. Wu, L. Li, L. Yu, K.H. Kim, Architecturally robust graphene-encapsulated MXene Ti_2CT_x@polyaniline composite for high-performance pouch-type asymmetric supercapacitor, ACS Appl. Mater. Interfaces 10 (2018) 34212-34221. https://doi.org/10.1021/acsami.8b10195

Supercapacitor Technology: Materials, Processes and Architectures   Materials Research Forum LLC
Materials Research Foundations **61** (2019) 171-222   https://doi.org/10.21741/9781644900499-8

Chapter 8

Carbon Nanoarchitectures for Supercapacitor Applications

T. Manovah David and Tom Mathews*

Materials Science Group, Indira Gandhi Centre for Atomic Research, Kalpakkam - 603 102, India

*tom@igcar.gov.in

Abstract

Presently, carbon nano-architectures have received an elevated position to be attractive candidates for advanced energy storage applications. Carbon has been a material of choice in its various forms because of inherent properties such as high surface-area, inter-linked pores, large electrical conductivity and superior wettability towards the electrolyte ions. This chapter summarizes carbon materials as supercapacitor electrodes based on their architecture. Carbon materials have been broadly classified into three distinct categories viz. activated materials, non-activated materials and graphene-structured materials. The discussion is confined only to pristine carbon nano-architectures that display electrical double layer capacitance.

Keywords

Carbon, Electrical Double Layer Capacitance, Activated Carbon, Fibres, Aerogels, Glassy Carbon, Carbon Black, Carbide Derived, Graphene, Fullerenes, Nanotubes, Nanowalls

Contents

1.   Introduction

1.1   Capacitors

A *capacitor* is a device that stores electrical energy in the form of an electrostatic field at the interface of the conductors (i.e. anode and cathode) separated by a dielectric material. In recent devices, the dielectric material is replaced with an electrolytic solution. In such cases an electric double layer is formed at the conductor electrolyte interface [1, 2]. The amount of energy stored, also known as capacitance is directly proportional to the surface area of the conducting plates and the dielectric constant of the plates, and is inversely proportional to the separation distance between the plates. Capacitance is measured in terms of Farad.

1.2   Supercapacitors

A supercapacitor, also known as *electrochemical capacitor* and *ultracapacitor,* is a capacitor where very high surface area conducting plates/electrodes like porous carbon are separated by an electrolyte. It has the ability to store electrical charge at the surface-electrolyte interface as a double layer. In other words, the charge is stored in the form of *static electricity* in supercapacitors [3]. In general the charge stored is around 10 to 100 times of that observed in normal capacitors due to the large electrode surface area [4].

Though the concept of supercapacitors originated in the late 19th century, the idea found its possible utility in commercial applications only during the year 1957 when H.I. Becker of *General Electric* first patented the electrical charge double layer based device made of using crude porous carbon electrode and an aqueous electrolyte [5, 6]. Subsequently, the years 1962 and 1970 became milestones with the patents of Robert A. Rightmire [5] and Donald L Boos [5, 7], respectively of *Standard Oil Company of Ohio.* They used a disc shaped device comprising carbon paste as electrodes separated by an ion permeable membrane soaked in an aqueous electrolyte [8]. These developments were the first fruits of the technology and formed the basis for the elevation of this technology. Till this time, the capacitors were using only aqueous electrolytes. In the year 1978, Panasonic came up with supercapacitors named as 'Goldcap' using a non-aqueous electrolyte operating at a higher cell voltage. In all these cases, carbon was used as the primary electrode at both the terminals. Later, the supercapacitor from the company Nec in the same year paved way for the production of complementary metal-oxide semiconductor (CMOS) batteries commonly used in computers to moderate volatile clock chips [9]. The first major change in the supercapacitor electrodes occurred in the year 1989 when the company Elit replaced the positive electrode with nickel oxy-hydroxide while retaining the negative carbon electrode. This resulted in the formation of the first asymmetric supercapacitor. Thereafter, the supercapacitor sector witnessed several contributions with the introduction of hybrid capacitors, battery coupled and solar powered supercapacitors. At present supercapacitors are able to operate heavy-load vehicles such as trucks and busses. Yet, as observed in the progress, carbon was inevitable throughout the development of supercapacitors.

1.3 Supercapacitors vs. batteries

The Ragone plot (power density [W kg^{-1}] vs. energy density [J kg^{-1}]) for various energy storage systems [10] notify that the supercapacitors occupy a position between conventional capacitors and batteries. As observed, it is also understood that the supercapacitors are 100 to 1000 times more powerful than the batteries. That is supercapacitors display a power density of 1 – 10 kW kg^{-1} and whereas the batteries exhibit only a mere 150 W kg^{-1} [11]. This excellent power density allows supercapacitors to be used in applications that require a 'power burst'. However, the energy density side of supercapacitors is far less compared to that of the batteries. This unique combination of higher power density along with higher specific energy defines the possibility of using the supercapacitor as a complement to other power sources [12]. The supercapacitors are now largely employed in electronic, memory back-up and industrial as well as power management systems [13]. More recently, the emergency exit doors of Airbus A380 have utilized the supercapacitors citing their reliability and safety [14]. Though, the terms of supercapacitors have resemblance to that of batteries, but both the energy systems differ entirely. The differences are summarized in Table 1.

1.4 Electrical energy storage in supercapacitors

Basically, two types of energy storage mechanisms are involved in supercapacitors, viz. electrical double layer capacitance (EDLC) and pseudocapacitance. The former involved majorly in electrode materials that are not electrochemically active, where there is no transfer of electrons between the electrode and the electrolyte. Charge stored electrostatically in the absence of electron transfer across the interface is called *non-Faradaic* charge storage. Carbon is one material that is not electrochemically active and therefore is supposed to generate EDLC type of capacitance. Materials that are electrochemically active (transferring electrons) undergo pseudocapacitance type capacitance. Since, the electrons are transferred across the interface, it is called as *Faradaic* charge storage. In both these systems, the electrodes at the terminals are similar and are called symmetric supercapacitors. Alternatively, if a supercapacitor is constructed with EDLC based electrode on one end and pseudocapacitance based electrode on the other end, it is called as *asymmetric* supercapacitor. If a battery-type electrode is connected to one terminal with EDLC type electrode on the other, the combination is called *hybrid* supercapacitor. In this section, EDLC is majorly discussed owing to its association with carbon based supercapacitors, while pseudocapacitance is deliberated for the sake of common understanding.

Table 1.   *List illustrating the differences between Capacitors and Batteries*

S.No	Comparison Parameter	Capacitor	Battery
1	Redox reactions	Do not occur: Non-Faradaic energy storage	Occur: Faradaic type energy storage
2	Energy storage	Direct form of energy storage: Electrostatic	Indirect form of energy storage: Electrochemical
3	Power density	Relatively higher	Relatively lower
4	Energy density	Relatively lower	Relatively higher
5	Chemical change	Do not take place	Does take place
6	Limitation	Redox kinetics, mass transport	Conductivity of electrolyte
7	Cyclability	Almost unlimited	limited cyclability (few thousand cycles)
8	Electron transfer	Electrons are not transferred at the interface	Electrons are transferred at the interface
9	Faraday's law	Positive and negative charges are separated by a dielectric: Faradays law do not apply	Faradays law apply: This is related to the electrode potential
10	Oxidation state of the electrode	Do not change: Electrode intact	changes: Electrode oxidation state changes
11	Involvement of electrons in	Electrons involved are localized conduction band electrons	Electrons are transferred from valance band [from orbitals]
12	Life time	Long (except for corrosion)	Poor (electrode degradation)
13	Charge density	0.18 electrons/atom	1 – 3 electrons/atom
14	Energy storage	Capacitance (F)	Capacity (C/W.h)
15	Gibbs energy	$G = QV / 2$	$G = Q.\Delta E$
		Q = Charge, V = Potential and ΔE = Difference in potential	

1.4.1 Electric double layer capacitance

Electric double layer capacitance arises when a charged material (usually a carbon electrode) is dipped into an electrolyte. At the insertion, there is an immediate arrangement of ionic charges along the surface to balance the excess or deficiency of electronic charges with the accumulation of counter ions (or adsorption of ions on the electrode surface). This phenomenon is typical of the electrostatic charge storage and several theories have been put forward to describe the activity at the interface between the solid and the liquid. Likewise, when a cell is constructed, both the electrodes contribute individually to the formation of EDLC with a minor change, that, one of the electrodes has excess of electrons whereas the other has a deficiency of electrons. The accumulation and withdrawal of electrons over electrodes, both anode and cathode strictly depend on the applied potential between them. The counter ions *arrange* on the outer surface in the electrolyte depending on the *status* of the electrode to provide electro-neutrality at the interface and therefore, this process is considered as mere physical charge storage. The thickness of the double layer is around 5-10 Å depending on the size of the ions being accumulated. The electric field at the interface is expected to be as high as 10^6 V cm^{-1} [15].

For optimal operation of EDLC the following points are to be satisfied (a) large surface area for improvement in capacitance, (b) better electrical conduction to reduce inherent resistance, (c) superior pore size distribution (on electrodes) to balance the anticipated ionic size of the electrolyte employed, (d) interconnecting pores for large accessibility of ions and (e) high surface wettability, thereby allowing the maximum utilization of the specific surface area of the electrode [12].

1.4.2 Pseudocapacitance

Non-carbonaceous materials such as metal oxides and conducting polymers are likely to undergo Faradaic charge-transfer processes resulting in pseudocapacitance. Here capacitance is due to the swift redox reaction occurring near the surface of the electrode. Interestingly, the electrical response displayed by the pseudocapacitance type material is similar to that of EDLC type material, where charge varies continuously with respect to the applied potential [12]. However, different mechanisms are involved in pseudocapacitance type charge storage, for instance underpotential deposition of 'H' adatoms, transition metal oxide based redox reactions, intercalation reactions in porous materials and doping/de-doping inconducting polymers [16]. Notably, the power performance of the pseudocapacitance type materials is slightly lower than that of EDLC type materials owing to the involvement of rate-determining Faradaic processes. Also, on considering the electrode stability, the pseudocapacitance type capacitors are less stable owing to the swelling and shrinking of electrode phases during the cycling process leading to a poor mechanical stability and poor cycle life. The capacitance exhibited by pseudocapacitance is about 10 to 100 times higher than that of EDLC. Therefore, the combinatorial effect of pseudocapacitance and EDLC in a single electrode is expected to largely improve the specific capacitance [17]. A comparison of EDLC and pseudocapacitance is presented in Table 2.

Table 2.   Comparison between EDLC and Pseudocapacitance

S. No	Comparison Parameter	Electrical double layer capacitance	Pseudocapacitance
1	Charging process	Non-Faradaic (Physical with no redox reactions)	Faradaic (Chemical with redox reactions)
2	Charge storage	20-50 $\mu F\ cm^{-2}$	200-500 $\mu F\ cm^{-2}$
3	Applied potential	Restricted (Based on electrolyte)	Restricted (Based on electrolyte)
4	Reversibility	Highly reversible over a million times	Reversibility better than batteries yet behind EDLC
5	Capacitance	Constant with potential	Constant with potential
6	Voltammogram	Mirror image	Mirror image
7	Power performance	Swift	Slower (Faradaic process)
8	Stability of electrode	Highly stable	Prone to become brittle owing to swelling / shrinking

1.5   Construction of a supercapacitor cell

The assembly of a supercapacitor much similar to that of a battery system, where the two electrodes are dipped into an electrolyte has been developed. Yet as a difference from batteries, the electrodes of supercapacitors were separated by an ion permeable membrane. The electrode-electrolyte *interface* was considered as *capacitor*. Therefore a cell consists of two capacitors in series separated by an electrolyte resistance. The general electrical equivalent circuit for a symmetrical capacitor cell is given as,

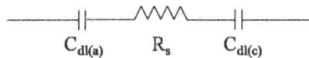

$$\dashv\vdash\!\!\!-\!\!\!\text{\Large\text{\textit{WW}}}\!\!\!-\!\!\vdash\vdash$$
$$C_{dl(a)} \qquad R_s \qquad C_{dl(c)}$$

where, the $C_{dl(a)}$ and $C_{dl(c)}$ are double layer capacitances of anode and cathode, respectively and R_s is the electrolyte resistance exerted.

The capacitance of the supercapacitor cell (C_{cell}) is given as [1],

$$1/\ C_{cell} = 1/C_{dl(a)} + 1/C_{dl(c)} \tag{1}$$

where, $C_{dl(a)}$ and $C_{dl(c)}$ are the double layer capacitances in series at anode and cathode terminals of the supercapacitor cell. If $C_{dl(a)} = C_{dl(c)}$ then the above equation could be written as

$$C_{cell} = C_{dl(a)}\ /\ 2 \tag{2}$$

Thereafter, the double layer capacitor is given to be as [18],

$$C_{dl} = \varepsilon A\ /\ 4\pi d \tag{3}$$

where, ε is the dielectric constant of the electrical double layer formed over the electrode, A denotes the surface area of the electrode and d represents the thickness of the double layer [19]. In order to achieve intensely high capacitance, thin charge separation in extreme large surface area is favored. Based on this information the energy and power density of the capacitor can be calculated using the formulae,

$$E = CV^2/2 \text{ or } QV/2 \qquad\qquad (4)$$

$$P = V^2/4R \qquad\qquad (5)$$

where, E, C, V, Q, P and R are energy, capacitance, applied voltage, charge, power density and equivalent series resistance, respectively [3, 18].

Two other factors that decide the efficiency of the supercapacitor material are the pore-size distribution and operating voltage. The pore-size distribution is a significant factor owing to the movement of ions within the pores. Since ions do not move in the same way as they do in the liquid. The movement is largely hindered if the size of the pore is very small and therefore smaller pores *do not* contribute to the double layer capacitance. In that case the relation between the specific surface area (obtained via Brunauer-Emmett-Teller [BET] surface area analysis) and the capacitance do not match [20]. In addition, the size of the ions to be adsorbed is expected to be rather very small to improve the capacitance. Similarly, the operating voltage plays a vital role in determining the specific energy and power of the supercapacitors. It is to be noted that the operating voltage is decided by the electrolyte employed. Usually, aqueous electrolyte involving acid (such as H_2SO_4) and basic solutions (such as KOH) have a very low decomposition voltage i.e. ~1.23 V [21]. This factor leads to quick disintegration of the electrolyte. Non-aqueous electrolytes such as propylene carbonate and acetonitrile have been utilized to improve the operating voltage up to 2.5 V [22].

1.6   Carbon as electrode

In general, carbon has four allotropes based on their bonding nature, viz. diamond (sp^3, σ bonds alone), graphite (sp^2), fullerene (distorted sp^2) and carbyne (sp^1) [12]. One of the primary requirements for carbon to be an electrode is that it should be conducting. Therefore, except diamond all the other hybridizations of carbon can be used as an electrode. As seen earlier, carbon materials strongly support energy storage with the formation of EDLC. The surface of the electrode utilized for the interfacial charge storage can get altered under the influence of surface groups. The presence of surface groups alters the wettability, electrical contact resistance, adsorption of ions and self-discharge characteristics [3, 23]. Graphitic carbon, with two different chemical surroundings is observed to have two different types of ion-adsorption sites, a) edge site and b) basal site. The edge sites with unpaired electrons are largely exposed to the electrolyte ions than the basal sites. These electrons are prone to interact with molecular oxygen thereby *oxidizing* the edge sites, primarily through reversible physisorption even below lower critical temperatures. Interaction with molecular oxygen at slightly higher temperatures, leads to the irreversible chemisorption process. Apart from oxygen, other molecular moieties such as hydrogen, nitrogen, sulfur and halogens can also contribute to the formation of surface functionalities [24]. The addition of these functionalities effectively contributes to the capacitance of a supercapacitor essentially through pseudocapacitance.

1.6.1 Change in electrical properties of carbon materials

Resistance: Carbon materials absolutely relate their structures to that of the electrical properties achieved. As observed from the allotropes of carbon, σ bonded carbon are highly insulating with resistivity greater than 10^{12} Ω cm [24]. This type of carbon is not suitable for being used in the capacitive applications. Therefore, the carbon with π bonds is necessary to act as capacitor electrode material. As discussed earlier, apart from graphite (the naturally obtained π bonded material), all the others have been synthesized. The conductivity of solid carbon, produced in the laboratory has been determined by the heat treatment, where the carbon was heated above 700 °C. The heat treatment paved way for the formation of π bonds that acted as charge carriers, thereby increasing conductivity. In this process, the separate entities of carbon were interconnected to form a conducting network [12, 25].

Equivalent series resistance: The intra-particle resistivity or the intrinsic resistance of the carbon material is largely dependent on the morphology and chemical structure. If the intra-particle contact is less than it would result in the poor conductivity [12]. The porosity of the carbon material also adds to the resistance where the current carrying path is altered [26]. Also, prolonged exposure of carbon to oxygen would aggravate the formation of surface oxygen functionalities preferentially at the edges, thereby increasing the resistivity [27]. The adverse effects of these could be reduced by making electrodes in the form of thin film, adding pressure to pack the synthesized carbon and heating above 1000 °C to remove any surface oxygen formation.

The physical and chemical properties such as high conductivity, large surface-area, superior corrosion resistance, effective thermal stability, defined pore-size distribution and utility as composite material contribute to the effective use as supercapacitor electrode.

With this background on supercapacitors and carbon, we shall look into the contributions, carbon has made in various forms towards supercapacitor applications.

2.   Electrochemical characterization of EDLC based carbon materials

The electrochemical properties of the carbon materials have been studied using cyclic voltammetry (CV), galvanostatic charge-discharge and electrochemical impedance spectroscopy (EIS). In general, CV measurements are carried out to estimate the EDLC performances of the carbon material based electrodes. galvanostatic charge-discharge is conducted to test the performance of the capacitors [28]. EIS helps in identifying the equivalent series resistance. General equation for analyzing the *specific capacitance* of the electrode material is given as

$$C_{specific} = 4 \times I \times t / \Delta V \times m \qquad (6)$$

where, I is the discharge current density (mA), t is time taken to discharge (s), ΔV denotes the difference in potential during discharge (V), m is the active mass of the material and $C_{specific}$ is the specific capacitance of the carbon electrode (F g^{-1}). Note, in all these the specific capacitance analysis would give varied results based on the electrode system used (either two or three electrode system).

2.1   Cyclic voltammetry (CV) studies of carbon electrode based supercapacitors

The voltammogram of a carbon material with EDLC properties should be rectangular owing to the *electrostatic* mechanism adopted. In common, the CV of an EDLC material shows pseudo-rectangular curves [Figure 1a]. At lower scan rate, typically more rectangular CV is obtained, whereas at a higher scan rate; the diffusion layers near the electrodes are narrowed down and as a result, higher currents are observed in the voltammogram with the formation of an oblique angle [Figure 1b]. The narrowing down is due to the larger ohmic resistance exhibited by the electrode at higher scan rates. Moreover, the difference in the two lines of the voltammogram loop denotes an ohmic drop while undergoing charging and discharging processes [Figure 1a]. The drop is also called as IR potential drop resulting due to anodic or cathodic polarization [3]. Broader the loop, greater is the capacitance indicating superior reversibility. The *capacitance* of the material is given by the formula

$$C = I / r \qquad (7)$$

where, C is the capacitance, I is the current density (middle point in the curve) and r is scan rate. The ΔV (difference in potential) depends on the nature of the electrolyte electrolyte used. If the current density is kept constant, the voltage varies linearly with time where the capacitance arising from EDLC is ideal. The linear relationship is then given by C = I (dV/dt), where, dV/dt is the slope.

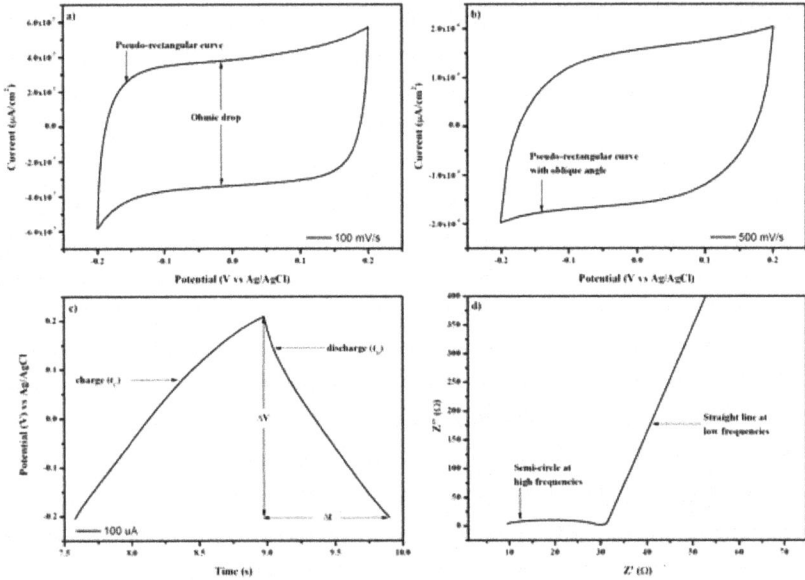

Figure 1. Cyclic voltammogram at a) lower scan rate, b) higher scan rate and model plots of c) galvanostatic charge-discharge and d) EIS

2.2   Galvanostatic charge/discharge for EDLC based carbon electrodes

The *specific capacitance* of a carbon electrode can be calculated employing galvanostatic charge-discharge (usually for two electrode cell configuration) using the formula,

$$C_{specific} = I \times \Delta t / A \times \Delta V \qquad (8)$$

where, I is the current density, Δt is the time taken to discharge from charged state, A is the active area of the electrode and ΔV is the potential difference [Figure 1c]. If the specific capacitance is estimated for powder samples then active surface area *A* given in the formula is replaced by *m*, denoting the active mass of the substance.

The *Columbic efficiency* (η) of a material can be determined using galvanostatic charge-discharge with the following equation [29],

$$\eta = t_D / t_C \times 100 \qquad (9)$$

where, t_D is the discharge time and t_C is the charging time, both usually measured in seconds [Figure 1c]. The Columbic efficiency is based on the cycle stability, where the first cycle is compared with the nth cycle for evaluation. The result is expressed in percentage.

Finally, the technique is also used to identify the *capacitance retention* of the carbon electrode material with the formula,

$$C_R = t_{Dn} / t_{D1} \times 100 \qquad (10)$$

where, C_R is the capacitance retention, t_{Dn} is the time taken to discharge for the nth cycle and t_{D1} is the time taken to discharge for the 1st cycle. The capacitance retention (obtained in percentage) of an electrode is held to be superior if it is intact for over 1 million cycles.

2.3   Electrochemical impedance spectroscopy (EIS) for carbon electrodes

Specific capacitance values have been determined using EIS. In EIS, Nyquist plots play a major role in defining the equivalent circuits of the system. The graph is plotted against Z_{real} (Z') vs. Z_{imag} (Z"), from higher to lower alternating current frequencies. Here Z_{real} and Z_{imag}, stand for impedance imaginary part and impedance real part, respectively. At high frequencies, a well-defined semi-circle appears and at lower frequencies a straight line appears [Figure 1d]. It is interesting to note that the diameter of the semi-circle denotes the interfacial charge-transfer Faradaic resistance. The semi-circle also provides input to estimate the *specific capacitance* value using the formula [30],

$$C_{specific} = 1 / m \times j \times 2\pi f \times Z_{imag} \qquad (11)$$

where, $C_{specific}$ is the specific capacitance, m is the active mass of the material, f is the frequency and Z_{imag} is the imaginary part impedance.

3.   Various architectures of *carbon* as supercapacitor electrodes

This chapter is intended to discuss the architectures of the carbon electrodes that undergo EDLC type of charge storage used in the supercapacitor applications as shown in Figure 2. The carbon materials are classified based on their *physical* features such as activated materials, non-activated materials and graphene-structured materials. Each section of the carbon classification is discussed here with respect to their supercapacitive properties.

3.1   Activated materials

3.1.1 Activated carbon

Activated carbons are generally prepared from the agricultural-waste biomass and hence these are economical, readily available, ecofriendly and renewable with the basic essence of the much needed *green* characteristics to safeguard the planet. In addition, utilization of biomass contributes to the agro-industrial waste mitigation [31]. Biomass; can be understood as renewable organic materials obtained either from plant kingdom or animal kingdom that have the ability to serve as an energy source. These biomass materials are also called *lignocellulosic*, since these are principally made up of hemicelluloses, cellulose and lignin [32]. These building blocks of the biomass comprise elements such as carbon, hydrogen, oxygen, nitrogen and traces of sulfur as well as chlorine. In general, the materials that are used to form activated carbons normally have suitability for EDLC formation and are prepared mainly from shells and husks of diverse *flora* and bones of various *fauna*. Preparation of activated carbons from the biomass is composed of two significant simultaneous processes, viz. carbonization and activation. The properties of the activated carbons such as surface area, pore structure and chemical polarity, largely depend on the precursor material and the activation process.

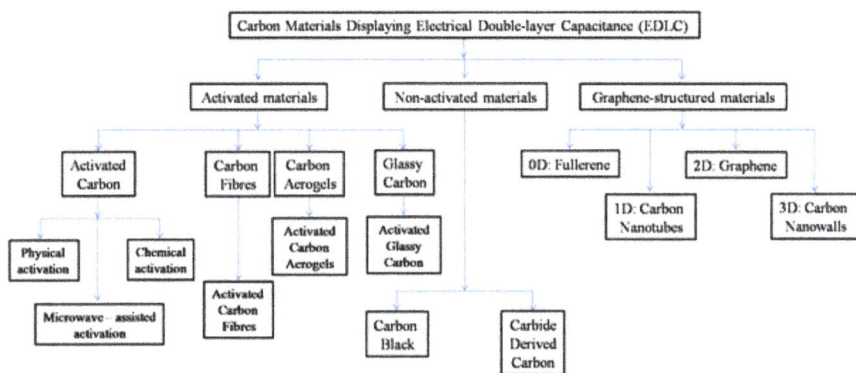

Figure 2. Classification of carbon nanomaterials pertaining to electrical double layer capacitance (EDLC) type supercapacitors

3.1.1.1 Carbonization

Carbonization is otherwise known as *charring*, where the organic (hydrocarbon) precursor turns out completely to carbon along with residues of heteroatom elements [33]. The process is conducted through a *conventional* heating procedure between 400 and 850 °C in an inert atmosphere devoid of oxygen [33]. The resultant product is of relatively low porosity, with the structure consisting of elementary crystallites with large number of disorganized carbon called as tars [12]. The tars block all the pores generated during the carbonization process.

3.1.1.2 Activation

Activation is a process where the pore-blocking tars are removed. At the end of activation, the existing pores are opened up and new pores are formed. The procedure defines the pore structure and subsequently the porosity increases. Usually, activation is carried out in a *convectional* fashion that leads to the thermal decomposition of the carbon materials [34]. Activation to form activated carbons can be carried out using three different methods namely, physical, chemical and microwave induced activation. *Physical activation* also called as controlled gasification. In this method, carbon dioxide (CO_2), air or steam (or combinations of these) is passed over carbonized materials at temperatures between 700 and 1100 °C [19, 35, 36]. In general, oxidizing atmosphere is used to increase the pore volume and CO_2 is used due to its cleanliness and easy-to-handle ability. As discussed earlier, activation improves the porosity of the activated carbons and it is accomplished with the gradual carbon burn-off and elimination of volatile matters. The quality of the generated activated carbons is influenced by two factors, e.g. gas flow and type of furnace.

Chemical activation is conducted using activating agents such as KOH, $ZnCl_2$, NaOH, K_2CO_3, H_3PO_4 and $FeCl_3$ [37-39]. Among these, KOH and $ZnCl_2$ are predominantly used. These agents are used as dehydrating agents and oxidants. This activation protocol is conducted at temperatures about 600 to 950 °C to influence pyrolytic decomposition. This process is confirmed to produce better porous structure with improved carbon yield as shown in Figure 3. This technique results in the generation of very high surface area. The demerit is the need of post-activation washing, for the removal of organic and inorganic residuals that are introduced into the activated carbons during activation. The removal of contaminants is largely tiresome, time-consuming and expensive [33]. *Microwave activation* is carried out through heating by conversion of microwave energy through dipole rotation and ionic conduction within the particles [40]. In this process, the thermal energy is introduced from the "inside" of the carbon materials. Whereas in the other activation processes (physical and chemical), thermal heating is introduced from the outside and the presence of thermal gradient is observed throughout the processes. The microwave activation requires short process time. The Table 3 briefly compares the activation procedures pertaining to the formation of the activated carbons.

3.1.1.3 Modifying the surface of activated carbons

The EDLC efficacy of the activated carbons is determined by their texture and surface properties. These two factors are reliant on the conditions and procedure of the chosen activation method. After activation, two changes take place, viz. increase in surface area with enhanced number of pores and surface heterogeneity. Primarily, at the completion of the activation process the activated carbons produce pores of different sizes. The pore-size determines the classification of pores, as micropores (< 2 nm), mesopores (2 – 50 nm) and macropores (> 50 nm) [34, 41, 42]. Amid these different pore sizes, the major contribution for EDLC is obtained from micropores [43]. However, for the statement to be true, these micropores should be available for the ions to interact for displaying improved EDLC, i.e. these pores should be at the proximity of the electrolyte ions to pass through. Therefore, this constraint is relieved by the utility of mesopores that efficiently contribute to the passage of ions into the bulk. Further, it is to be understood that specific capacitance is not directly proportional to specific surface-area [11, 36, 44, 45]. The fact is that the entire pores are not accessible to the electrolyte ions [46]. In general, capacitance obtained from micropores (C_{micro}) is likely to become half of the anticipated value, because one ion occupies both the sides of the walls of the micropores as in the case of organic electrolytes because of larger size. Therefore, organic electrolytes demonstrate lesser capacitance compared to aqueous electrolytes with smaller ions. The larger ions offer also hindrance, in terms of being irreversibly trapped at micropores. This phenomenon is called *sieving effect* of the micropores. On the whole, the EDLC depends on the size of the electrolyte ions and not on the BET surface area of the activated carbons. Still, the researchers have not come to a conclusion in defining the optimal pore size for displaying excellent EDLC. Yet, for the sake of understanding, the reports convey that pores of size 0.3 to 0.7 nm are suitable for aqueous electrolyte ions and 0.8 nm pores are effective for organic electrolyte ion [36, 47]. Maximum EDLC is expected when the ion size of the electrolyte matches well with the pore size of the electrode [14, 36]. Alternatively, the mesopores provide the ions with good electrolyte accessibility and excellent intra and inter-particle conductivity.

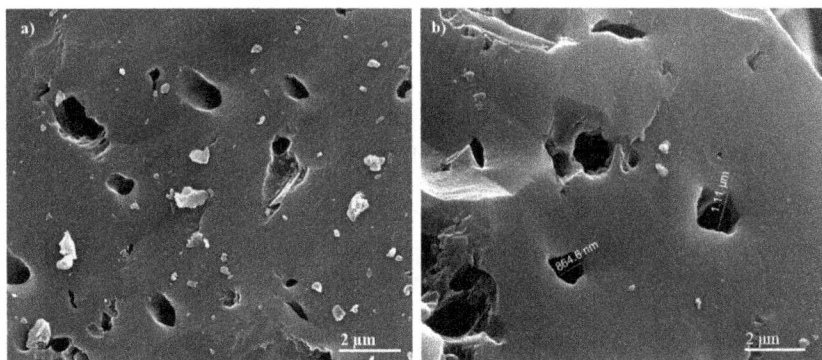

Figure 3. Activated carbon obtained after chemical activation using KOH from a) orange peel and b) groundnut shell

Supercapacitor Technology: Materials, Processes and Architectures   Materials Research Forum LLC
Materials Research Foundations **61** (2019) 171-222   https://doi.org/10.21741/9781644900499-8

Table 3. Comparison of the activation protocols

Property	Activation Method		
	Physical	Chemical	Microwave
Activating agents	CO_2, Air, Steam	KOH, $ZnCl_2$, NaOH, K_2CO_3, H_3PO_4 and $FeCl_3$	Microwave energy (dipole rotation and ionic conduction)
Process time	Long	Long	Short
Energy consumption	Huge	Huge	Less
Heating type	Thermal gradient – Heat transfer	Thermal gradient – Heat transfer	Volumetric heating – Energy transfer
Heating procedure	From the outside	From the outside	From the inside
Homogeneity	Inhomogeneous	Inhomogeneous	Homogeneous
Heating rate (Time-taken)	More	More	Less
Energy saving	Not supported	Not supported	High support
Advantages	Carbon burn-off	High surface area formation	Obvious problems of Physicochemical activations: avoided
Disadvantages	High temperature operation	Removal of organic and inorganic residues	Sophistication required

The electrolytes also have their obvious fingerprint in terms of capacitance, where organic electrolyte provides larger potential window and the aqueous electrolyte provides comparatively lower potential window. However, for a similar potential window, aqueous electrolyte has better capacitance than that of organic electrolyte because of the ionic size. Both the scenarios are dependent on the size of the ion. Further, in terms of EIS, the higher the interaction of the ions with the carbon surfaces, the better is the frequency response behavior of the electrode i.e. energy to be extracted from the electrode at higher frequencies

On the other hand, surface heterogeneity contributes to the existing capacitance of the activated carbons. Major contribution for the heterogeneity is from adsorbed oxygen and hydrogen, also, a minor contribution from nitrogen, sulfur, halogen and phosphorus. These heteroatoms are introduced onto the activated carbons either from the precursor or deliberately added during activation. Apart from these gases, the other activating agents that contribute to the heterogeneity are HNO_3, H_2SO_4 and H_2O_2, respectively [33]. The surface heterogeneity displays pseudocapacitance and it adds up to the existing EDLC, thereby improving the overall capacitance of the activated carbons. The combinatorial effect owing to surface-area and surface-heterogeneity, results with improved wettability, capacitance, electrical conductivity and low or zero self discharge of the activated carbons.

3.1.1.4 Supercapacitance exhibited by activated carbons

At present, activated carbons are commonly used in commercial supercapacitors. Though aqueous electrolytes display better cyclability and higher specific capacitance up to 300 F g^{-1}, non-aqueous electrolyte based activated carbons supercapacitors are more favored for commercial applications. This is owing to the fact that organic electrolytes allow supercapacitors to reach a maximum operating voltage of 2.7 V [48, 49]. Moreover, these supercapacitors (involving organic electrolytes) exhibit a specific capacitance ranging between 100 and 120 F g^{-1} or 60 F cm^{-3} in case of non-planar and planar formats, respectively [21, 49, 50]. The use of organic electrolyte faces challenges with the possibility of its decomposition and increment in the series resistance with the presence of active surface oxides and moisture. This event results in the instability of the electrode thereby preventing its effective utility [12]. The supercapacitors with surface-active activated carbon electrodes are reported to undergo speedy aging. Hence, supercapacitors with activated carbon electrodes without heteroatoms are preferred for commercial applications. Table 4 lists the specific capacitances of activated carbons obtained from various biomass materials using different activation techniques. The list includes activated carbon electrodes reported either in two or three electrode configurations used for testing the respective *capacitance*.

3.1.2 Carbon fibres

Carbon fibres (CFs) are generally obtained from two different procedures as *activated CFs* and *exfoliated CFs*. CFs are prepared from cellulose (also known as rayon), phenolic resins, polyacrylonitrile and pitch (tar) based materials [24]. *Activated CFs* are synthesized by electro-spinning, where the precursor solution is taken in a spinnerette and the extrudant is drawn as a thin fibre onto a substrate. The obtained product is stabilized in the temperature range 200 - 400 °C. Thereafter, the stabilized material is carbonized in the temperature range 800 - 1500 °C. This procedure results in the formation of CFs. The CFs are then activated by annealing in oxidizing atmosphere at 400 – 900 °C under inert atmosphere at graphitizing temperature above 3000 °C [51]. The quality of the activated CFs largely depends on two factors viz. the precursor employed and the manufacturing process. Among the precursors mentioned above, soft precursor (such as pitch) derived activated CFs exhibit superior electrical properties compared to that of hard precursors (such as polyacrilonitrile) [12, 46]. Remarkably, almost every activated CFs has a diameter of around 10 μm and possesses a very narrow pore-size distribution with the pores of size below 2 nm. Therefore, quick adsorption or desorption of electrolyte ions is highly possible in activated CFs [46, 52]. The morphologies of the activated CFs ranged from bundles, chopped fibers, thread to cloth [12, 35]. The activated CFs bundles can be woven into cloth fabrics with a possibility to form a large surface area up to 2500 m^2g^{-1}.

Table 4. Specific capacitance of activated carbons reported in literature

Material	Activation Method	BET Surface Area ($m^2 g^{-1}$)	Specific Capacitance ($F\ g^{-1}$)	Electrolyte	Cell Design	Ref.
Fibres of oil palm empty fruit bunches	Chemical (KOH)	1704	150 4.297 Wh kg^{-1}	1 M H_2SO_4	2E	[130]
Waste coffee grounds	Chemical (ZnCl$_2$)	1000	368 20 Wh kg^{-1}	1 M H_2SO_4	3E	[38]
Apricot shell	Chemical (NaOH)	2335	348	6 M KOH	2E	[39]
Bamboo species	Chemical (KOH)	850-1100	161	2 M H_2SO_4	2E	[131]
Casaval peel waste	Chemical (KOH)	1352	264.08	0.5 M H_2SO_4	3E	[132]
Sunflower seed shell	Chemical (KOH)	899-1163	311	1 M H_2SO_4	2E	[133]
Coffee endocarp	Chemical (KOH)	89 to 1050	176		3E	[134]
Argan seed shells	Chemical (KOH)	2100	355	1 M H_2SO_4	3E	[135]
Biochar	Chemical (KOH)	2959	260	6 M KOH	2E	[136]
Waste tea-leaves	Chemical (KOH)	2245 - 2841	330	2 M KOH	3E	[137]
Furfurol	Physical (Steam)	1040	111	1Me$_3$BuImBF$_4$	2E	[138]
Firwood	Physical (Steam)	1130	142	1 M HNO_3	3E	[139]
Sucrose	Physical (CO$_2$)	2100	163	1 M H_2SO_4	2E	[140]
Sucrose	Physical (CO$_2$)	1940	148	EMImBF$_4$	2E	[141]
Peanut shell and rice husk	Microwave (ZnCl$_2$)	1527-1634	99	Et$_4$NBF$_4$/PC	2E	[142]

Table 5. Specific capacitance of carbon fibres

Material	Activation Method	BET Surface Area $(m^2 g^{-1})$	Specific Capacitance $(F g^{-1})$	Electrolyte	Cell Design	Ref.
Activated carbon fibres from banana stems	Chemical treatment with KOH or $ZnCl_2$	1097	74	1 M Na_2SO_4	3E	[143]
Textile carbon fibres	Synthesized from organic polymers	-	88	$H_4[W_{12}SiO_{40}]$, H_3PO_4 and 85% H_2O	3E	[144]
Multi-walled carbon nanotube fibres	Chemical vapor deposition and electrospinning	-	13.31 3.87 kW kg^{-1} 2.84 Wh kg^{-1}	H_3PO_4 and PVA	3E	[145]
Carbon microfibers coated multi-walled carbon nanotubes	Electrospinning depsotion	-	11.1 3.7 μW cm^{-1} 0.7 μWh cm^{-1}	H_3PO_4 and PVA	3E	[146]
Activated carbon fibres	Electrospinning / carbonisation	1400	344 2.98 kW kg^{-1} 8.1 Wh kg^{-1}	6 M KOH	1E	[147]
Activated carbon cloth	Exfoliation by modified Hummers	61.2	8.8	1 M H_2SO_4	3E	[148]
Activated carbon fibres web	Electrospinning / carbonisation	940-2100	175	30 wt% KOH	2E	[149]
Activated carbon fibres	Chemical treatment	2000	340 (acidic), 270 (basic)	4 M H_2SO_4 7 M KOH	3E	[48]
Activated carbon fibres	Carbonization / chemical activation	448-1520	255 (acidic) 202 (basic) 87 (organic)	2 M H_2SO_4, 1 M KOH, 1 M Et_4NBF_4	2E	[150]
Activated carbon fibres	Carbonization / steam activation	1200-3223	280	1 M H_2SO_4	3E	[151]

Activated carbons are largely different from activated CFs, primarily due to their pore structure. In activated CFs, the micropores are directly exposed on the surface, subsequently providing good accessibility to the active sites, whereas in activated carbon micropores are within the walls of mesopores and macropores [52]. Moreover, activated CFs allow the control of pore sizes and pore lengths, whereas it is difficult in activated carbons. In addition, activated carbons have very high BET surface area, yet activated

CFs of the same value perform better by displaying excellent electrical properties [46]. In general, the specific capacitance demonstrated by activated CFs is about 100 F g^{-1}. Under other conditions, the *exfoliated CFs* exhibit large surface area owing to the exfoliation process conducted over the precursor carbon fibres. Compared to the activated CFs, the pores in exfoliated CFs are larger. Exfoliated CFs have the capacitance of about 400 F g^{-1}. The high specific capacitance may be due to the involvement of pseudocapacitance in addition to EDLC. The electrolyte ions have the tendency to intercalate into the graphite gallery of exfoliated CFs [53, 54]. Table 5 lists the specific capacitances exhibited by various activated CFs and exfoliated CFs.

Table 6. Specific capacitance of carbon aerogels

Material	Activation Method	BET Surface Area (m^2 g^{-1})	Specific Capacitance (F g^{-1})	Electrolyte	Cell Design	Ref.
Carbon aerogels	Pyrolysis	1873	302	6 M KOH	3E	[152]
Carbon aerogels	Pyrolysis	403-587	171	30% H$_2$SO$_4$	3E	[153]
Carbon aerogels	Polycondensation	-	110.06	6 M KOH	3E	[154]
Carbon aerogels	Pyrolysis	660-1400	110 29 Wh kg^{-1}	0.8 M Et$_4$NBF$_4$-PC	2E	[58]
Carbon aerogels	Pyrolysis	600	40 (H$_2$SO$_4$) 35 (KOH)	3 M H$_2$SO$_4$ 4 M KOH	2E	[155]
Carbon aerogels/Fe$_3$O$_4$ composite	Hydrothermal	-	333.1	6 M KOH	3E	[156]

3.1.3 Carbon aerogels

Organic aerogels such as formaldehyde [55, 56] and resorcinol [56] subjected to pyrolysis result in the formation of carbon aerogels. The final product as solid powders is obtained via two steps, sol-gel process (poly-condensation of the organic aerogels takes place) followed by pyrolysis. Manipulating the sol-gel process, the physical properties such as shape, size, porosity and density of the obtained carbon aerogels can be controlled [57]. At the completion of the pyrolysis process, the electrical conductivity of the carbon aerogels significantly improved [12]. Solid powder carbon aerogels form an interconnected network of intraparticle pores, consequently enhancing the BET surface area up to 400 – 1000 m^2 g^{-1}. Further, the technique provides large porosity and uniformity in the pore sizes. The sizes of the primary carbon aerogels particles were in the range of 4 and 9 nm [58, 59] and the pore sizes obtained largely fell in the mesoporous range, i.e. between 2 and 50 nm. The electrodes based on carbon aerogels can be prepared as binder loaded carbon aerogel powders [59], binder-free carbon aerogel monolithic [60, 61] and stable carbon aerogel thin films [60]. In all the cases, EDLC mode of energy storage operates. The electrochemical studies revealed that the capacitance was large when the mesopores predominant. Moreover, carbon aerogels with pore sizes in the range of 3 to 13 nm displayed excellent capacitance properties [61]. Notably, BET surface-area played only a minor role in the capacitance determination, however, the capacitance matched well with the mesoporous surface-area.

3.1.3.1 Activated carbon aerogels

Physical [56, 58] activation was conducted on the carbon aerogels to improve their supercapacitive properties. Activating carbon aerogels in CO_2 at higher temperatures increased the BET surface area from ~650 $m^2\,g^{-1}$ to ~2500 $m^2\,g^{-1}$. Yet the capacitance of the carbon aerogels did not proportionally improved as the capacitance decreased up to 8 $\mu F\,cm^{-2}$ from the original value of 18 $\mu F\,cm^{-2}$. This phenomenon is owing to the fact that the increment has happened only in the degree of pores, yet most of them are inaccessible to the electrolyte anions. Therefore, the maximum capacitance experienced in this type of materials with surface area around 1000 - 2500 $m^2\,g^{-1}$ has been only up to 50 F cm^{-3} [56]. In another case, with an increase in the microporous surface area from a value of 530 $m^2\,g^{-1}$ to 1290 $m^2\,g^{-1}$ along with an increase in the mesoporous surface area from a value of 170 $m^2\,g^{-1}$ to 530 $m^2\,g^{-1}$, the capacitance amplified from 112 to 171 F g^{-1} [62]. Table 6 lists the specific capacitances exhibited by carbon aerogels.

3.1.4 Glassy carbon

Carbon that has a vitreous (or glassy) appearance with polymeric chain is called glassy carbon. This type of carbon is usually produced by thermally degrading the selected polymeric resins. Common resins that are employed for the purpose are, phenolic resins and furfuryl alcohol [63, 64]. Glassy carbons are produced through three stages, a) curing of the polymeric resin, b) slow carbonization and c) annealing at elevated temperatures. Among the three steps, the final step decides the physical property of the synthesized glassy carbons. The heating temperature can range between 600 ℃ and 3000 ℃. However, the literature reports revealed that the temperature around 1800 ℃ has been optimal for producing glassy carbon with excellent physical properties [63]. Glassy carbons has been reported to have a pore diameters of 1 to 5 nm [65]. Notably, the obtained glassy carbons have *closed* pores that inhibit the complete access of the surface to that of the incoming electrolyte ions leading to the poor contribution to EDLC. Moreover, the structure of glassy carbon is a cumulative of intertwined graphene sheets that on the other hand supports better conductivity. Upon comparing the density and tensile strength of glassy carbon with graphite, the density was lower and the tensile strength was higher than that of graphite. The lower density is attributed to the more presence of pores and the higher tensile strength is ascribed to the architecture of glassy carbon with intertwined graphene sheets. Glassy carbon is reported to have very low internal resistance of the value of about 3 to 8×10^{-6} Ω m [24]. This inherent quality with desired low internal resistance of the glassy carbons makes them highly suitable for high-power enabled supercapacitors [66].

3.1.4.1 Activated glassy carbons

One of the hitches of glassy carbons is to have close-ended pores that have an innate quality to suppress the effective supercapacitive behavior. The close-ended pores cause the induction of isolated porosity that does not contribute to EDLC formation onto the surface of the glassy carbon material [65]. Therefore, thermally activating the glassy carbon is expected to open up the pores subsequently increasing the surface area suitable for EDLC type charge storage. Interestingly, thermal activation controls both growth and thickness of the glassy carbon with the controlled diffusion of the oxidant into the active material. As a result, activated glassy carbon produces a well interconnected outer surface and the carbon layer beneath, in terms of both mechanical strength and electrical conductivity. Another way of activation is electrochemical oxidation which causes discrepancies in the charge storage process as it adopts augmented ionic resistance when organic electrolytes are employed [67]. Therefore in comparison, thermal activation well suits for glassy carbon over electrochemical activation. The thermal activation results in the double layer capacitance of approximately 20 μF cm^{-2}. The supercapacitor properties displayed by glassy carbons are presented in Table 7.

Table 7. Specific capacitance of glassy carbons

Material	Preparation Method	BET Surface Area (m^2g^{-1})	Areal Capacitance	Electrolyte	Cell Design	Ref.
Glassy carbon	Thermochemical oxidation	900	-	-	-	[64]
Glassy carbon	Thermochemical oxidation	-	20 μF/cm^2	3 M H$_2$SO$_4$	3E	[66]
Glassy carbon	Thermal oxidation	1000	12 μF/cm^2	3 M H$_2$SO$_4$	3E	[63]
Glassy carbon	Anodic oxidation at 1.95 V	-	200 mF/cm^2	1 M H$_2$SO$_4$	3E	[65]

3.2 Non-activated materials

3.2.1 Carbon blacks

Carbon blacks in colloidal size have nearly spherical structures. The structure carbon black is said to be *nearly spherical* owing to the fact that while being synthesized the tiny carbon particles of colloidal size coalesce together to form aggregates or agglomerates of any shape. These materials are produced either by partial combustion or by thermal decomposition of hydrocarbons in the gaseous phase [12]. Carbon black generally assumes fine particle size, aggregated spherical structure and large amount of pores [12, 67]. In general, efficiently conducting carbon blacks are highly branched with open-ended structure and oxygen-free chemically-clean surface. This type of carbon blacks have superior conductivity in the range between 10 and 10^4 S m^{-1} [68]. The electron conduction in carbon blacks is possible through two probable pathways, either jumping between closely associated aggregates or though graphitic manner when the agglomerates are *in touch* with others.

Table 8. Specific capacitance of carbon blacks

Material	Preparation Method	BET Surface Area (m^2g^{-1})	Specific Capacitance $(F\,g^{-1})$	Electrolyte	Cell Design	Ref.
Carbon black	Inkjet printing	1800	115	25% KOH	3E	[71]
Graphene nanosheet/Carbon black	Insitu reduction method	586	175	6 M KOH	3E	[69]
Carbon black/ RuO$_2$.xH$_2$O	Precipitation and mixing Carbon black	90	768 10 kW kg^{-1} 26.7 Wh kg^{-1}	H$_2$SO$_4$	3E	[157]
Graphite/Carbon black /polytetrafluoroethylene	Cold pressing	2692	213 - 279	7 M H$_2$SO$_4$	3E	[70]
Poly vinylidene fluoride gel/ Carbon	Cold pressing	-	4.1	PVdF gel electrolyte	2E	[158]

Carbon black is employed in supercapacitor application in two distinct modes viz. as a) conductive fillers in composites [69, 70] and b) primary electrode material [71]. The *conductive filler* form, the carbon black is added to aid the conductivity of the active material. Though the filling process plays a vital role in improving the conductivity of the active material, it has to be filled slightly above a critical loading. Below the critical loading level, the particles of carbon black are far away from each other and therefore

Supercapacitor Technology: Materials, Processes and Architectures   Materials Research Forum LLC
Materials Research Foundations **61** (2019) 171-222   https://doi.org/10.21741/9781644900499-8

influencing the conductivity of the active material is not possible. However, if the particles are available above the critical loading, then the composite becomes highly conducting rather than behaving as a supercapacitor. Hence, optimal loading above critical loading is essential to completely utilize the composite surface. The highly branched nature of the carbon black materials make them suitable to be employed as fillers and electrolyte reservoirs in the carbon-based composites that are suitable to be employed in supercapacitors [69]. Furthermore, carbon black also paves way to be used as a *primary electrode* material [71]. Commercially available carbon black is reported to have less packing density resulting in poor volumetric capacitance [72]. To overcome the set-back, the carbon black is printed on a conventional current collector or on the separator membrane using the inkjet printer as thin film of 1 μm thickness. Here, the diameters of the particles were brought down to 8 nm and the agglomerates were guarded to be around 100 nm. The double layer capacitance achieved though this procedure was around 800 μF [71]. In most cases, carbon black has attained a maximum specific capacitance of about 250 F g^{-1}. Table 8 lists the specific capacitances of carbon blacks where they have been used as fillers and as primary electrode materials.

3.2.2 Carbide-derived carbons

Carbide-derived carbons are commonly termed as CDCs. This type of nanoporous carbon is likely to be prepared from metal carbides such as titanium carbide (TiC) [73-76], zirconium carbide (ZrC) [77], silicon carbide (SiC) [78, 79], molybdenum carbide (Mo_2C), aluminum carbide (Al_4C), boron carbide (B_4C) [80] and titanium-aluminum alloy carbide (Ti_2AlC) [81]. In general, two different synthesis routes are followed viz. vacuum decomposition [77] and chlorination of metal carbides at high temperature [82]. Interesting phenomenon of the CDCs is that the synthesis procedure allows a better control over the porous morphology and architecture involving the surface functional groups [83]. In other words, the porosity tuning of the CDCs is greatly possible by altering the synthesis temperature. It is to be noted that the metal carbides mentioned in the above list produce desired microporous morphology based on two primary factors a) the atomic number of the metal in the precursor carbide and b) the temperature involved in the synthesis process between 400 and 1200 °C with the continuous flow of Cl_2 [78]. Normally, a precursor material with lower atomic number of the metal element in the carbide (e.g. TiC) is prone to produce narrower pore-size distribution than carbide containing a metal with higher atomic number (e.g. SiC). The phenomenon is witnessed when two different carbides are treated at the very same temperature. Conversely, within a carbide precursor, increase in the chlorination temperature proportionally increases the pore-size distribution [77, 78]. CDCs grown with TiC as the starting material are reported to have a large specific capacitance of about 220 F g^{-1}. Also, the size of pores plays a

194

vital role in improving the capacitance of the CDCs. Notably, micropores (< 2 nm) of the CDCs largely contribute to the double layer capacitance than mesopores (> 2 nm) [77]. Table. 9 illustrates the specific capacitances displayed by CDCs.

Table 9. Specific capacitance of carbide derived carbons

Material	Carbide Precursor	Preparation Method	BET Surface Area (m^2g^{-1})	Specific Capacitance $(F\ g^{-1})$	Electrolyte	Cell Degn.	Ref.
Nanoporous carbon	Titanium carbide	Chlorination	1450	100-130	1M Et_3MeNBF_4, propylene carbonate and acetonitrile	3E	[74]
Nanoporous carbon	Titanium carbide	Thermo-chemical etching	1200	130	1.5 M $(C_2H_5)_4N(BF_4)$ in acetonitrile	2E	[73]
Nanoporous carbon	TiC and ZrC	Leaching of metals	2000	150	1 M H_2SO_4	2E	[77]
Nanoporous carbon	TiC, α-SiC, Mo_2C, Al_4C_3 and B_4C	Chlorination	1085, 1505, 1490, 1525 and 1470	16.3 to 98.3	1 M $(C_2H_5)_3CH_3NBF_4$ in acetonitrile	3E	[80]
Nanoporous carbon	SiC ceramics	Chlorination	-	-	-	-	[159]
Nanoporous carbon	TiC and AlC, Ti_2AlC and B_4C	Chlorination	1550	175 and 147	1 M H_2SO_4	2E	[81]
Nanoporous carbon	NEt_4BF_4 in acetonitrile and propylene carbonate	Chlorination	1625	-	-		[75]

3.3   Graphene-structured materials

Graphene is derived from the naturally available mineral graphite [84]. In fact, a monolayer of graphite is called as graphene and it has superior physical, chemical and mechanical properties. It assumes a basic 2D structure completely made up of sp^2 carbons. Graphene has excellent in-plane electrical as well as thermal conductivity, robust mechanical strength of about 1 T Pa, highly manipulate-able surface area, approximately up to 2600 $m^2\ g^{-1}$ and effective chemical stability [85]. The single layer of graphene has the capacity to display a double layer capacitance of about 21 μF cm^{-1} and upon utilization of the entire surface it is expected to exhibit a specific capacitance of about 55 F g^{-1}. However, this value has not attained by any of the graphene-based supercapacitors so far, owing to the fact that the graphene layers undergo serious agglomeration. As a consequence, not all the surfaces are utilized leading to much lower capacitance than that is anticipated. Graphene is generally prepared through two distinct routes, by a) *physical means* via vapor depositions [86, 87], mechanical cleavage using atomic force microscope cantilever over graphite [88], plasma-based synthesis [89] and b) *chemical means* via chemical exfoliation of graphene from graphite using organic solvents [90] and reduction of graphite oxide to form graphene [91]. Graphene is also known as the mother of all graphitic forms such as fullerenes, carbon nanotubes, layered graphene and few-layer graphene [92]. The following section deals with the exhibition of supercapacitor properties, by all the architectures of graphene viz. zero-dimensional (0D): *fullerene*, one dimensional (1D): *carbon nanotubes* (CNTs), two dimensional (2D): *graphene* and *graphene oxides* and three dimensional (3D): *vertical graphene nanowalls*.

3.3.1 Fullerenes: 0D

Fullerenes are one of the most remarkable allotropes of carbon and are also commonly called as carbon dots. They are basically classified as single-shell fullerenes [93], multi-shell fullerenes (otherwise known as carbon onions) [94] or fullerene whiskers [95]. They are employed for the supercapacitor applications as both pristine materials and as composites with polyaniline [96], graphene [97], metal oxides [98] or metals [99]. The fullerenes are in general termed as C_{60} [93] or C_{70} [100, 101] based on the number of carbons they possess in a single carbon dot. C_{60} is reported to have 60 vertices and 32 faces [93]. Though fullerenes have been identified and synthesized quite long ago, yet the investigations on their capacitive properties remained very poor for a lengthy period. However, recently fullerenes have caught the attention of the scientific community owing to their possibility to form electrodes of micro-supercapacitors with lofty power density. These materials have capacity to reversibly hold up to six electrons into their architectures [102]. Fullerenes are procured as such [93] or synthesized through electron beam irradiation, condensation of carbon vapor and from carbon black while burning ghee in direct flame [98] and vacuum annealing of nanodiamond [94]. On the whole, the two latter techniques are commonly utilized for the preparation of carbon onions owing to their ease in synthesis.

The molecular ratio of C_{60} in an aromatic solvent (such as m-xylene) has been found to bring the change in the lattice parameters. The ratio of m-xylene to C_{60} with 0.83, 0.39 and 0.36 moles resulted in the formation of hollow-long cylindrical, hollow-long square and thick solid-short forms of fullerenes, respectively [93]. Each of the three showed variance in their lattice parameters, crystal structure and van der Waals forces compared to that of the C_{60} (bucky ball) owing to the incorporation of solvent, into the C_{60} architecture, during synthesis. Large variance in the lattice parameters was observed for hollow-long cylindrical C_{60}, that change principally contributed to a superior supercapacitor behavior [93]. Similarly, C_{70} can be investigated for its double layer capacitance [100]. The specific capacitances of C_{60} was found to be about 145.5 F g^{-1} [101]. Secondly, C_{60} has been used to make composites with conducting polymer such as polyaniline emeraldine base (C_{60}-PANI-EB) [96].

Table 10. Specific capacitance of fullerenes

Material	Preparation Method	BET Surface Area (m^2g^{-1})	Specific Capacitance	Electrolyte	Cell Design	Ref.
Fullerene crystals (C_{60})	-	-	36 μF cm^{-2}	-	3E	[93]
Onion-like carbon	Thermochemical gas phase oxidation	397-579	-	1 M H$_2$SO$_4$ and organic	3E	[94]
Fullerene-polyaniline emeraldine base	Covalent bonding functionalized fullerene	1000	776 μF g^{-1} 36.595 kW kg^{-1} 64 Wh kg^{-1}	3 M H$_2$SO$_4$	3E	[63]
Fullerene C_{60} whisker and polyaniline emeraldine based composite	-	-	813 F g^{-1}	1 M H$_2$SO$_4$	3E	[65]
Mesoporous graphitic carbon microtubes derived from fullerene C_{70} tubes	Ultrasonic liquid–liquid interfacial precipitation method	609	212.2 F g^{-1}	1 M H$_2$SO$_4$	3E	[160]
Fullerenes (C_{60})/graphene composite	Facile solution method	-	135.36 F g^{-1}	6 M KOH	3E	[161]
Fullerene-Capped gold nanoparticles on graphene	Seeded growth method	-	197 F g^{-1}	0.5 M KCl	3E	[162]
1D C_{70} Microstructures	KOH activation	1391.8	362.0 F g^{-1}	1 M H$_2$ SO$_4$	2E	[163]

Apart from this, the fullerenes have superiority over their counterparts such as activated carbon and other porous carbons. In common, activated carbons have been added into carbon blacks of size approximately 40 nm to improve the electrical properties. On the other hand, addition of C_{60} of sizes 5 – 10 nm to activated carbon has effectively improved the electrical properties [103]. Likewise, porous carbons have been reported to show better double layer capacitance behavior, in spite of the fact that the pathways for the electrolyte ions are largely blocked after a certain limit into the porous channels. In contrast, C_{60} exhibited better electrochemical behavior owing to the use of outer surface alone while forming the double layer. This opportunity allowed electrolyte ions to interact with C_{60} only at the outer facets and facilitates several cycles of double layers at high scan rates leaving little or no defects onto its surface [104]. The capacitance and other electrochemical features demonstrated by C_{60} fullerenes are given in Table. 10.

3.3.2 Carbon Nanotubes: 1D

Carbon nanotubes (CNTs) are considered as a material of choice for constructing a high-power electrode owing to their good electrical conductivity and effective accessible outer surface area [14]. Their outer surface is largely exposed and consists only of basal planes instead of edge planes. The occurrence leads to the formation of a nearly ideal surface suitable for a large potential window [46]. In spite of this superior effect, the inner surface area of the CNTs is mostly left unused. CNTs prepared in different architectures such as single-walled (SWCNTs) [105-107], double (DWCNTs) [108] or multi-walled (MWCNTs) [109] and vertically aligned (VACNTs) [110] have been employed for supercapacitor applications. In every case, the external surface of the CNTs is involved majorly in the formation of the double layer capacitance. However, the phase purity of the CNTs and their orientation largely contribute in enhancing the achievable capacitance. CNTs grown as a forest, directly over the substrate are effective in generating a superior capacitance as the need of a binder is nullified. Each form of the CNTs and their role in supercapacitors are briefly discussed here.

SWCNTs have been largely synthesized by techniques like, hydrogen arc-discharge of graphite [107, 111] and water-assisted chemical vapor deposition [105, 112] The latter method, also known as 'super growth' technique is said to produce highly pristine and poorly bundled SWCNTs thereby exposing 80 % of the surface. As a result, the ensuing SWCNTs produced are expected to display capacitance better than that of activated carbons. Yet, the improvement has been made only on the utilization of the outer surface of SWCNTs. Therefore, in order to utilize the inner surface of the SWCNTs, an oxidation process has been tried to induce tube openings in the tube walls [46]. The process of tube-opening allows the entry of electrolyte ions into the tube wall to contribute to the

EDLC. In this case almost two times increase in capacitance was achieved. Though the excess oxidation can damage the surface, optimal oxidation enhances the capacitance. The other architectures (DWCNTs and MWCNTs) have been widely reported owing to the ease in synthesizing. However, the capacitance values did not exceed that of SWCNTs due to the lack of effective surface area. On the whole, CNTs majorly linger with specific capacitances around 200 F g^{-1} [112]. In certain architectures of CNTs, capacitance was found to surpass 500 F g^{-1} [109]. Table. 11 illustrates the capacitances displayed by various forms of CNTs.

Table 11. Specific capacitance of carbon nanotubes

Material	Preparation Method	BET Surface Area (m^{2}g^{-1})	Specific Capacitance (F g^{-1})	Electrolyte	Cell Design	Ref.
SWCNTs	Opened carbon nanotubes	2200	114 98.9 kW kg^{-1} 24.7 Wh kg^{-1}	TEABF$_4$ in propylene carbonate	3E	[164]
SWCNTs	Pure SWCNTs	1100	34	1 M TEABF$_4$ in propylene carbonate	3E	[112]
SWCNTs	Electrochemical doping of pure SWCNTs	1100	25	1 M TEABF$_4$ in propylene carbonate	2E	[106]
SWCNTs	SWCNTs modified electrochemically	109	56	6 M KOH	2E	[111]
SWCNTs	HiPco buckytubes	500	45	1 M LiClO$_4$ in propylene carbonate	3E	[165]
SWCNTs	Films of SWCNTs	-	283	0.1 M C$_{16}$H$_{36}$F$_6$NP	3E	[166]
DWCNTs	Vertically aligned	-	44	1.96 M TEABF$_4$ in propylene carbonate	2E	[108]
MWCNTs	Aligned carbon nanotube electrodes	400	440 110 kW kg^{-1} 52 Wh kg^{-1}	[EMIM] [Tf$_2$N]	3E	[167]
MWCNTs	Carbon-nanotube aerogel electrodes	1059	524	5 M KOH	2E	[168]
MWCNTs	Multi walled carbon nanotubes	1315	66	1 M TEABF$_4$ mixed in acetonitrile	Two	[169]

3.3.3 Graphene sheets: 2D

Two-dimensional graphene is a sheet of one-atom thickness completely made up of sp^2 hybridized carbon atoms [Figure 4a]. The crystal lattice is polyaromatic with honeycomb structure [113]. The electrochemical properties such as rate capability, cycle capability and enhanced capacitance make them suitable for use as high-efficiency energy storage systems. Graphene also has ability to swiftly conduct electrons and 'store charges' on its surfaces. Moreover, it has good structural flexibility, tunable thickness, possible transparency, lightweight and impermeability [85, 114]. Other properties include high mechanical strength and chemical stability. All these advantages make graphene materials very desirable. Graphene is produced via spin-coating, layer-by-layer deposition and vacuum filtration [114]. While processing, graphene undergoes strong agglomeration and restacking [Figure 4b]. This phenomenon is owing to the π-π interaction and the van der Waals forces [115]. The agglomeration can greatly reduce the surface area and in turn hinders the diffusion of the electrolyte thereby potentially reducing the capacitance. This bottleneck has been overcome by (a) the addition of spacers (such as carbonaceous materials, metals or metal oxides), (b) by template-assisted growth of graphene and (c) crumpling [116-118].

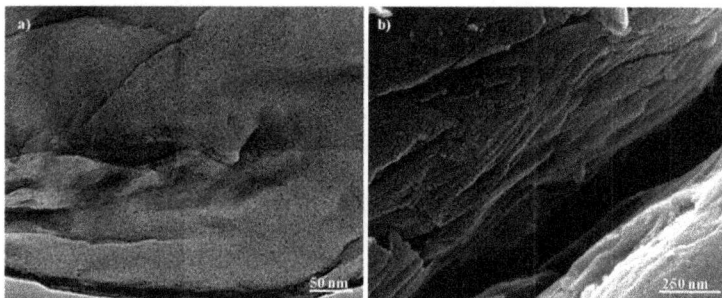

Figure. 4: a) TEM image of graphene sheets and b) FESEM image of rGO

Graphene has exhibited the specific capacitance of about 135 F g^{-1} in aqueous media and about 100 F g^{-1} in organic electrolytes [83]. Another close associate of the 2D graphene is the graphene oxide that is also considered to be its derivative. In general, graphene oxides prepared from natural graphite by modified Hummer's method have exhibited layered structure and suitable electrochemical properties [119, 120]. Based on the synthesis technique, functional groups containing oxygen such as C=O (carboxyl), C–OH (hydroxyl) and C–O (epoxy) are formed on the surface of graphene and behave as scaffolding agents between sheets [85]. The oxide functionalities are reduced by means of a reducing agent and the resultant product is called as the reduced graphene oxide

[Figure 4b]. The reduced form of graphene oxide exhibits supercapacitive properties up to a capacitance of around 135 F g^{-1}. Table 12 lists the specific capacitances of various 2D graphene-based materials.

Table 12. Specific capacitance of graphene

Material	Preparation Method	BET Surface Area (m^{2}g^{-1})	Specific Capacitance (F g^{-1})	Power Density (kW kg^{-1})	Energy Density (Wh kg^{-1})	Electrolyte	Cell Design	Ref.
Graphene nanosheets	Alkali modification	492	136 18.9 Wh kg^{-1}	-		1M Na$_2$SO$_4$	3E	[170]
Reduced graphene oxide	Hummer's method	-	348 (H$_2$SO$_4$) 158 (BMIPF$_6$)	-	-	1 M H$_2$SO$_4$ and BMIPF$_6$	3E	[171]
Non-covalently synthesized graphene	Chemical reduction	-	148	-	-	-	3E	[172]
Graphene oxide nanostructures	Hydrothermal process	-	352	27	5.1	alkaline	3E	[173]
Functionalized graphene	Solvothermal process	-	276	34.5	20	1 M H$_2$SO$_4$	3E	[174]
Graphene	-	-	100-250	9.838	136	EMIMBF$_4$	2E	[175]
Graphene	Chemical reduction	320	205	10	28.5	30 wt% KOH	2E	[176]
Graphene	Mechanical exfoliation	-	-	-	-	NaF, BMIMPF$_6$	2E / 3E	[177]
Graphene	Chemical vapour deposition		80 µFcm^{-1}			PVA/H$_3$PO$_4$	2E	[178]
Graphene oxide	Chemical reduction	-	394 µFcm^{-1}	-	-	PVA/H$_3$PO$_4$	2E	[178]
Graphene films	Vacuum filtration	-	135	7.2	15.4	2M KCl	3E	[179]
Graphene hydrogels	Hydrothermal reduction	780-960	220	30	5.7	5 M KOH	2E	[180]

Supercapacitor Technology: Materials, Processes and Architectures   Materials Research Forum LLC
Materials Research Foundations **61** (2019) 171-222   https://doi.org/10.21741/9781644900499-8

3.3.4 Vertical Graphene Nanowalls: 3D

Vertical graphene nanowalls (VGNs) have been uniformly grown as oriented 3D nanosheets with excellent special arrangement [Figure 5]. They have the capability to outperform all the other non-oriented graphene nanostructures [121]. VGNs were prepared primarily from thermal-induced chemical vapor deposition [122] or plasma-induced chemical vapor deposition of carbon generated from a mixture of methane and nitrogen or argon [123]. The plasma was generated using microwave [124] and radio frequencies [121]. Among the two types, the latter (i.e.) the radio frequency based chemical vapor deposition has been reported to produce denser carbon plane edges [125]. VGNs with effective unique features such as sturdy vertical orientation, exposed plane edges and inter-sheet open channels have been suitable for double layer capacitance. VGNs also render the possibility of large area and conformal coatings over its surface [126, 127]. In addition, VGNs do not undergo agglomeration as observed in the case of 2D graphene. Interestingly, the pore voids in VGNs contribute to the capacitive behavior as electrolyte ion reservoirs and every single sheet or the plane edges form a nanoelectrode [128]. However, the maximum performance ability of the VGNs depends on the height of growth, density of the graphene sheet, surface availability and the electrolyte used. Notably, apart from electrolyte used all the other needed parameters rest on the grown VGNs. The edge planes of VGNs are capable of storing of about $50 - 60$ μF $cm^{-2}$and displaying minimized distribution of charge-storage=. On the whole, the combinatorial effect of VGNs minimizes the series resistance thereby contributing much to the supercapacitive property [125]. The reported areal capacitance of VGNs is has been around 2 mF cm^{-2} [124][B1-304] in pristine format. Interestingly, VGNs not only form the basis for EDLC formation but also provide skeleton for functionalization [129]. The various areal capacitances displayed by VGNs are presented in Table 13.

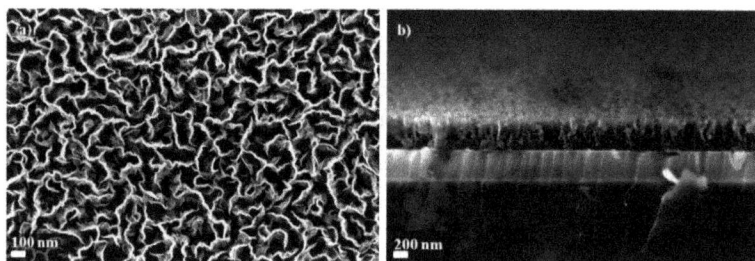

Figure. 5: FESEM images of VGNs, a) top-view and b) cross-section

Table 13. Specific capacitance of carbon nanowalls

Material	Preparation Method	Plasma Treatment	Areal Capacitance	Electrolyte	Cell Design	Ref.
Vertically oriented graphene	Inductively coupled plasma CVD	350 W, 350 Pa	713.6 μFcm^{-2}	6 M KOH	2E	[181]
Vertical graphene nanosheets	Microwave plasma enhanced CVD	300 W, 1.2×10^{-3} mbar	3.31 mFcm^{-2}	1 M tetra ethyl ammonium tetrafluoroborate in H$_2$SO$_4$	3E	[126]
Oriented graphene sheets	Plasma induced in-built electric field	375 W, 1×10^{-6} mbar	95 μFcm^{-2}	Aqueous H$_2$SO$_4$	3E	[182]
Vertical graphene nanosheets	Plasma enhanced CVD	100, 300, 600 W, 1.2×10^{-3} mbar	1.7 mFcm^{-2}	1 M KOH	3E	[124]
Vertical graphene nanosheets	Electron cyclotron resonance plasma enhanced CVD	200 W, 1×10^{-3} mbar	197 μFcm^{-2} (KOH) 188 μFcm^{-2} (H$_2$SO$_4$) 43.8 μFcm^{-2} (Na$_2$SO$_4$)	neutral (Na$_2$SO$_4$), alkaline (KOH), and acidic (H$_2$SO$_4$)	3E	[128]
Vertically oriented graphene	Plasma enhanced CVD	1000 V	Specific capacitance 156 F g^{-1} 112.6 kW kg^{-1} 4.98 Wh kg^{-1}	6 M KOH	2E	[123]
Vertical graphene nanosheets	Electron cyclotron resonance plasma CVD	375 W, 2×10^{-3} mbar	158 μFcm^{-2}	1 M KOH	3E	[183]

Summary

In this chapter, a concise overview is presented with focus on the various carbon nano-architectures as electrodes for supercapacitors. The general synthesis procedures and the indication of those architectures pertaining to their use in EDLC dependant supercapacitors are briefly discussed. The merits and the demerits of each carbon material pertaining to their pore structure, passivity of electrolyte ion into the carbon material based electrode and surface morphology have been discussed. The specific capacitances of the pristine carbon nanomaterials ranged between 100 and 500 F g^{-1}, a change that is witnessed solely on the change in their architecture. Correlation to the BET surface area and the capacitance varied largely owing to the obstruction of the electrolyte ions to occupy the entire surface. Notably, the materials that did not involve the use of binders displayed considerably improved capacitance owing to the decrement in the series resistance. The future of the carbon-based supercapacitors has large opening in terms of optimizing these surface architectures suitable for hybrid and asymmetric forms of supercapacitors.

Acknowledgement

T. Manovah David gratefully acknowledges the award of 'Research Associateship' received from Indira Gandhi Centre for Atomic Research, Kalpakkam, Department of Atomic Energy (DAE), Government of India. The authors wish to thank P.A. Manoj Kumar, Jayanthi Karthikeyan, Rebeca Gopu, Mahesh Ravichandran, Dhanavel Swaminathan and Vadivel Mani for their timely help in providing valuable inputs.

List of Abbreviations

0D:	zero dimensional
1D:	one dimensional
2D:	two dimensional
3D:	three dimensional
BET:	Brunauer-Emmett-Teller
CFs:	carbon fibres
CMOS:	complementary metal-oxide semiconductor
CNTs:	carbon nanotubes
CV:	cyclic voltammetery
DWCNTs:	double walled carbon nanotubes
EDLC:	electrical double layer capacitance
EIS:	electrochemical impedance spectroscopy
SWCNTs:	single walled carbon nanotubes
MWCNTs:	multi walled carbon nanotubes
VACNTs:	vertically aligned carbon nanotubes
VGNs:	vertical graphene nanowalls

References

[1] M. Winter, R.J. Brodd, What are batteries, fuel cells, and supercapacitors?, Chem. Rev. 104 (2004) 4245-4269. https://doi.org/10.1021/cr020730k

[2] X. Li, B. Wei, Supercapacitors based on nanostructured carbon, Nano Energy 2 (2013) 159-173. https://doi.org/10.1016/j.nanoen.2012.09.008

[3] B.E. Conway, Electrochemical supercapacitors: Scientific fundamentals and technological applications, Springer Science & Business Media, 2013.

[4] B. Conway, V. Birss, J. Wojtowicz, The role and utilization of pseudocapacitance for energy storage by supercapacitors, J. Power Sources 66 (1997) 1-14. https://doi.org/10.1016/S0378-7753(96)02474-3

[5] J. Miller, A brief history of supercapacitors, Battery Energ. Storage Technol. (2007) 61.

[6] H.I. Becker, Low voltage electrolytic capacitor, United States Patent, 1957.

[7] D. Boos, S. Argade, International Seminar on Double Layer Supercapacitors and Similar Energy Storage Devices, Florida Educational Seminars, Deerfield Beach, FL, 1991, pp. 1.

[8] D.L. Boos, Electrolytic capacitor having carbon paste electrodes, United States Patent, 1970.

[9] M. Endo, T. Takeda, Y. Kim, K. Koshiba, K. Ishii, High power electric double layer capacitor (EDLC's); from operating principle to pore size control in advanced activated carbons, Carbon Sci. 1 (2001) 117-128. https://doi.org/10.7209/tanso.2001.14

[10] T. Christen, M.W. Carlen, Theory of Ragone plots, J. Power Sources 91 (2000) 210-216. https://doi.org/10.1016/S0378-7753(00)00474-2

[11] G. Wang, L. Zhang, J. Zhang, A review of electrode materials for electrochemical supercapacitors, Chem. Soc. Rev. 41 (2012) 797-828. https://doi.org/10.1039/C1CS15060J

[12] A. Pandolfo, A. Hollenkamp, Carbon properties and their role in supercapacitors, J. Power Sources 157 (2006) 11-27. https://doi.org/10.1016/j.jpowsour.2006.02.065

[13] J.R. Miller, A.F. Burke, Electrochemical capacitors: challenges and opportunities for real-world applications, Electrochem. Soc. Interface 17 (2008) 53.

[14] L.L. Zhang, X. Zhao, Carbon-based materials as supercapacitor electrodes, Chem. Soc. Rev. 38 (2009) 2520-2531. https://doi.org/10.1039/b813846j

[15] S. Faraji, F.N. Ani, The development supercapacitor from activated carbon by electroless plating—A review, Renew. Sust. Energ. Rev. 42 (2015) 823-834. https://doi.org/10.1016/j.rser.2014.10.068

[16] B. Conway, W. Pell, Double-layer and pseudocapacitance types of electrochemical capacitors and their applications to the development of hybrid devices, J. Solid State Electrochem. 7 (2003) 637-644. https://doi.org/10.1007/s10008-003-0395-7

[17] C.M. Chuang, C.W. Huang, H. Teng, J.M. Ting, Effects of carbon nanotube grafting on the performance of electric double layer capacitors, Energ. Fuel 24 (2010) 6476-6482. https://doi.org/10.1021/ef101208x

[18] X. Andrieu, Energy Storage Syst., Electron. New Trends Electrochem. Technol. 1 (2000) 521.

[19] D. Qu, H. Shi, Studies of activated carbons used in double-layer capacitors, J. Power Sources 74 (1998) 99-107. https://doi.org/10.1016/S0378-7753(98)00038-X

[20] P. Sharma, T. Bhatti, A review on electrochemical double-layer capacitors, Energ. Convers. Manage. 51 (2010) 2901-2912. https://doi.org/10.1016/j.enconman.2010.06.031

[21] J. Fernández, T. Morishita, M. Toyoda, M. Inagaki, F. Stoeckli, T.A. Centeno, Performance of mesoporous carbons derived from poly (vinyl alcohol) in electrochemical capacitors, J. Power Sources 175 (2008) 675-679. https://doi.org/10.1016/j.jpowsour.2007.09.042

[22] X. Du, W. Zhao, Y. Wang, C. Wang, M. Chen, T. Qi, C. Hua, M. Ma, Preparation of activated carbon hollow fibers from ramie at low temperature for electric double-layer capacitor applications, Bioresource Technol. 149 (2013) 31-37. https://doi.org/10.1016/j.biortech.2013.09.026

[23] F.M. Delnick, Proceedings of the Symposium on Electrochemical Capacitors II, The Electrochemical Society, 1997.

[24] K. Kinoshita, Carbon: Electrochemical and physicochemical properties, Wiley Interscience, New York, 1988.

[25] S. Biniak, A. Swiatkowski, M. Pakula, L. Radovic, Electrochemical studies of phenomena at active carbon-electrolyte solution interfaces, Marcel Dekker, New York, 2001. https://doi.org/10.1002/chin.200117228

[26] A. Espinola, P.M. Miguel, M.R. Salles, A.R. Pinto, Electrical properties of carbons—resistance of powder materials, Carbon 24 (1986) 337-341. https://doi.org/10.1016/0008-6223(86)90235-6

[27] K. Radeke, K. Backhaus, A. Swiatkowski, Electrical conductivity of activated carbons, Carbon 29 (1991) 122-123. https://doi.org/10.1016/0008-6223(91)90103-P

[28] X. Li, W. Xing, S. Zhuo, J. Zhou, F. Li, S.Z. Qiao, G.Q. Lu, Preparation of capacitor's electrode from sunflower seed shell, Bioresource Technol. 102 (2011) 1118-1123. https://doi.org/10.1016/j.biortech.2010.08.110

[29] V. Khomenko, E. Raymundo-Pinero, F. Béguin, Optimisation of an asymmetric manganese oxide/activated carbon capacitor working at 2 V in aqueous medium, J. Power Sources 153 (2006) 183-190. https://doi.org/10.1016/j.jpowsour.2005.03.210

[30] S.M. Chen, R. Ramachandran, V. Mani, R. Saraswathi, Recent advancements in electrode materials for the high-performance electrochemical supercapacitors: A review, Int. J. Electrochem. Sci. 9 (2014) 4072-4085.

[31] D. Adinata, W.M.A.W. Daud, M.K. Aroua, Preparation and characterization of activated carbon from palm shell by chemical activation with K_2CO_3, Bioresource Technol. 98 (2007) 145-149. https://doi.org/10.1016/j.biortech.2005.11.006

[32] L. Zhang, C.C. Xu, P. Champagne, Overview of recent advances in thermo-chemical conversion of biomass, Energ. Convers Manage. 51 (2010) 969-982. https://doi.org/10.1016/j.enconman.2009.11.038

[33] A.M. Abioye, F.N. Ani, Recent development in the production of activated carbon electrodes from agricultural waste biomass for supercapacitors: A review, Renew. Sust. Energ. Rev. 52 (2015) 1282-1293. https://doi.org/10.1016/j.rser.2015.07.129

[34] J.L. Figueiredo, M. Pereira, M. Freitas, J. Orfao, Modification of the surface chemistry of activated carbons, Carbon 37 (1999) 1379-1389. https://doi.org/10.1016/S0008-6223(98)00333-9

[35] M. Endo, T. Maeda, T. Takeda, Y. Kim, K. Koshiba, H. Hara, M. Dresselhaus, Capacitance and pore-size distribution in aqueous and nonaqueous electrolytes using various activated carbon electrodes, J. Electrochem. Soc. 148 (2001) A910-A914. https://doi.org/10.1149/1.1382589

[36] E. Raymundo-Pinero, K. Kierzek, J. Machnikowski, F. Béguin, Relationship between the nanoporous texture of activated carbons and their capacitance properties in different electrolytes, Carbon 44 (2006) 2498-2507. https://doi.org/10.1016/j.carbon.2006.05.022

[37] R. Farma, M. Deraman, A. Awitdrus, I. Talib, E. Taer, N. Basri, J. Manjunatha, M. Ishak, B. Dollah, S. Hashmi, Preparation of highly porous binderless activated carbon electrodes from fibres of oil palm empty fruit bunches for application in supercapacitors, Bioresource Technol. 132 (2013) 254-261. https://doi.org/10.1016/j.biortech.2013.01.044

[38] T.E. Rufford, D. Hulicova-Jurcakova, K. Khosla, Z. Zhu, G.Q. Lu, Microstructure and electrochemical double-layer capacitance of carbon electrodes prepared by zinc chloride activation of sugar cane bagasse, J. Power Sources 195 (2010) 912-918. https://doi.org/10.1016/j.jpowsour.2009.08.048

[39] B. Xu, Y. Chen, G. Wei, G. Cao, H. Zhang, Y. Yang, Activated carbon with high capacitance prepared by NaOH activation for supercapacitors, Mater. Chem. Phys. 124 (2010) 504-509. https://doi.org/10.1016/j.matchemphys.2010.07.002

[40] H. Deng, G. Li, H. Yang, J. Tang, J. Tang, Preparation of activated carbons from cotton stalk by microwave assisted KOH and K_2CO_3 activation, Chem. Eng. J. 163 (2010) 373-381. https://doi.org/10.1016/j.cej.2010.08.019

[41] B.D. Zdravkov, J.J. Cermak, M. Sefara, J. Janku, Pore classification in the characterization of porous materials: A perspective, Cent. Eur. J. Chem. 5 (2007) 385-395. https://doi.org/10.2478/s11532-007-0017-9

[42] S. Porada, R. Zhao, A. Van Der Wal, V. Presser, P. Biesheuvel, Review on the science and technology of water desalination by capacitive deionization, Prog. Mat. Sci. 58 (2013) 1388-1442. https://doi.org/10.1016/j.pmatsci.2013.03.005

[43] E. Frackowiak, Carbon materials for supercapacitor application, Phys. Chem. Chem. Phys. 9 (2007) 1774-1785. https://doi.org/10.1039/b618139m

[44] G. Salitra, A. Soffer, L. Eliad, Y. Cohen, D. Aurbach, Carbon electrodes for double-layer capacitors I. Relations between ion and pore dimensions, J. Electrochem. Soc. 147 (2000) 2486-2493. https://doi.org/10.1149/1.1393557

[45] O. Barbieri, M. Hahn, A. Herzog, R. Kötz, Capacitance limits of high surface area activated carbons for double layer capacitors, Carbon 43 (2005) 1303-1310. https://doi.org/10.1016/j.carbon.2005.01.001

[46] M. Inagaki, H. Konno, O. Tanaike, Carbon materials for electrochemical capacitors, J. Power Sources 195 (2010) 7880-7903. https://doi.org/10.1016/j.jpowsour.2010.06.036

[47] C.O. Ania, V. Khomenko, E. Raymundo-Piñero, J.B. Parra, F. Beguin, The large electrochemical capacitance of microporous doped carbon obtained by using a zeolite template, Adv. Funct. Mater. 17 (2007) 1828-1836. https://doi.org/10.1002/adfm.200600961

[48] K. Babel, K. Jurewicz, KOH activated carbon fabrics as supercapacitor material, J. Phys. Chem. Solids 65 (2004) 275-280. https://doi.org/10.1016/j.jpcs.2003.08.023

[49] K. Jurewicz, C. Vix-Guterl, E. Frackowiak, S. Saadallah, M. Reda, J. Parmentier, J. Patarin, F. Béguin, Capacitance properties of ordered porous carbon materials prepared by a templating procedure, J. Phys. Chem. Solids 65 (2004) 287-293. https://doi.org/10.1016/j.jpcs.2003.10.024

[50] P. Simon, A. Burke, Nanostructured carbons: Double-layer capacitance and more, Electrochem. Soc. Interface 17 (2008) 38.

[51] A. Yoshida, I. Tanahashi, A. Nishino, Effect of concentration of surface acidic functional groups on electric double-layer properties of activated carbon fibers, Carbon 28 (1990) 611-615. https://doi.org/10.1016/0008-6223(90)90062-4

[52] M. Inagaki, Pores in carbon materials-importance of their control, New Carbon Mater. 24 (2009) 193-232. https://doi.org/10.1016/S1872-5805(08)60048-7

[53] Y. Soneda, J. Yamashita, M. Kodama, H. Hatori, M. Toyoda, M. Inagaki, Pseudo-capacitance on exfoliated carbon fiber in sulfuric acid electrolyte, Appl. Phys. A 82 (2006) 575-578. https://doi.org/10.1007/s00339-005-3395-x

[54] M. Toyoda, Y. Tani, Y. Soneda, Exfoliated carbon fibers as an electrode for electric double layer capacitors in a 1 mol/dm^3 H$_2$SO$_4$ electrolyte, Carbon 42 (2004) 2833-2837. https://doi.org/10.1016/j.carbon.2004.06.022

[55] R. Pekala, Organic aerogels from the polycondensation of resorcinol with formaldehyde, J. Mater. Sci. 24 (1989) 3221-3227. https://doi.org/10.1007/BF01139044

[56] H. Pröbstle, M. Wiener, J. Fricke, Carbon aerogels for electrochemical double layer capacitors, J. Porous Mat. 10 (2003) 213-222. https://doi.org/10.1023/B:JOPO.0000011381.74052.77

[57] U. Fischer, R. Saliger, V. Bock, R. Petricevic, J. Fricke, Carbon aerogels as electrode material in supercapacitors, J. Porous Mater. 4 (1997) 281-285. https://doi.org/10.1023/A:1009629423578

[58] B. Fang, B. Wei, K. Maruyama, M. Kumagai, High capacity supercapacitors based on modified activated carbon aerogel, J. Appl. Electrochem. 35 (2005) 229-233. https://doi.org/10.1007/s10800-004-3462-6

[59] C. Lin, J.A. Ritter, B.N. Popov, Correlation of double-layer capacitance with the pore structure of sol-gel derived carbon xerogels, J. Electrochem. Soc. 146 (1999) 3639-3643. https://doi.org/10.1149/1.1392526

[60] C. Schmitt, H. Pröbstle, J. Fricke, Carbon cloth-reinforced and activated aerogel films for supercapacitors, J. Non-Cryst. Sol. 285 (2001) 277-282. https://doi.org/10.1016/S0022-3093(01)00467-7

[61] R. Petričević, M. Glora, J. Fricke, Planar fibre reinforced carbon aerogels for application in PEM fuel cells, Carbon 39 (2001) 857-867. https://doi.org/10.1016/S0008-6223(00)00190-1

[62] H. Wang, Q. Gao, Synthesis, characterization and energy-related applications of carbide-derived carbons obtained by the chlorination of boron carbide, Carbon 47 (2009) 820-828. https://doi.org/10.1016/j.carbon.2008.11.030

[63] A. Braun, M. Bärtsch, B. Schnyder, R. Kötz, O. Haas, H.-G. Haubold, G. Goerigk, X-ray scattering and adsorption studies of thermally oxidized glassy carbon, J. Non-Cryst. Solids 260 (1999) 1-14. https://doi.org/10.1016/S0022-3093(99)00571-2

[64] A. Braun, M. Bärtsch, B. Schnyder, R. Kötz, O. Haas, A. Wokaun, Evolution of BET internal surface area in glassy carbon powder during thermal oxidation, Carbon 40 (2002) 375-382. https://doi.org/10.1016/S0008-6223(01)00114-2

[65] D. Alliata, P. Häring, O. Haas, R. Kötz, H. Siegenthaler, In situ atomic force microscopy of electrochemically activated glassy carbon, Electrochem. Solid-State Lett. 2 (1999) 33-35. https://doi.org/10.1149/1.1390725

[66] A. Braun, M. Bärtsch, O. Merlo, B. Schnyder, B. Schaffner, R. Kötz, O. Haas, A. Wokaun, Exponential growth of electrochemical double layer capacitance in glassy carbon during thermal oxidation, Carbon 41 (2003) 759-765. https://doi.org/10.1016/S0008-6223(02)00390-1

[67] M. Noked, A. Soffer, D. Aurbach, The electrochemistry of activated carbonaceous materials: past, present, and future, J. Solid State Electrochem. 15 (2011) 1563. https://doi.org/10.1007/s10008-011-1411-y

[68] A. Clague, J. Donnet, T. Wang, J. Peng, A comparison of diesel engine soot with carbon black, Carbon 37 (1999) 1553-1565. https://doi.org/10.1016/S0008-6223(99)00035-4

[69] J. Yan, T. Wei, B. Shao, F. Ma, Z. Fan, M. Zhang, C. Zheng, Y. Shang, W. Qian, F. Wei, Electrochemical properties of graphene nanosheet/carbon black composites as electrodes for supercapacitors, Carbon 48 (2010) 1731-1737. https://doi.org/10.1016/j.carbon.2010.01.014

[70] M. Toupin, D. Bélanger, I.R. Hill, D. Quinn, Performance of experimental carbon blacks in aqueous supercapacitors, J. Power Sources 140 (2005) 203-210. https://doi.org/10.1016/j.jpowsour.2004.08.014

[71] P. Kossyrev, Carbon black supercapacitors employing thin electrodes, J. Power Sources 201 (2012) 347-352. https://doi.org/10.1016/j.jpowsour.2011.10.106

[72] F. Beck, M. Dolata, E. Grivei, N. Probst, Electrochemical supercapacitors based on industrial carbon blacks in aqueous H_2SO_4, J. Appl. Electrochem. 31 (2001) 845-853. https://doi.org/10.1023/A:1017529920916

[73] R. Dash, J. Chmiola, G. Yushin, Y. Gogotsi, G. Laudisio, J. Singer, J. Fischer, S. Kucheyev, Titanium carbide derived nanoporous carbon for energy-related applications, Carbon 44 (2006) 2489-2497. https://doi.org/10.1016/j.carbon.2006.04.035

[74] L. Permann, M. Lätt, J. Leis, M. Arulepp, Electrical double layer characteristics of nanoporous carbon derived from titanium carbide, Electrochim. Acta 51 (2006) 1274-1281. https://doi.org/10.1016/j.electacta.2005.06.024

[75] R. Lin, P.-L. Taberna, J. Chmiola, D. Guay, Y. Gogotsi, P. Simon, Microelectrode study of pore size, ion size, and solvent effects on the charge/discharge behavior of microporous carbons for electrical double-layer capacitors, J. Electrochem. Soc. 156 (2009) A7-A12. https://doi.org/10.1149/1.3002376

[76] J. Chmiola, G. Yushin, Y. Gogotsi, C. Portet, P. Simon, P.-L. Taberna, Anomalous increase in carbon capacitance at pore sizes less than 1 nanometer, Science 313 (2006) 1760-1763. https://doi.org/10.1126/science.1132195

[77] J. Chmiola, G. Yushin, R. Dash, Y. Gogotsi, Effect of pore size and surface area of carbide derived carbons on specific capacitance, J. Power Sources 158 (2006) 765-772. https://doi.org/10.1016/j.jpowsour.2005.09.008

[78] Y. Gogotsi, A. Nikitin, H. Ye, W. Zhou, J.E. Fischer, B. Yi, H.C. Foley, M.W. Barsoum, Nanoporous carbide-derived carbon with tunable pore size, Nature Mater. 2 (2003) 591. https://doi.org/10.1038/nmat957

[79] D.A. Ersoy, M.J. McNallan, Y. Gogotsi, Carbon coatings produced by high temperature chlorination of silicon carbide ceramics, Mater. Res. Innov. 5 (2001) 55-62. https://doi.org/10.1007/s100190100136

[80] A. Jänes, L. Permann, M. Arulepp, E. Lust, Electrochemical characteristics of nanoporous carbide-derived carbon materials in non-aqueous electrolyte solutions,

Electrochem. Commun. 6 (2004) 313-318.
https://doi.org/10.1016/j.elecom.2004.01.009

[81] J. Chmiola, G. Yushin, R.K. Dash, E.N. Hoffman, J.E. Fischer, M.W. Barsoum, Y. Gogotsi, Double-layer capacitance of carbide derived carbons in sulfuric acid, Electrochem. Solid-State Lett. 8 (2005) A357-A360.
https://doi.org/10.1149/1.1921134

[82] A. Kravchik, J.A. Kukushkina, V. Sokolov, G. Tereshchenko, Structure of nanoporous carbon produced from boron carbide, Carbon 44 (2006) 3263-3268.
https://doi.org/10.1016/j.carbon.2006.06.037

[83] A. González, E. Goikolea, J.A. Barrena, R. Mysyk, Review on supercapacitors: Technologies and materials, Renew. Sust. Energ. Rev. 58 (2016) 1189-1206.
https://doi.org/10.1016/j.rser.2015.12.249

[84] L.L. Zhang, R. Zhou, X. Zhao, Graphene-based materials as supercapacitor electrodes, J. Mater. Chem. 20 (2010) 5983-5992. https://doi.org/10.1039/c000417k

[85] Q. Ke, J. Wang, Graphene-based materials for supercapacitor electrodes–A review, J. Materiomics 2 (2016) 37-54. https://doi.org/10.1016/j.jmat.2016.01.001

[86] Q. Yu, J. Lian, S. Siriponglert, H. Li, Y.P. Chen, S.S. Pei, Graphene segregated on Ni surfaces and transferred to insulators, Appl. Phys. Lett. 93 (2008) 113103.
https://doi.org/10.1063/1.2982585

[87] Y. Zhu, S. Murali, W. Cai, X. Li, J.W. Suk, J.R. Potts, R.S. Ruoff, Graphene and graphene oxide: synthesis, properties, and applications, Adv. Mater. 22 (2010) 3906-3924. https://doi.org/10.1002/adma.201001068

[88] K.S. Novoselov, A.K. Geim, S.V. Morozov, D. Jiang, Y. Zhang, S.V. Dubonos, I.V. Grigorieva, A.A. Firsov, Electric field effect in atomically thin carbon films, Science 306 (2004) 666-669. https://doi.org/10.1126/science.1102896

[89] A. Dato, V. Radmilovic, Z. Lee, J. Phillips, M. Frenklach, Substrate-free gas-phase synthesis of graphene sheets, Nano Lett. 8 (2008) 2012-2016.
https://doi.org/10.1021/nl8011566

[90] Y. Hernandez, V. Nicolosi, M. Lotya, F.M. Blighe, Z. Sun, S. De, I. McGovern, B. Holland, M. Byrne, Y.K. Gun'Ko, High-yield production of graphene by liquid-phase exfoliation of graphite, Nature Nanotechnol. 3 (2008) 563.
https://doi.org/10.1038/nnano.2008.215

[91] K. Zhang, L.L. Zhang, X. Zhao, J. Wu, Graphene/polyaniline nanofiber composites as supercapacitor electrodes, Chem. Mater. 22 (2010) 1392-1401. https://doi.org/10.1021/cm902876u

[92] A.K. Geim, K.S. Novoselov, The rise of graphene, Nanoscience and Technology: A Collection of Reviews from Nature Journals, World Scientific, Oxford, 2010, pp. 11-19. https://doi.org/10.1142/9789814287005_0002

[93] E. Bae, N.D. Kim, B.K. Kwak, J. Park, J. Lee, Y. Kim, K. Choi, J. Yi, The effects of fullerene (C_{60}) crystal structure on its electrochemical capacitance, Carbon 48 (2010) 3676-3681. https://doi.org/10.1016/j.carbon.2010.06.007

[94] J.K. McDonough, A.I. Frolov, V. Presser, J. Niu, C.H. Miller, T. Ubieto, M.V. Fedorov, Y. Gogotsi, Influence of the structure of carbon onions on their electrochemical performance in supercapacitor electrodes, Carbon 50 (2012) 3298-3309. https://doi.org/10.1016/j.carbon.2011.12.022

[95] H. Wang, X. Yan, G. Piao, A high-performance supercapacitor based on fullerene C_{60} whisker and polyaniline emeraldine base composite, Electrochim. Acta 231 (2017) 264-271. https://doi.org/10.1016/j.electacta.2017.02.057

[96] S. Xiong, F. Yang, H. Jiang, J. Ma, X. Lu, Covalently bonded polyaniline/fullerene hybrids with coral-like morphology for high-performance supercapacitor, Electrochim. Acta 85 (2012) 235-242. https://doi.org/10.1016/j.electacta.2012.08.056

[97] J. Ma, Q. Guo, H.L. Gao, X. Qin, Synthesis of C_{60}/graphene composite as electrode in supercapacitors, Fuller. Nanotub. Carbon 23 (2015) 477-482. https://doi.org/10.1080/1536383X.2013.865604

[98] M.V.K. Azhagan, M.V. Vaishampayan, M.V. Shelke, Synthesis and electrochemistry of pseudocapacitive multilayer fullerenes and MnO_2 nanocomposites, J. Mater. Chem. A 2 (2014) 2152-2159. https://doi.org/10.1039/C3TA14076H

[99] K. Winkler, E. Grodzka, F. D'Souza, A.L. Balch, Two-component films of fullerene and palladium as materials for electrochemical capacitors, J. Electrochem. Soc. 154 (2007) K1-K10. https://doi.org/10.1149/1.2434683

[100] P. Bairi, R.G. Shrestha, J.P. Hill, T. Nishimura, K. Ariga, L.K. Shrestha, Mesoporous graphitic carbon microtubes derived from fullerene C_{70} tubes as a high performance electrode material for advanced supercapacitors, J. Mater. Chem. A 4 (2016) 13899-13906. https://doi.org/10.1039/C6TA04970B

[101] S. Zheng, H. Ju, X. Lu, A High-Performance Supercapacitor Based on KOH Activated 1D C_{70} Microstructures, Adv. Energ. Mater. 5 (2015) 1500871. https://doi.org/10.1002/aenm.201500871

[102] Q. Xie, E. Perez-Cordero, L. Echegoyen, Electrochemical detection of C_{60}^{6-} and C_{70}^{6-}: Enhanced stability of fullerides in solution, J. Am. Chem. Soc. 114 (1992) 3978-3980. https://doi.org/10.1021/ja00036a056

[103] C. Portet, G. Yushin, Y. Gogotsi, Electrochemical performance of carbon onions, nanodiamonds, carbon black and multiwalled nanotubes in electrical double layer capacitors, Carbon 45 (2007) 2511-2518. https://doi.org/10.1016/j.carbon.2007.08.024

[104] J. Huang, B.G. Sumpter, V. Meunier, G. Yushin, C. Portet, Y. Gogotsi, Curvature effects in carbon nanomaterials: Exohedral versus endohedral supercapacitors, J. Mater. Res. 25 (2010) 1525-1531. https://doi.org/10.1557/JMR.2010.0195

[105] D.N. Futaba, K. Hata, T. Yamada, T. Hiraoka, Y. Hayamizu, Y. Kakudate, O. Tanaike, H. Hatori, M. Yumura, S. Iijima, Shape-engineerable and highly densely packed single-walled carbon nanotubes and their application as super-capacitor electrodes, Nature Mater. 5 (2006) 987. https://doi.org/10.1038/nmat1782

[106] O. Kimizuka, O. Tanaike, J. Yamashita, T. Hiraoka, D.N. Futaba, K. Hata, K. Machida, S. Suematsu, K. Tamamitsu, S. Saeki, Electrochemical doping of pure single-walled carbon nanotubes used as supercapacitor electrodes, Carbon 46 (2008) 1999-2001. https://doi.org/10.1016/j.carbon.2008.08.026

[107] C.y. Liu, A.J. Bard, F. Wudl, I. Weitz, J.R. Heath, Electrochemical characterization of films of single-walled carbon nanotubes and their possible application in supercapacitors, Electrochem. Solid-State Lett. 2 (1999) 577-578. https://doi.org/10.1149/1.1390910

[108] Y. Honda, M. Takeshige, H. Shiozaki, T. Kitamura, K. Yoshikawa, S. Chakrabarti, O. Suekane, L. Pan, Y. Nakayama, M. Yamagata, Vertically aligned double-walled carbon nanotube electrode prepared by transfer methodology for electric double layer capacitor, J. Power Sources 185 (2008) 1580-1584. https://doi.org/10.1016/j.jpowsour.2008.09.020

[109] T. Bordjiba, M. Mohamedi, L.H. Dao, New class of carbon-nanotube aerogel electrodes for electrochemical power sources, Adv. Mater. 20 (2008) 815-819. https://doi.org/10.1002/adma.200701498

[110] M. Chhowalla, K. Teo, C. Ducati, N. Rupesinghe, G. Amaratunga, A. Ferrari, D. Roy, J. Robertson, W. Milne, Growth process conditions of vertically aligned carbon

nanotubes using plasma enhanced chemical vapor deposition, J. Appl. Phys. 90 (2001) 5308-5317. https://doi.org/10.1063/1.1410322

[111] C.G. Liu, H.T. Fang, F. Li, M. Liu, H.M. Cheng, Single-walled carbon nanotubes modified by electrochemical treatment for application in electrochemical capacitors, J. Power Sources 160 (2006) 758-761. https://doi.org/10.1016/j.jpowsour.2006.01.072

[112] O. Tanaike, D.N. Futaba, K. Hata, H. Hatori, Supercapacitors using pure single-walled carbon nanotubes, Carbon Lett. 10 (2009) 90-93. https://doi.org/10.5714/CL.2009.10.2.090

[113] Z.S. Wu, G. Zhou, L.C. Yin, W. Ren, F. Li, H.M. Cheng, Graphene/metal oxide composite electrode materials for energy storage, Nano Energ. 1 (2012) 107-131. https://doi.org/10.1016/j.nanoen.2011.11.001

[114] K.S. Novoselov, V. Fal, L. Colombo, P. Gellert, M. Schwab, K. Kim, A roadmap for graphene, Nature 490 (2012) 192. https://doi.org/10.1038/nature11458

[115] T. Chen, L. Dai, Carbon nanomaterials for high-performance supercapacitors, Mater. Today 16 (2013) 272-280. https://doi.org/10.1016/j.mattod.2013.07.002

[116] Z. Lei, N. Christov, X. Zhao, Intercalation of mesoporous carbon spheres between reduced graphene oxide sheets for preparing high-rate supercapacitor electrodes, Energ. Environ. Sci. 4 (2011) 1866-1873. https://doi.org/10.1039/c1ee01094h

[117] J.L. Vickery, A.J. Patil, S. Mann, Fabrication of graphene-polymer nanocomposites with higher-order three-dimensional architectures, Adv. Mater. 21 (2009) 2180-2184. https://doi.org/10.1002/adma.200803606

[118] G. Wang, X. Sun, F. Lu, H. Sun, M. Yu, W. Jiang, C. Liu, J. Lian, Flexible pillared graphene-paper electrodes for high-performance electrochemical supercapacitors, Small 8 (2012) 452-459. https://doi.org/10.1002/smll.201101719

[119] M.D. Stoller, S. Park, Y. Zhu, J. An, R.S. Ruoff, Graphene-based ultracapacitors, Nano Lett. 8 (2008) 3498-3502. https://doi.org/10.1021/nl802558y

[120] A. Nishino, Capacitors: operating principles, current market and technical trends, J. Power Sources 60 (1996) 137-147. https://doi.org/10.1016/S0378-7753(96)80003-6

[121] H. Yang, J. Yang, Z. Bo, S. Zhang, J. Yan, K. Cen, Edge effects in vertically-oriented graphene based electric double-layer capacitors, J. Power Sources 324 (2016) 309-316. https://doi.org/10.1016/j.jpowsour.2016.05.072

[122] D. Wang, H. Tian, Y. Yang, D. Xie, T.-L. Ren, Y. Zhang, Scalable and direct growth of graphene micro ribbons on dielectric substrates, Sci. Rep. 3 (2013) 1348. https://doi.org/10.1038/srep01348

[123] Z. Bo, W. Zhu, W. Ma, Z. Wen, X. Shuai, J. Chen, J. Yan, Z. Wang, K. Cen, X. Feng, Vertically oriented graphene bridging active-layer/current-collector interface for ultrahigh rate supercapacitors, Adv. Mater. 25 (2013) 5799-5806. https://doi.org/10.1002/adma.201301794

[124] G. Sahoo, S. Polaki, S. Ghosh, N. Krishna, M. Kamruddin, K.K. Ostrikov, Plasma-tuneable oxygen functionalization of vertical graphenes enhance electrochemical capacitor performance, Energ. Storage Mater. 14 (2018) 297-305. https://doi.org/10.1016/j.ensm.2018.05.011

[125] J.R. Miller, R. Outlaw, B. Holloway, Graphene double-layer capacitor with ac line-filtering performance, Science 329 (2010) 1637-1639. https://doi.org/10.1126/science.1194372

[126] S. Ghosh, G. Sahoo, S. Polaki, N.G. Krishna, M. Kamruddin, T. Mathews, Enhanced supercapacitance of activated vertical graphene nanosheets in hybrid electrolyte, J. Appl. Phys. 122 (2017) 214902. https://doi.org/10.1063/1.5002748

[127] S. Ghosh, K. Ganesan, S.R. Polaki, T. Ravindran, N.G. Krishna, M. Kamruddin, A. Tyagi, Evolution and defect analysis of vertical graphene nanosheets, J. Raman Spectrosc. 45 (2014) 642-649. https://doi.org/10.1002/jrs.4530

[128] S. Ghosh, T. Mathews, B. Gupta, A. Das, N.G. Krishna, M. Kamruddin, Supercapacitive vertical graphene nanosheets in aqueous electrolytes, Nano-Struct. Nano-Objects 10 (2017) 42-50. https://doi.org/10.1016/j.nanoso.2017.03.008

[129] S. Ghosh, B. Gupta, K. Ganesan, A. Das, M. Kamruddin, S. Dash, A. Tyagi, MnO_2-vertical graphene nanosheets composite electrodes for energy storage devices, Mater. Today: Proc. 3 (2016) 1686-1692. https://doi.org/10.1016/j.matpr.2016.04.060

[130] R. Farma, M. Deraman, A. Awitdrus, I. Talib, E. Taer, N. Basri, J. Manjunatha, M. Ishak, B. Dollah, S. Hashmi, Preparation of highly porous binderless activated carbon electrodes from fibres of oil palm empty fruit bunches for application in supercapacitors, Bioresour.Technol. 132 (2013) 254-261. https://doi.org/10.1016/j.biortech.2013.01.044

[131] P. González-García, T. Centeno, E. Urones-Garrote, D. Ávila-Brande, L. Otero-Díaz, Microstructure and surface properties of lignocellulosic-based activated carbons, Appl. Surf. Sci 265 (2013) 731-737. https://doi.org/10.1016/j.apsusc.2012.11.092

[132] A.E. Ismanto, S. Wang, F.E. Soetaredjo, S. Ismadji, Preparation of capacitor's electrode from cassava peel waste, Bioresour.Technol. 101 (2010) 3534-3540. https://doi.org/10.1016/j.biortech.2009.12.123

[133] X. Li, W. Xing, S. Zhuo, J. Zhou, F. Li, S.-Z. Qiao, G.-Q. Lu, Preparation of capacitor's electrode from sunflower seed shell, Bioresour.Technol. 102 (2011) 1118-1123. https://doi.org/10.1016/j.biortech.2010.08.110

[134] J.V. Nabais, J.G. Teixeira, I. Almeida, Development of easy made low cost bindless monolithic electrodes from biomass with controlled properties to be used as electrochemical capacitors, Bioresour. Technol. 102 (2011) 2781-2787. https://doi.org/10.1016/j.biortech.2010.11.083

[135] A. Elmouwahidi, Z. Zapata-Benabithe, F. Carrasco-Marín, C. Moreno-Castilla, Activated carbons from KOH-activation of argan (Argania spinosa) seed shells as supercapacitor electrodes, Bioresour.Technol. 111 (2012) 185-190. https://doi.org/10.1016/j.biortech.2012.02.010

[136] H. Jin, X. Wang, Z. Gu, J. Polin, Carbon materials from high ash biochar for supercapacitor and improvement of capacitance with HNO_3 surface oxidation, J. Power Sources 236 (2013) 285-292. https://doi.org/10.1016/j.jpowsour.2013.02.088

[137] C. Peng, X.-b. Yan, R.-t. Wang, J.-w. Lang, Y.-j. Ou, Q.-j. Xue, Promising activated carbons derived from waste tea-leaves and their application in high performance supercapacitors electrodes, Electrochim. Acta 87 (2013) 401-408. https://doi.org/10.1016/j.electacta.2012.09.082

[138] K. Denshchikov, M. Izmaylova, A. Zhuk, Y. Vygodskii, V. Novikov, A. Gerasimov, 1-Methyl-3-butylimidazolium tetraflouroborate with activated carbon for electrochemical double layer supercapacitors, Electrochim. Acta 55 (2010) 7506-7510. https://doi.org/10.1016/j.electacta.2010.03.065

[139] F.-C. Wu, R.-L. Tseng, C.-C. Hu, C.-C. Wang, Physical and electrochemical characterization of activated carbons prepared from firwoods for supercapacitors, J. Power Sources 138 (2004) 351-359. https://doi.org/10.1016/j.jpowsour.2004.06.023

[140] L. Wei, G. Yushin, Electrical double layer capacitors with activated sucrose-derived carbon electrodes, Carbon 49 (2011) 4830-4838. https://doi.org/10.1016/j.carbon.2011.07.003

[141] L. Wei, G. Yushin, Electrical double layer capacitors with sucrose derived carbon electrodes in ionic liquid electrolytes, J. Power Sources 196 (2011) 4072-4079. https://doi.org/10.1016/j.jpowsour.2010.12.085

[142] X. He, P. Ling, J. Qiu, M. Yu, X. Zhang, C. Yu, M. Zheng, Efficient preparation of biomass-based mesoporous carbons for supercapacitors with both high energy density and high power density, J. Power Sources 240 (2013) 109-113. https://doi.org/10.1016/j.jpowsour.2013.03.174

[143] V. Subramanian, C. Luo, A.M. Stephan, K. Nahm, S. Thomas, B. Wei, Supercapacitors from activated carbon derived from banana fibers, J. Phys. Chem. C 111 (2007) 7527-7531. https://doi.org/10.1021/jp067009t

[144] K. Jost, D. Stenger, C.R. Perez, J.K. McDonough, K. Lian, Y. Gogotsi, G. Dion, Knitted and screen printed carbon-fiber supercapacitors for applications in wearable electronics, Energ. Environ. Sci. 6 (2013) 2698-2705. https://doi.org/10.1039/c3ee40515j

[145] J. Ren, L. Li, C. Chen, X. Chen, Z. Cai, L. Qiu, Y. Wang, X. Zhu, H. Peng, Twisting carbon nanotube fibers for both wire-shaped micro-supercapacitor and micro-battery, Adv. Mater. 25 (2013) 1155-1159. https://doi.org/10.1002/adma.201203445

[146] V.T. Le, H. Kim, A. Ghosh, J. Kim, J. Chang, Q.A. Vu, D.T. Pham, J.H. Lee, S.W. Kim, Y.H. Lee, Coaxial fiber supercapacitor using all-carbon material electrodes, ACS Nano 7 (2013) 5940-5947. https://doi.org/10.1021/nn4016345

[147] S. Hu, S. Zhang, N. Pan, Y.L. Hsieh, High energy density supercapacitors from lignin derived submicron activated carbon fibers in aqueous electrolytes, J. Power Sources 270 (2014) 106-112. https://doi.org/10.1016/j.jpowsour.2014.07.063

[148] G. Wang, H. Wang, X. Lu, Y. Ling, M. Yu, T. Zhai, Y. Tong, Y. Li, Solid-state supercapacitor based on activated carbon cloths exhibits excellent rate capability, Adv. Mater. 26 (2014) 2676-2682. https://doi.org/10.1002/adma.201304756

[149] C. Kim, Y.O. Choi, W.J. Lee, K.S. Yang, Supercapacitor performances of activated carbon fiber webs prepared by electrospinning of PMDA-ODA poly(amic acid) solutions, Electrochim. Acta 50 (2004) 883-887. https://doi.org/10.1016/j.electacta.2004.02.072

[150] V. Barranco, M. Lillo-Rodenas, A. Linares-Solano, A. Oya, F. Pico, J. Ibáñez, F. Agullo-Rueda, J.M. Amarilla, J. Rojo, Amorphous carbon nanofibers and their activated carbon nanofibers as supercapacitor electrodes, J. Phys. Chem. C 114 (2010) 10302-10307. https://doi.org/10.1021/jp1021278

[151] Z. Jin, X. Yan, Y. Yu, G. Zhao, Sustainable activated carbon fibers from liquefied wood with controllable porosity for high-performance supercapacitors, J. Mater. Chem. A 2 (2014) 11706-11715. https://doi.org/10.1039/C4TA01413H

[152] G. Zu, J. Shen, L. Zou, F. Wang, X. Wang, Y. Zhang, X. Yao, Nanocellulose-derived highly porous carbon aerogels for supercapacitors, Carbon 99 (2016) 203-211. https://doi.org/10.1016/j.carbon.2015.11.079

[153] R. Saliger, U. Fischer, C. Herta, J. Fricke, High surface area carbon aerogels for supercapacitors, J. Non Cryst. Solids 225 (1998) 81-85. https://doi.org/10.1016/S0022-3093(98)00104-5

[154] J. Li, X. Wang, Q. Huang, S. Gamboa, P. Sebastian, Studies on preparation and performances of carbon aerogel electrodes for the application of supercapacitor, J. Power Sources 158 (2006) 784-788. https://doi.org/10.1016/j.jpowsour.2005.09.045

[155] S.J. Kim, S. Hwang, S. Hyun, Preparation of carbon aerogel electrodes for supercapacitor and their electrochemical characteristics, J. Mater. Sci. 40 (2005) 725-731. https://doi.org/10.1007/s10853-005-6313-x

[156] X.L. Wu, T. Wen, H.L. Guo, S. Yang, X. Wang, A.-W. Xu, Biomass-derived sponge-like carbonaceous hydrogels and aerogels for supercapacitors, ACS Nano 7 (2013) 3589-3597. https://doi.org/10.1021/nn400566d

[157] J. Zheng, T. Jow, High energy and high power density electrochemical capacitors, J. Power Sources 62 (1996) 155-159. https://doi.org/10.1016/S0378-7753(96)02424-X

[158] T. Osaka, X. Liu, M. Nojima, Acetylene black/poly(vinylidene fluoride) gel electrolyte composite electrode for an electric double-layer capacitor, J. Power Sources 74 (1998) 122-128. https://doi.org/10.1016/S0378-7753(98)00043-3

[159] D.A. Ersoy, M.J. McNallan, Y. Gogotsi, Carbon coatings produced by high temperature chlorination of silicon carbide ceramics, Mater. Res. Innovations 5 (2001) 55-62. https://doi.org/10.1007/s100190100136

[160] P. Bairi, R.G. Shrestha, J.P. Hill, T. Nishimura, K. Ariga, L.K. Shrestha, Mesoporous graphitic carbon microtubes derived from fullerene C70 tubes as a high performance electrode material for advanced supercapacitors, J. Mater. Chem. A 4 (2016) 13899-13906. https://doi.org/10.1039/C6TA04970B

[161] J. Ma, Q. Guo, H.-L. Gao, X. Qin, Synthesis of C60/graphene composite as electrode in supercapacitors, Fullerenes, Nanotubes and Carbon Nanostructures 23 (2015) 477-482. https://doi.org/10.1080/1536383X.2013.865604

[162] V. Yong, H.T. Hahn, Synergistic effect of fullerene-capped gold nanoparticles on graphene electrochemical supercapacitors, (2013). https://doi.org/10.4236/anp.2013.21001

[163] S. Zheng, H. Ju, X. Lu, A High-performance supercapacitor based on KOH activated 1D C70 microstructures, Adv. Energy Mater. 5 (2015) 1500871. https://doi.org/10.1002/aenm.201500871

[164] T. Hiraoka, A. Izadi-Najafabadi, T. Yamada, D.N. Futaba, S. Yasuda, O. Tanaike, H. Hatori, M. Yumura, S. Iijima, K. Hata, Compact and light supercapacitor electrodes from a surface-only solid by opened carbon nanotubes with 200 $m^2 g^{-1}$ surface area, Adv. Funct. Mater. 20 (2010) 422-428. https://doi.org/10.1002/adfm.200901927

[165] S. Shiraishi, H. Kurihara, K. Okabe, D. Hulicova, A. Oya, Electric double layer capacitance of highly pure single-walled carbon nanotubes (HiPco™ Buckytubes™) in propylene carbonate electrolytes, Electrochem. Commun. 4 (2002) 593-598. https://doi.org/10.1016/S1388-2481(02)00382-X

[166] C.y. Liu, A.J. Bard, F. Wudl, I. Weitz, J.R. Heath, Electrochemical characterization of films of single-walled carbon nanotubes and their possible application in supercapacitors, Electrochem. Solid-State Lett. 2 (1999) 577-578. https://doi.org/10.1149/1.1390910

[167] W. Lu, L. Qu, K. Henry, L. Dai, High performance electrochemical capacitors from aligned carbon nanotube electrodes and ionic liquid electrolytes, J. Power Sources 189 (2009) 1270-1277. https://doi.org/10.1016/j.jpowsour.2009.01.009

[168] T. Bordjiba, M. Mohamedi, L.H. Dao, New class of carbon-nanotube aerogel electrodes for electrochemical power sources, Adv. mater. 20 (2008) 815-819. https://doi.org/10.1002/adma.200701498

[169] C. Emmenegger, P. Mauron, P. Sudan, P. Wenger, V. Hermann, R. Gallay, A. Züttel, Investigation of electrochemical double-layer (ECDL) capacitors electrodes based on carbon nanotubes and activated carbon materials, J. Power Sources 124 (2003) 321-329. https://doi.org/10.1016/S0378-7753(03)00590-1

[170] Y. Li, M. Van Zijll, S. Chiang, N. Pan, KOH modified graphene nanosheets for supercapacitor electrodes, J. Power Sources 196 (2011) 6003-6006. https://doi.org/10.1016/j.jpowsour.2011.02.092

[171] Y. Chen, X. Zhang, D. Zhang, P. Yu, Y. Ma, High performance supercapacitors based on reduced graphene oxide in aqueous and ionic liquid electrolytes, Carbon 49 (2011) 573-580. https://doi.org/10.1016/j.carbon.2010.09.060

[172] S. Bose, T. Kuila, A.K. Mishra, N.H. Kim, J.H. Lee, Preparation of non-covalently functionalized graphene using 9-anthracene carboxylic acid, Nanotechnology 22 (2011) 405603. https://doi.org/10.1088/0957-4484/22/40/405603

[173] N. Li, S. Tang, Y. Dai, X. Meng, The synthesis of graphene oxide nanostructures for supercapacitors: a simple route, J. Mater. Sci. 49 (2014) 2802-2809. https://doi.org/10.1007/s10853-013-7986-1

[174] Z. Lin, Y. Liu, Y. Yao, O.J. Hildreth, Z. Li, K. Moon, C. Wong, Superior capacitance of functionalized graphene, J. Phys. Chem. C 115 (2011) 7120-7125. https://doi.org/10.1021/jp2007073

[175] C. Liu, Z. Yu, D. Neff, A. Zhamu, B.Z. Jang, Graphene-based supercapacitor with an ultrahigh energy density, Nano Lett. 10 (2010) 4863-4868. https://doi.org/10.1021/nl102661q

[176] Y. Wang, Z. Shi, Y. Huang, Y. Ma, C. Wang, M. Chen, Y. Chen, Supercapacitor devices based on graphene materials, J. Phys. Chem. C 113 (2009) 13103-13107. https://doi.org/10.1021/jp902214f

[177] J. Xia, F. Chen, J. Li, N. Tao, Measurement of the quantum capacitance of graphene, Nat. Nanotechnol. 4 (2009) 505. https://doi.org/10.1038/nnano.2009.177

[178] J.J. Yoo, K. Balakrishnan, J. Huang, V. Meunier, B.G. Sumpter, A. Srivastava, M. Conway, A.L. Mohana Reddy, J. Yu, R. Vajtai, Ultrathin planar graphene supercapacitors, Nano Lett. 11 (2011) 1423-1427. https://doi.org/10.1021/nl200225j

[179] A. Yu, I. Roes, A. Davies, Z. Chen, Ultrathin, transparent, and flexible graphene films for supercapacitor application, Appl. Phys. Lett. 96 (2010) 253105. https://doi.org/10.1063/1.3455879

[180] L. Zhang, G. Shi, Preparation of highly conductive graphene hydrogels for fabricating supercapacitors with high rate capability, J. Phys. Chem. C 115 (2011) 17206-17212. https://doi.org/10.1021/jp204036a

[181] H. Yang, J. Yang, Z. Bo, S. Zhang, J. Yan, K. Cen, Edge effects in vertically-oriented graphene based electric double-layer capacitors, J. Power Sources 324 (2016) 309-316. https://doi.org/10.1016/j.jpowsour.2016.05.072

[182] S. Ghosh, S. Polaki, M. Kamruddin, S.M. Jeong, K.K. Ostrikov, Plasma-electric field controlled growth of oriented graphene for energy storage applications, J. Phys.D: Appl. Phys. 51 (2018) 145303. https://doi.org/10.1088/1361-6463/aab130

Supercapacitor Technology: Materials, Processes and Architectures Materials Research Forum LLC
Materials Research Foundations **61** (2019) 171-222 https://doi.org/10.21741/9781644900499-8

[183] G. Sahoo, S. Ghosh, S. Polaki, T. Mathews, M. Kamruddin, Scalable transfer of vertical graphene nanosheets for flexible supercapacitor applications, Nanotechnology 28 (2017) 415702. https://doi.org/10.1088/1361-6528/aa8252

Supercapacitor Technology: Materials, Processes and Architectures   Materials Research Forum LLC
Materials Research Foundations **61** (2019) 223-232   https://doi.org/10.21741/9781644900499-9

Chapter 9

Photo-Supercapacitor

S. Vadivel*[1], S. Hariganesh[1], Pothu Ramyakrishna[2], Rajender Boddula[3]

[1]Department of Chemistry, PSG College of Technology, Coimbatore-641004, India

[2]College of Chemistry and Chemical Engineering, Hunan University, Changsha 410082, PR China

[3]CAS Key Laboratory of Nanosystem and Hierarchical Fabrication, National Center for Nanoscience and Technology, Beijing 100190, PR China

*vlvelu7@gmail.com

Abstract

In this modern technology-based world, the consumption of electric energy has increased as most part of our day to day life is connected with electricity simply starting from telecommunication devices to laptops and computers. As energy demand increased the energy production needs to be increased but unfortunately as the fossil fuels are being depleted we are in the situation to turn to more renewable energy sources, especially solar energy could be utilised and transformed to electric energy via solar powered cells. However, the problems arise related to the fluctuation of sunlight and its unmatched energy production. Hence, energy produced by solar cells should be stored using energy storage devices. This thought sparked the idea of integrating a solar cell with a supercapacitor, where energy produced by the solar cell will be stored by the capacitor and so it was termed as photo-supercapacitor. In this chapter we discussed basic working, progressive timeline and the future scope in this fascinating field.

Keywords

Photosupercapacitor, Dye Sentisied Solar Cell, Supercapacitor, Pervoskite Solar Cell

Contents

Supercapacitor Technology: Materials, Processes and Architectures   Materials Research Forum LLC
Materials Research Foundations **61** (2019) 223-232   https://doi.org/10.21741/9781644900499-9

1.   Introduction

The relentless quest for never-ending energy production can only be succeeded by proper and complete utilisation of the energy sources that are renewable like solar energy, wind energy and hydro-powered energy. Especially solar power, unimaginable amount of energy can be extracted where the sun supplies continuous energy of about 3×10^{24} J per year that seems to be way more than enough for the human consumption [1-4]. Based on solar power to electric power conversion, the solar cells were developed. Solar cells were extensively researched for the past 50 years and from the first generation (gen) silicon solar cell, it has grown to second generation solar cells which used semiconducting thin films on solar cells that in case changed the cost of the solar cell construction and now it is new third-gen solar cells comprising of dye sensitised solar cells (DSSC), polymer-based solar cells, perovskite-based solar cells and quantum dot sensitized solar cells [5-7]. First generation silicon solar cell achieved efficiency of 25% by extensive research which in case seems to be higher than the new third-gen solar cells, which is only concerned about the cost of production, design, transparency, flexibility and simple method of construction [8-10]. As the solar to electric conversion efficiency has improved as well as the size, shape also improved regarding the solar cells but they couldn't store the converted electric energy. Energy storing devices such as batteries, capacitors and ultra/supercapacitors are also researched extensively and even some of these were into commercial applications [11,12]. Especially supercapacitors with high power density, fast charge-discharge mechanism and prolonging stability seem to be the better energy storage device on comparison. Thus, to overcome charge storage problem in the solar cells, energy storage devices like supercapacitors were coupled with solar cell to form a device called photo- supercapacitor which plays dual role where the solar powered cell transforms the solar energy to electric energy and the supercapacitor part internally stores the produced electrical energy [13-15]. The field of photo-supercapacitor was explored for the first time by Miyasaka et al. in 2004, where they fabricated light-driven self-charging capacitor and named it as photocapacitor [16].

2.   Mechanism of photo-supercapacitor

A photo-supercapacitor device consists of two parts, one is DSSC and other is supercapacitor, the DSSC has a photoanode coated with the light absorbing semiconductor material, preferably TiO_2 on transparent glass substrate, a liquid electrolyte, a counter electrode like porous activated carbon or Pt and a dye as a

Supercapacitor Technology: Materials, Processes and Architectures   Materials Research Forum LLC
Materials Research Foundations **61** (2019) 223-232   https://doi.org/10.21741/9781644900499-9

sensitizer. Nowadays mostly ruthenium-based dyes are used. On the other hand, the supercapacitor consists of electrodes and an electrolyte placed between them [17-19].

In a photo-supercapacitor, charging is done with the help of solar energy whereas in the normal supercapacitor the charging is done directly by an external electrical energy supply. In case of photo-supercapacitors, on irradiation with solar energy, the light induces the electrons in the dye molecule which then introduces the electron into the conduction band (CB) of the semiconducting material, simultaneously the holes produced at the dye molecule relocates to the counter electrode. The redox ions present in the electrolyte play a big role where they balance the photo induced electron circulation. The electrons in the electrolyte help in dye regeneration and simultaneously the electrons from the counter electrode transferred to fill the vacant sites in the electrolyte [20,21]. The following equations provide the ideal working mechanism of a photo supercapacitor [1,22].

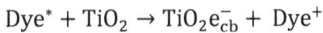

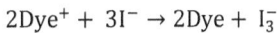

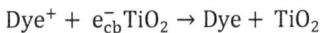

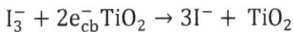

$$Dye + hv \rightarrow Dye^*$$

$$Dye^* + TiO_2 \rightarrow TiO_2 e_{cb}^- + Dye^+$$

$$Dye^* \rightarrow Dye$$

$$2Dye^+ + 3I^- \rightarrow 2Dye + I_3^-$$

$$Dye^+ + e_{cb}^- TiO_2 \rightarrow Dye + TiO_2$$

$$I_3^- + 2e^-(catalyst) \rightarrow 3I^-$$

$$I_3^- + 2e_{cb}^- TiO_2 \rightarrow 3I^- + TiO_2$$

In the following sections, we will discuss the basic understandings, working of various types of photo-supercapacitor, recent findings and advancements that had taken place in this field, concluding with its future scope.

The photo-supercapacitors can be categorized into four types based on the new third-gen solar cell which was used for the construction [23].

    i) DSSC type photo-supercapacitor

    ii) Polymer solar cells type photo-supercapacitor

    iii) Quantum dot sensitized solar cells type photo-supercapacitor

    iv) Perovskite solar cells type photo-supercapacitor

Supercapacitor Technology: Materials, Processes and Architectures   Materials Research Forum LLC
Materials Research Foundations **61** (2019) 223-232   https://doi.org/10.21741/9781644900499-9

3.   Research progress on photo-supercapacitors

Miyasaka et al. were the first to develop a model for the dye-sensitized photocapacitor. In 2004, they fabricated a photocapacitor with two electrodes and then in 2005, they fabricated a three-electrode system for a photocapacitor. In the two electrode system, ruthenium complex dye was used for sensitization which formed heterojunction with TiO_2 nanocrystals coated on transparent conductive oxide (TCO) coated glass, here the TCO is transparent conductive F-doped SnO_2. Along with immobilized activated carbon acting as photoelectrode and activated carbon layered on Pt-coated TCO glass substrate acting as counter electrode, $15wt\%(CH_3CH_2)_4NBF_4$ in propylene carbonate was used as an electrolyte. LiI in acetonitrile(0.1 mol L^{-1}) was used to cover the surface of TiO_2 to prevent the direct contact of dye molecule with the activated carbon. LiI acted as a hole trapping intermediate and helped in the regeneration of the dye. Visible light source of 500W xenon arc lamp was used and the cut off filters were used to eliminate the ultraviolet and infrared radiation. This device yielded specific capacitance of 0.69 F cm^{-2} [16]. The three electrode system consists of dye-sensitized mesoporous TiO_2 electrode as working electrode and one carbon coated electrode as counter electrode, the third electrode which was also a carbon electrode was inserted between the two electrodes working and counter electrodes where redox electrolyte with I^-/I_3^- was filled in the gap between working electrode and intermediate electrode and the non-redox electrolyte was filled in between intermediate and counter electrodes. Thus, the intermediate electrode has a dual role of conducting redox electron at one half whereas storing the charge on the other half. Due to the introduction of the third electrode, the achieved energy density of about 47μ W h cm^{-2} showed five-fold increase on comparing with the two electrode system. The capacitance value was about 0.65 F cm^{-2} and the estimated internal resistance was 330 Ω, whereas the two electrode system exhibited higher resistance of about 2.6 KΩ showing that the three electrode system facilitated more conduction [24]. In 2010, Chen et al. reported a three-electrode photo-supercapacitor system, N719 dye was used as the sensitizer, the photoactive TiO_2 electrode of 11μm thickness was made by electrophoretic deposition followed by mechanical compression at 100 Mpa which facilitated the TiO_2 to adhere well on the plastic substrate and PEDOT, a conducting polymer was used as the charge storage material in the supercapacitor. This flexible photo-supercapacitor exhibited excellent capacitance of about 0.52F cm^{-2}, the other photovoltaic parameters which proved the excellent property of this facile TiO_2 film deposition method were efficiency of power conversion, density of short circuit and fill factor of 4.37%, 8.38 mA cm^{-2} and 0.71 respectively [25]. Similarly, Hsu et al. assembled three-electrode photo-supercapacitor with N3-dye adsorbed TiO_2 as photoactive working electrode and Pt electrodes were used as counter and intermediate electrodes. The PProDOT-ET$_2$ polymer

Supercapacitor Technology: Materials, Processes and Architectures   Materials Research Forum LLC
Materials Research Foundations **61** (2019) 223-232   https://doi.org/10.21741/9781644900499-9

film was used in the supercapacitor for charge storing purpose where the electrolyte was 0.5M LiClO$_4$/MPN solution. This device showed specific capacitance of about 0.39 F cm^{-2}, energy density of about 22 μ W h/cm^2 and power density of about 0.6 m w/cm^2 [26]. In 2011, Wee et al. made-up a printable, solid photo-supercapacitor using a carbon nanotube as common base to integrate organic photovoltaic and supercapacitor of the photo-supercapacitor. In the work, organic photovoltaic part comprises PEDOT:PSS followed by P3HT:PCBM coated ITO as anode and Al cathode and supercapacitor having CNT as the electrode and polyvinyl alcohol with 1 M H$_3$PO$_4$ as electrolyte. The introduction of CNT has reduced the size and weight of the device where it became < 0.6 mm thinner and < 1 g in weight and also reduced the internal resistance by 43%. Finally, it yielded specific capacitance of about 28 F g^{-1} and could go upto 80 F g^{-1} if couple of organic photo voltaic were connected in series [27]. In 2013, yang et al. reported free standing photo-supercapacitors using multi-walled carbon nanotube (MWCNT) films as the electrodes with TiO$_2$ as photoactive electrode alongside with N719 dye as sensitizer. PVA-H$_3$PO$_4$ electrolyte was used between two MWCNT electrodes in supercapacitor and gel electrolyte having LiI as one of the constituents was used as the electrolyte in the photovoltaic cell. When light falls on the dye, the photo electrons generated from the dye migrate to the conduction band of TiO$_2$ which is the photoactive electrode, then they flow to the MWCNT via an external circuit connection as the surface area was large for MWCNT it exhibited better performance. The specific capacitance delivered by this integrated device was about 48 F g^{-1} and the charge storage efficiency was about ~84%. The same device exhibited specific capacitance of about 208 F g^{-1} when the bare MWCNT electrode was replaced by modified polyaniline (PANI) incorporated MWCNT [28]. In 2013, skunik-Nuckowska et al. made a solid state photo-supercapacitor using the dye (D35) (E)-3-(5-(4-(Bis(20,40-dibutoxybiphenyl-4-yl)amino)phenyl)thiophen-2-yl)-2-cyanoacrylic acid with TiO$_2$ as photoactive electrode. A polymer electrolyte was employed for conduction in the solar cell part, whereas ruthenium oxide layers were used as supercapacitor electrodes between which Nafion was used as conducting membrane, a silver electrode was inserted as the intermediate electrode between the solar cell and supercapacitor parts. This hybrid device exhibited Coulombic efficiency of 88%, specific capacitance of 407 F g^{-1}, energy density of 0.17 mWh cm^{-2} and power density of 0.34mW cm^{-2} [29]. In 2014, Bagheri et al. constructed a photo-supercapacitor with DSSC having dye sensitised TiO$_2$ and cobalt polypyridl complex as electrolyte, and a nickel foil coated on one side with PEDOT was used as an intermediate electrode, the main novelty here is in the supercapacitor part where they have made a asymmetric supercapacitor with nickel–cobalt oxide as the positive electrode and activated carbon as the negative electrode. This facile asymmetric supercapacitor integrated liquid dye sensitized solar cell gave specific capacitance of 32 F g^{-1}, energy density 2.3 Wh kg^{-1} and energy conversion efficiency of about 4.9%. The Couloumbic efficiency of the device was 54% and the total charge conversion and charge storage efficiency was about 0.6% [30].

In 2016, Zhou et al. reported a perovskite solar cell type photo-supercapacitor in which they used ($CH_3NH_3PbI_{3-x}Cl_x$) perovskite layered c-TiO_2 glass electrode, MoO_3/Au/MoO_3 acted as the integrating electrode that combined the perovskite solar cell and the electrochromic capacitor consisting WO_3 electrodes. The co-anode type showed areal capacitance of about 286.6 F m^{-2}, where co-cathode showed 430.7 F m^{-2}, the energy density of co-anode was 13.4 mW h m^{-2} where co cathode gave 24.5 mW h m^{-2} and the power density of co-anode typae was 187.6 mW m^{-2} with co cathode of about 377 mW m^{-2}. The main advantage of this perovskite solar cells type photo-supercapacitor was that the device can even work at low operating power and has a prolonged life time [31]. In 2018, Das et al, prepared quantum dots sensitized solar cell type photo-supercapacitor using cadmium sulfide quantum dots/hibiscus dye co-sensitised TiO_2 as photoactive electrode for the solar-powered cell, and PEDOP@MnO_2 served as common electrode for both the solar cell and the supercapacitors. Polymeric ion conducting electrode containing PMMA dissolved in ([BuMeIm$^+$] [$CF_3SO_3^-$])/propylene carbonate (PC) as the electrolyte for the supercapacitor and redox electrolyte used for solar cell was the solution of 1M Na_2S and 1 M S. This hybrid device yielded specific capacitance of about 183 F g^{-1}, with energy density of about 13.2 Wh Kg^{-1} and power density of about 360 W Kg^{-1} at 1 Ag^{-1} current density. It also showed good power conversion efficiency of 6.11% [32]. Recently Ng et al, made a photo-supercapacitor based on cesium lead halide inorganic perovskite solar cell combined with an asymmetric supercapacitor. The perovskite material was $CsPbBr_{3-x}I_x$ along with mesoporous TiO_2 and Spiro-OMeTAD, 1M $LiClO_4$/acetonitrile was used as redox electrolyte. The researchers analyzed the photo-supercapacitance activity of the device which exhibited about areal capacitance of 30 m F cm^{-2}. The advantage of the cesium lead halide based perovskite solar cells was that the cells retained 33% of their power conversion efficiency even after 25 h which showed their high stability at high level of humidity of about < 80%, 1.6 times better than the methylammonium based solar cells [33]. During the past 15 years, lot of technical advancement, innovative materials and developments have taken place in the field of photo-supercapacitors.

Summary and Future Scope

In this fascinating field of photo-supercapacitor, still lots of questions have remained unanswered and lots of details need to be explored especially in connection with the theoretical approaches of photo-supercapacitors. Recently Lechene et al. published an article about the theoretical analysis and characterization of the basic workings of photo-supercapacitors, with emphasis that more such researches are needed to gain good basic understanding of this field [34]. Unfortunately, most of the reported devices yielded very low capacitance value as well as low energy conversion efficiency. More focus must be given to prepare highly efficient solar cells and supercapacitors and then both the high efficient devices should be integrated. Research must be narrowed down towards the commercialization of such devices wherein the near future we might be able to wear such flexible photo-supercapacitors, thus preferentially research on solid electrolytes, flexible and light weighted substrates and cost-effective materials will ease us the way for such future.

References

[1]   C.H. Ng, H.N. Lim, S. Hayase, I. Harrison, A. Pandikumar, N.M. Huang, Potential active materials for photo-supercapacitor: A review, J. Power Sources. 296 (2015) 169–185. https://doi.org/10.1016/j.jpowsour.2015.07.006.

[2]   N.S. Lewis, Introduction: Solar energy conversion, Chem. Rev. 115 (2015) 12631–12632. https://doi.org/10.1021/acs.chemrev.5b00654.

[3]   Q. Zhang, G. Cao, Nanostructured photoelectrodes for dye-sensitized solar cells, Nano Today 6 (2011) 91–109. https://doi.org/10.1016/j.nantod.2010.12.007.

[4]   J. Gong, J. Liang, K. Sumathy, Review on dye-sensitized solar cells (DSSCs): Fundamental concepts and novel materials, Renew. Sustain. Energy Rev. 16 (2012) 5848–5860. https://doi.org/10.1016/j.rser.2012.04.044.

[5]   M. Grätzel, Solar cells to dye for, Nature 421 (2003) 586. https://doi.org/10.1038/421586a.

[6]   E.W. McFarland, J. Tang, A photovoltaic device structure based on internal electron emission, Nature. 421 (2003) 616. https://doi.org/10.1038/nature01316.

[7]   M.A. Green, K. Emery, Y. Hishikawa, W. Warta, E.D. Dunlop, Solar cell efficiency tables (Version 45), Prog. Photovoltaics Res. Appl. 23 (2015) 1–9. https://doi.org/10.1002/pip.2573.

[8]   P.J. Kulesza, M. Skunik-Nuckowska, K. Grzejszczyk, N. Vlachopoulos, L. Yang, L. Häggman, A. Hagfeldt, Development of solid-state photo-supercapacitor by coupling dye-sensitized solar cell utilizing conducting polymer charge relay with proton-conducting membrane based electrochemical capacitor, ECS Trans. 50 (2013) 235–244. https://doi.org/10.1149/05043.0235ecst .

[9]   Q. Wang, Y. Xie, F. Soltani-Kordshuli, M. Eslamian, Progress in emerging solution-processed thin film solar cells – Part I: Polymer solar cells, Renew. Sustain. Energy Rev. 56 (2016) 347–361. https://doi.org/10.1016/j.rser.2015.11.063.

[10]   A. Fakharuddin, R. Jose, T.M. Brown, F. Fabregat-Santiago, J. Bisquert, A perspective on the production of dye-sensitized solar modules, Energy Environ. Sci. 7 (2014) 3952–3981. https://doi.org/10.1039/C4EE01724B.

[11]   S. Vadivel, B. Saravanakumar, M. Kumaravel, D. Maruthamani, N. Balasubramanian, A. Manikandan, G. Ramadoss, B. Paul, S. Hariganesh, Facile solvothermal synthesis of BiOI microsquares as a novel electrode material for supercapacitor applications, Mater. Lett. 210 (2018) 109–112. https://doi.org/10.1016/j.matlet.2017.08.137.

[12]   M. Yassine, D. Fabris, Performance of commercially available supercapacitors, Energies . 10 (2017) 1340-1352. https://doi.org/10.3390/en10091340.

[13]   Z. Gao, C. Bumgardner, N. Song, Y. Zhang, J. Li, X. Li, Cotton-textile-enabled flexible self-sustaining power packs via roll-to-roll fabrication, Nat. Commun. 7 (2016) 11586. https://doi.org/10.1038/ncomms11586.

[14]   B. Liu, B. Liu, X. Wang, X. Wu, W. Zhao, Z. Xu, D. Chen, G. Shen, Memristor-integrated voltage-stabilizing supercapacitor system, Adv. Mater. 26 (2014) 4999–5004. https://doi.org/10.1002/adma.201401017.

[15]   S.C. Lau, H.N. Lim, T.B.S.A. Ravoof, M.H. Yaacob, D.M. Grant, R.C.I. MacKenzie, I. Harrison, N.M. Huang, A three-electrode integrated photo-supercapacitor utilizing graphene-based intermediate bifunctional electrode, Electrochim. Acta. 238 (2017) 178–184. https://doi.org/10.1016/j.electacta.2017.04.003.

[16]   T. Miyasaka, T.N. Murakami, The photocapacitor: An efficient self-charging capacitor for direct storage of solar energy, Appl. Phys. Lett. 85 (2004) 3932–3934. https://doi.org/10.1063/1.1810630.

[17]   N.-G. Park, K.M. Kim, M.G. Kang, K.S. Ryu, S.H. Chang, Y.-J. Shin, Chemical sintering of nanoparticles: A methodology for low-temperature fabrication of dye-sensitized TiO_2 films, Adv. Mater. 17 (2005) 2349–2353. https://doi.org/10.1002/adma.200500288.

[18]   Y. Kijitori, M. Ikegami, T. Miyasaka, Highly efficient plastic dye-sensitized photoelectrodes prepared by low-temperature binder-free coating of mesoscopic titania pastes, Chem. Lett. 36 (2007) 190–191. https://doi.org/10.1246/cl.2007.190.

[19]   A. Mishra, M.K.R. Fischer, P. Bäuerle, Metal-free organic dyes for dye-sensitized solar cells: From structure, property relationships to design rules, Angew. Chemie Int. Ed. 48 (2009) 2474–2499. https://doi.org/10.1002/anie.200804709.

[20]   N. Heo, Y. Jun, J.H. Park, Dye molecules in electrolytes: new approach for suppression of dye-desorption in dye-sensitized solar cells, Sci. Rep. 3 (2013) 1712. https://doi.org/10.1038/srep01712.

[21]   J. van de Lagemaat, N.-G. Park, A.J. Frank, Influence of electrical potential distribution, charge transport, and recombination on the photopotential and photocurrent conversion efficiency of dye-sensitized nanocrystalline TiO_2 solar cells: A study by electrical impedance and optical modulation techniques, J. Phys. Chem. B. 104 (2000) 2044–2052. https://doi.org/10.1021/jp993172v.

[22]   P. Calandra, G. Calogero, A. Sinopoli, P.G. Gucciardi, Metal Nanoparticles and carbon-based nanostructures as advanced materials for cathode application in dye-sensitized solar cells, Int. J. Photoenergy 2010 (2010). https://doi.org/10.1155/2010/109495

[23]   Y. Sun, X. Yan, Recent advances in dual-functional devices integrating solar cells and supercapacitors, Solar RRL. 1 (2017) 1700002. https://doi.org/10.1002/solr.201700002.

[24]   T.N. Murakami, N. Kawashima, T. Miyasaka, A high-voltage dye-sensitized photocapacitor of a three-electrode system, Chem. Commun. (2005) 3346–3348. https://doi.org/10.1039/B503122B.

[25]   H.-W. Chen, C.-Y. Hsu, J.-G. Chen, K.-M. Lee, C.-C. Wang, K.-C. Huang, K.-C. Ho, Plastic dye-sensitized photo-supercapacitor using electrophoretic deposition and compression methods, J. Power Sources. 195 (2010) 6225–6231. https://doi.org/10.1016/j.jpowsour.2010.01.009.

[26]   C.-Y. Hsu, H.-W. Chen, K.-M. Lee, C.-W. Hu, K.-C. Ho, A dye-sensitized photo-supercapacitor based on PProDOT-Et2 thick films, J. Power Sources. 195 (2010) 6232–6238. https://doi.org/10.1016/j.jpowsour.2009.12.099.

[27]   G. Wee, T. Salim, Y.M. Lam, S.G. Mhaisalkar, M. Srinivasan, Printable photo-supercapacitor using single-walled carbon nanotubes, Energy Environ. Sci. 4 (2011) 413–416. https://doi.org/10.1039/C0EE00296H.

[28]   Z. Yang, L. Li, Y. Luo, R. He, L. Qiu, H. Lin, H. Peng, An integrated device for both photoelectric conversion and energy storage based on free-standing and aligned

carbon nanotube film, J. Mater. Chem. A. 1 (2013) 954–958.
https://doi.org/10.1039/C2TA00113F.

[29]   M. Skunik-Nuckowska, K. Grzejszczyk, P.J. Kulesza, L. Yang, N. Vlachopoulos, L. Häggman, E. Johansson, A. Hagfeldt, Integration of solid-state dye-sensitized solar cell with metal oxide charge storage material into photoelectrochemical capacitor, J. Power Sources. 234 (2013) 91–99. https://doi.org/10.1016/j.jpowsour.2013.01.101.

[30]   N. Bagheri, A. Aghaei, M.Y. Ghotbi, E. Marzbanrad, N. Vlachopoulos, L. Häggman, M. Wang, G. Boschloo, A. Hagfeldt, M. Skunik-Nuckowska, P.J. Kulesza, Combination of asymmetric supercapacitor utilizing activated carbon and nickel oxide with cobalt polypyridyl-based dye-sensitized solar cell, Electrochim. Acta. 143 (2014) 390–397. https://doi.org/10.1016/j.electacta.2014.07.125.

[31]   F. Zhou, Z. Ren, Y. Zhao, X. Shen, A. Wang, Y.Y. Li, C. Surya, Y. Chai, Perovskite photovoltachromic supercapacitor with all-transparent electrodes, ACS Nano. 10 (2016) 5900–5908. https://doi.org/10.1021/acsnano.6b01202.

[32]   A. Das, S. Deshagani, R. Kumar, M. Deepa, Bifunctional photo-supercapacitor with a new architecture converts and stores solar energy as charge, ACS Appl. Mater. Interfaces. 10 (2018) 35932–35945. https://doi.org/10.1021/acsami.8b11399.

[33]   C.H. Ng, H.N. Lim, S. Hayase, Z. Zainal, S. Shafie, H.W. Lee, N.M. Huang, Cesium lead halide inorganic-based perovskite-sensitized solar cell for photo-supercapacitor application under high humidity condition, ACS Appl. Energy Mater. 1 (2018) 692–699. https://doi.org/10.1021/acsaem.7b00103.

[34]   B.P. Lechene, R. Clerc, A.C. Arias, Theoretical analysis and characterization of the energy conversion and storage efficiency of photo-supercapacitors, Sol. Energy Mater. Sol. Cells. 172 (2017) 202–212. https://doi.org/10.1016/j.solmat.2017.07.034.

Supercapacitor Technology: Materials, Processes and Architectures   Materials Research Forum LLC
Materials Research Foundations **61** (2019) 233-262   https://doi.org/10.21741/9781644900499-10

Chapter 10

Novel Bimetal Oxides/Sulfides Composites Electrodes for Electrochemical Supercapacitors

Shuhua Liu, Zongyu Huang, and Xiang Qi*

Hunan Key Laboratory of Micro-Nano Energy Materials and Devices, Laboratory for Quantum Engineering and Micro-Nano Energy Technology and School of Physics and Optoelectronic, Xiangtan University, Hunan 411105, P.R. China

* xqi@xtu.edu.cn

Abstract

Supercapacitors (SCs) possess a bright future because of their high power supply and superior stability. Pseudocapacitive oxides of transition metals ($NiCo_2O_4$, $FeCo_2O_4$, Zn_2SnO_4 etc.) have been widely investigated for SCs ascribed to their high theoretical capacitances (>1000 F/g), environmental benignity, and low cost. In Bimetal sulfides (BMSs) such as MCo_2S_4 (M = Ni, Fe, and Cu) the efficient synergistic effect between two different metal ions makes them have more abundant oxidation reducing states, and their conductivity is qualitatively improved (four-order magnitude). This paper is hoped to provide a scientific basis to investigate the SCs technology.

Keywords

Supercapacitors, Pseudocapacitors, Bimetal Sulfides, Transition Metals, Energy-Storage

Contents

1.   Introduction

After the industrial revolution broke out, the demand for energy in countries around the world continued to increase, a large amount of non-renewable fossil fuels are consumed, at the same time, due to the excessive combustion of fossil fuels and harmful gases such as SO_X, CO, etc., and greenhouse gases CO_2 are discharged in large quantities, causing the earth's ecological environment to be seriously damaged [1-3]. Up to now, governments have invested a lot of manpower and resources in the research of renewable clean energy represented by wind energy, solar energy and ocean tides, and have shown good development momentum. Since new energy sources are indirectly supplied, we need to convert them into electricity and store them to ensure a stable and efficient supply

[4, 5]. Besides, energy conversion efficiency also plays a key role in energy storage [6]. To achieve high efficiency, non-pollution, and rapid energy conversion has become a research hotspot of excellent performance and green safe energy storage materials [7, 8]. Electrochemical (EC) storage systems are the best solution so far, using batteries, supercapacitors (SCs) and fuel cells to store energy in the mode of chemical energy, and via briefly reversing the EC reaction, releasing it when needed [9, 10].

Lithium-ion batteries, have higher energy density (100-200 W h k/g), but their power density (500 W k/g) is not satisfactory, so there cannot meet the requirements of cranes (slope) and large excavators (quickly start) [11]. In addition, the almost irreversible redox reaction and the energy storage process of lithium-ion cross-materialization inside the material will destroy the internal structure of the material, further affecting the life of the lithium-ion battery, which is greatly limited in practical applications [7]. Fortunately, the interesting feature of SCs is their ultra-high power density, as shown in Figure 1, it is significantly higher than fuel cells and lithium-ion batteries. The energy density of SCs is 10-100 times that of conventional capacitors. Besides that, the SCs require a very short charging time (high current charging), but can complete the instantaneous high current discharge (Table 1). Not only that, the cycle charge and discharge can reach more than 500,000 times, which has obvious advantages in emergency safety doors and rail transit [11-13]. According to their energy storage mechanism (such as Faradaic or non-Faradaic reactions) and configurations (symmetric or asymmetric systems), SCs have two types: ① EDLCs (electrochemical double-layer capacitors), ② PCs (pseudocapacitors) [2, 14-16]. Moreover, SCs have been applied in hybrid electric vehicles, power back-ups, fuel cell starting power, electronic equipment sudden power supply and other aspects [17, 18]. Excitingly, SCs have been successfully applied in buses, for-example, in Shenzhen, capacitor buses have spread all over the city. Nevertheless, the energy densities of commercial SCs are generally lower than batteries, which hinder the application to the portable and flexible devices [5]. Many studies have been actively carried out increase the gravimetric energy density of SCs while reducing the manufacturing cost of SCs without sacrificing the high power transmission and cycle life [4, 19]. A promising and effective strategy is to synthesize nanostructured materials with multiple components containing metals, to design the architecture (core-shell [20], nanowires (NWs) [21], and nanoflower (NF) [22]).

Currently, lot of MOs/MSs and BMOs/BMSs have been investigated for the SCs response [23-25], and in advanced SCs materials, BMOs/BMSs such as $FeCo_2S_4$ etc. are believed to be the most potential materials for next-generation of SCs. Hence, this chapter opens up the survey of SCs properties of MOs/ MSs and BMOs/BMSs via main methods. Finally, the trends, challenges, and future tasks in SCs will be discussed.

Supercapacitor Technology: Materials, Processes and Architectures   Materials Research Forum LLC
Materials Research Foundations **61** (2019) 233-262   https://doi.org/10.21741/9781644900499-10

Table 1: *Comparsion of some characteristics of the supercapacitors and batteries.*

Characteristic	Supercapacitors	Batteries
Cycle life	>100000	500~2000
Charge time	1~30s	1~5h
Discharge time	1~30s	0.3~3h
Energy density (W h kg^{-1})	1~10	20~180
Power density (W kg^{-1})	1000~2000	50~300
Charge and discharge efficiency	0.9~0.95	0.7~0.85

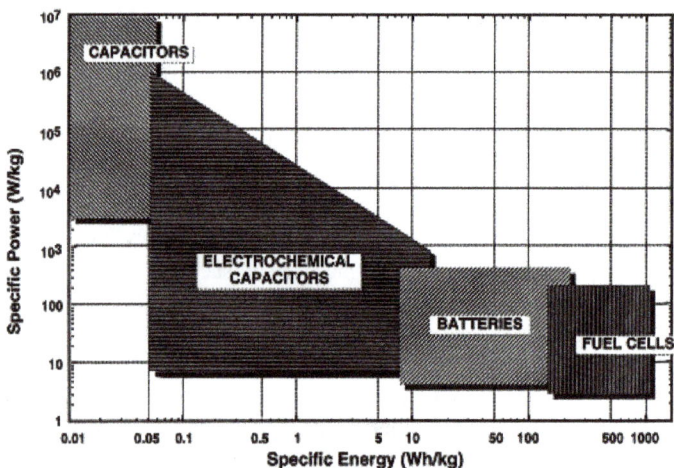

Figure 1 Ragone diagram of various energy storage conversion devices[26].

2. Fundamentals of electrochemical supercapacitors

SCs are energy storage devices that are similar to batteries in terms of test parameters and design. Meanwhile, SCs can perform charge and discharge quickly. The SCs includes electrodes, electrolyte and separator (Figure 2) [26]. Electrode material is an integral part of SCs, as well as it is made of nanomaterial that have a high specific surface area and a high porosity. Currently, SCs can be classified as follows:

According to the electrolyte, SCs can be divided into two categories. The more common are the aqueous electrolytes. Usually, electrolyte is ultrapure water containing appropriate amount of H_2SO_4, Na_2SO_4, NaOH or KOH to configure as an acidic, neutral, alkaline electrolyte. The other type is the organic electrolyte. In general, the difference between these is that the voltage window of the water-based electrolyte SCs is significantly lower compared to the organic electrolyte SCs, at the same time, the energy density is affected by different electrolytes.

According to electrode materials and their corresponding energy storage mechanisms [27], SCs classified as PDCs and EDLCs.

When classified by structure, they can be divided into asymmetric SCs and symmetric SCs.

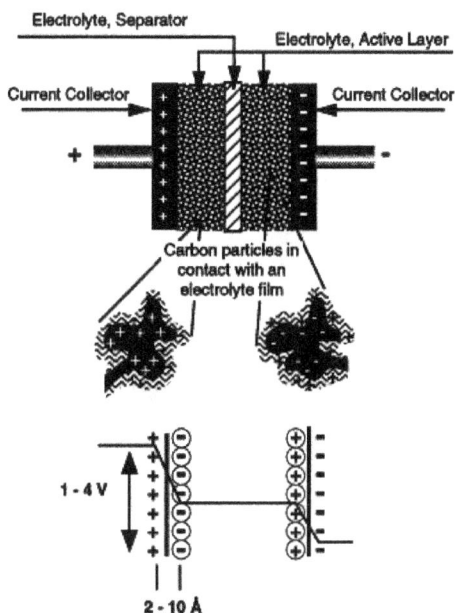

Figure 2. Schematic diagram of the structure of each part of the capacitor[26].

2.1   Electric double-layer capacitance (EDLC)

The EDLCs has a similar energy storage mechanism as the conventional capacitor (Figure 3) [27], when we conduct an EC (Electrochemical) test, a suitable voltage is applied to the electrodes to begin the charging process of the system. At this time, the anion/cation in the solution starts to move toward the two electrodes, the positive charge is attracted by the negative plate and diffuses to the surface, and the negative charge is attracted by the positive plate to diffuse to the surface. The accumulation of electric charge is caused by the electrostatic force, and a stable EDL structure is constituted at the interface between the electrode and the electrolyte to complete charging and energy storage. When it is used for work, the charge undergoes a reverse motion to pass the energy out. Since the EDLC does not undergo a redox reaction in this process, it is a physical process in which a simple charge is adsorbed and desorbed under an electrostatic field, and the process is extremely stable and has no pollution to the environment. The ability of many batteries to store their charge after long-term use is irreversible due to the destruction of their internal structure. Moreover, since ion adsorption-desorption is generally not affected by the environment, such as working in high temperature or low-temperature environments (-20 °C ~ +60 °C), this is very difficult for many energy storage devices. However, the energy storage achieved by simply adsorbing the charge has a relatively small specific capacitance and limited space for improvement.

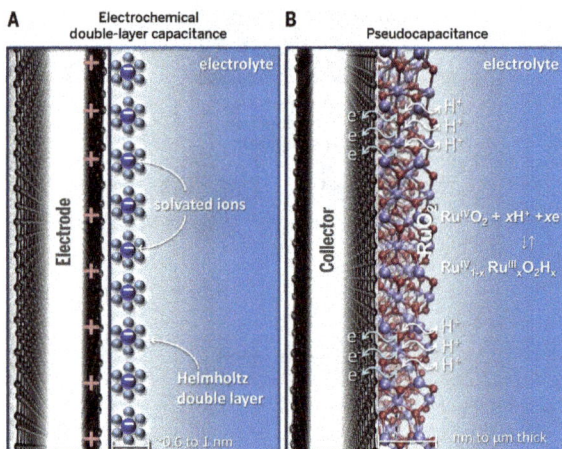

Figure 3. Two different supercapacitors and their corresponding energy storage mechanism diagram[27].

2.2   Pseudocapacitors

Pseudocapacitors (PDCs), whose main working principle is to use electroactive material to carry out the deposition of underpotential (the potential should be within the specified range). A high degree of chemical adsorption-desorption or redox reaction is performed at the surface position and surrounding position of the electrode to store energy [28]. Importantly, these reactions are almost always highly reversible. Therefore, the PDCs are worse than the EDLC in terms of output power and cycle stability. Nevertheless, their ability to store charge and energy density are much stronger than EDLC. Under the same conditions (electrode area or mass of active material), the specific capacity is 10 to 100 times that of EDLC. Moreover, PDCs energy storage is not limited to the surface of materials and electrolytes. In many cases, it can penetrate deep into the material and charge/discharge in a very small period of time. The energy storage mechanism of PDCs is different for different materials. For rare metals such as ruthenium, the energy storage largely relies on the surface hydrogen adsorption of active materials, while increase of the materials such as group VI compounds and hydroxides, the energy storage mainly relies on highly reversible redox reactions. But in general, PDCs energy storage is not only through the above method, but also has more or less EDLC system energy storage in its energy storage process, but it does not occupy a dominant position.

2.3   Symmetric supercapacitors

Symmetrical SCs use the identical active material for both negative and positive plates [29], including EDLC material and PDCs material, and the energy storage mechanism of the two electrodes is similar. It is noteworthy that although the negative and positive electrodes of the symmetric SCs use the same SCs type of material, the quality of two electrodes is different due to the absorption of different electrolyte ions on the negative and positive electrodes during discharging and charging processes.

2.4   Asymmetric supercapacitors

Asymmetric SCs also called hybrid SCs, not only have a voltage window higher than the symmetry type, but also show great improvement in energy density. The two electrode materials of the asymmetric SCs have different characteristic properties, and one of the positive electrodes converts and stores energy through a redox reaction generated by the active material, and the active material is usually made of a conductive polymer, or various oxides (PDCs materials). However, the carbon material (EDLC material) makes the negative electrode more efficient for storage and release of energy. Asymmetric SCs can combine the advantages of two electrodes with different materials. First of all, in terms of the voltage window, the positive and the negative potentials can be reasonably coordinated, so that the overall voltage window has a great widening because it covers the positive and negative potentials. As the voltage window increases, the energy density will also be greatly improved [30]. In order to achieve a perfect balance of the asymmetric capacitor bipolar plate charge, it is necessary to select the appropriate active material and its corresponding quality for a reasonable match. In addition, since the entire capacitor system combines the advantages of EDLC and PDCs, the corresponding charge-discharge cycle stability, rate performance, and energy power density are greatly improved. To assemble an asymmetric SCs device, Tang et al. [8] adopted the $FeCo_2S_4$

Supercapacitor Technology: Materials, Processes and Architectures   Materials Research Forum LLC
Materials Research Foundations **61** (2019) 233-262   https://doi.org/10.21741/9781644900499-10

nanotubes (NTs) as the positive electrode, the 3D porous nitrogen-doped graphene gel as the negative electrode, and the PVA/KOH solid electrolyte as a diaphragm. In addition, the working voltage is 1.6 V, the ultra-high power density is 755 W k/g, and the ultra-high energy density is 76.1 W h k/g, the remaining capacity is 82% after 1000 cycles of the constant current charge and discharge cycle, and only use one device can light up a LED bulb. Huang et al. [31] successfully combined $NiCo_2S_4$@NiO core-shell structure, then the $NiCo_2S_4$@NiO (positive electrode), the activated carbon (negative electrode), and the electrolyte were used to assemble the asymmetric SCs device. Furthermore, the power density reaches 0.335 KW Kg^{-1}, when the energy density reaches 22.4 W h Kg^{-1}, and the stability is extremely superior. At present, the more common asymmetric SCs on the market are composed of activated carbon and $Ni(OH)_2$, and Shanghai Aowei Company has been applied it in electric buses.

3.   Electrode materials

The energy storage characteristics of SCs are basically determined by the characteristics of their corresponding energy storage mechanisms and the electrode materials. Therefore, in practical applications, electrode materials with different characteristics have their corresponding characteristics, as described below.

EDLC materials: Carbon materials have been the most successful electrode materials in industrialization [32]. Among materials, carbon particle composite, carbon fiber composite and carbon gel have been widely used in the market. However, people have gradually discovered that their potential is limited, and with the deepening of research, carbon NTs and graphene have dominated many carbon materials owing to their distinctive physical and chemical characteristics [33]. In terms of capacitive materials, the conductivity and stability of carbon materials are undoubtedly the highest. In addition, the larger specific surface area of carbon materials is 2500 m^2/g, so it's the best choice for EDLC [34]. Furthermore, its green and non-polluting characteristics make it popular with researchers. Carbon NTs (CNTs) were discovered in 1991, can be prepared via CVD, arc discharge, and laser ablation, etc. [35]. We used CNTs as electrode material for SCs and found them superior. Compared with other carbon materials, CNTs have greatly improved conductivity, rate performance, and power density. After the discovery of graphene by researchers in 2004 [36], the focus of research on carbon materials was gradually shifted from traditional carbon materials and carbon NTs to graphene. The electrode material of pure graphene showed possess specific capacity of 135 F/g in aqueous solution [37]. For-example, Liu et al. [38] reduced GO (graphene oxide) by EC deposition and deposit it onto a foamed nickel (Ni) substrate, after testing, the specific capacity reached 131.6 F/g at the sweep rate of 10 mv/s. However, the specific capacity of simple graphene electrode is limited compared to PDCs materials. Therefore, for further to improve its EC properties, different methods to dope and combine graphene in various ways to form binary or ternary electrode materials have been used, thus the advantages and disadvantages of each material were combined, and the EC performance of the graphene-based electrode material was qualitatively improved. Shen et al. [39] reported the $FeNi_2S_4/MoSe_2$/graphene with outstanding EC performance, at a current density of 2 A g^{-1}, the capacity of $FeNi_2S_4/MoSe_2$/graphene was 1700 F g^{-1} and a specific capacity was 106% of the initial capacity, after 4000 cycles. This can be attribute to the ions in the electrolyte are embedded in the interlayer of graphene during the long-term constant current discharge and charge processes, and continuously contacting the graphene electrode material at a deep distance from the electrolyte to complete the activation of the electrode material. In the charge and discharge process, its capacity will continue to increase after cycling, which is the advantage of graphene electrodes [40].

PDCs materials: Compared to the EDLC material, which relies on the adsorption and desorption of ions to store and release energy, the most EC characteristic of PDCs is to store or release energy through redox reaction [41]. The most typical material is RuO_2, which belongs to transition metal oxides. Its energy storage mechanism is the redox reaction in aqueous electrolytes through continuous electron transfer [42]. The valence state of ruthenium element changes from Ru^{2+}, Ru^{3+}, and Ru^{4+} leads to a sharp electronic transition, resulting in extremely high conductivity and specific capacity [43]. According to the reports [44], the specific capacity of RuO_2 in acidic electrolyte can reach 980 F g^{-1}. However, Ru metal is extremely expensive, and there is only one tenth of the content in the earth's crust. Although it has excellent EC performance, its inability to be widely used has limited its development and application in the future. In order to find a substitute for RuO_2, researchers gradually focused on MnO_2, which is extremely low-cost and green without pollution. For-example, the specific capacity of flaky MnO_2 increased by 8.1% after 200 cycles of constant current discharge and charge (100 mA/cm^2) was reported by Yan et al. [45] . However, the biggest drawback of MnO_2 is that its conductivity is not ideal, which limits its application in energy storage materials [46]. The best solution is to combine it with other materials of better conductivity, such as graphene [47-49]. In addition, the conductive polymer is also an excellent PDCs material with superior conductivity. The principle of energy storage mainly relies on doping and de-doping for redox reaction, and its energy storage process can fully penetrate the active material, so it also has a good specific capacity [50]. Figure 4 displays the specific capacitance comparison of the reported EDLC and PDCs materials.

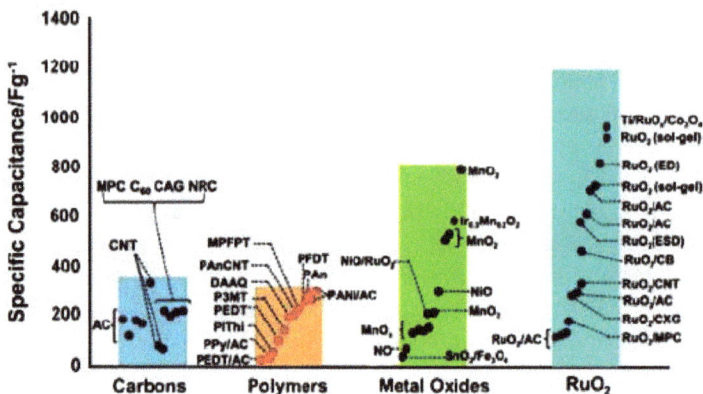

Figure 4. Comparison of specific capacitances of double layer capacitors materials and pseudocapacitance materials that have been reported[104].

3.1   Metal oxides/sulfides

Cobalt, iron, nickel, copper and other elements are extremely rich in reserves on the earth [51-53]. According to recent literature reports [54-56], the oxides and sulfides formed by the combination of elements such as oxygen and sulfur of the group VI have good PDCs properties, and thus have received widespread attention.

Co_3O_4 is an AB_2O_4-shaped spinel structure. As a PDCs material, Co_3O_4 showed highly reversible redox reactions in an alkaline solution, following reaction formula which can be expressed as [57]:

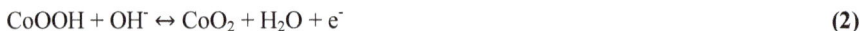

$$Co_3O_4 + OH^- + H_2O \leftrightarrow CoOOH + e^- \tag{1}$$

$$CoOOH + OH^- \leftrightarrow CoO_2 + H_2O + e^- \tag{2}$$

During the oxidation reaction, when the charging process is performed, Co_3O_4 first reacts with OH^- in the alkaline electrolyte to form an intermediate product CoOOH, which also emits an electron, and then CoOOH continues to react, and at the same time give out an electron to generate the final product CoO_2. When the discharge process is carried out, CoO_2 generates electrons by reversible redox reaction to form Co_3O_4. Therefore, through the above reactions, Co_3O_4 exhibits both the specific capacity and the cycle stability extremely well. Meher et al. [58] reported a porous Co_3O_4 with specific capacity of 548 F

g^{-1}, and it can be maintained at 98.5 % after 2000 cycles and it has 100% coulomb efficiency, showing excellent EC performance.

3.1.1 Nickel oxide (NiO) based supercapacitors

NiO has been prepared via sol-gel [59], electrodeposited [60], and electrostatic spray depostion methods [61]. Ni was deposited on the chemically etched Si NWs and then annealed to obtain highly ordered NiO-coated Si NWs [62]. In addition, the constructed electrode (NiO) has excellent electrical conductivity and a more active chemical reaction site per unit area, and thus has good cycle stability, low internal resistance, and high specific volume. For discharge current of 2.5 mA, the specific capacity was 787.5 F g $^{-1}$, which decreased slightly after 500 circles, with a loss of 4.039%. The anodic and cathodic peaks at 0.54 and 0.21 eV were due to the oxidation of NiO to NiOOH and opposite process. The redox process ascribed to the quasi-reversible PDC of the NiO/Si NWs electrodes in an alkaline solution, happened according to the equation [63]:

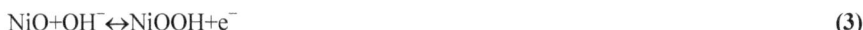

$$NiO+OH^- \leftrightarrow NiOOH+e^- \tag{3}$$

3.1.2 Nickel sulfide (Ni_3S_2) based supercapacitors

The EC performance and hydrothermal prepared Ni_3S_2 in SCs have been reported [64-66], meanwhile, the Ni_3S_2 electrode exhibted a porous, rough and unique nanostructured morphology. Huang et al. have been reported that the three-dimensional (3D) framework of Ni_3S_2 with high specific capacitance of 710.4 F/g at the current density of 2 A/g [67]. Ni_3S_2 with different morphologies can be obtained by adjust different reaction temp and reaction time. The mechanism of faradaic redox reaction in alkaline electrolyte is as follows:

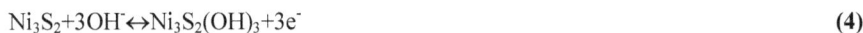

$$Ni_3S_2+3OH^- \leftrightarrow Ni_3S_2(OH)_3+3e^- \tag{4}$$

Meanwhile, the Ni_3S_2/Ni foam and Ni_3S_2/graphene have been reported [64, 68], the specific capacitances of Ni_3S_2/Ni foam and Ni_3S_2/graphene were 1293 F/g, 1420 F/g (at current density of 5 mA cm^{-2}, 2 A g^{-1}), respectively.

3.1.3 Zinc oxide (ZnO) based supercapacitors

There have been many reports on ZnO oxide based SCs electrodes [53, 69, 70]. For instance, the ZnO/graphene NSs (GNS) composites were obtained by one-pot green electrodeposition [71], the ZnO/GNS with a specific capacitance of 291 F g^{-1} at current density of 3 A g^{-1}. During the EC test, ZnO/GNS used as the working electrode, Pt foil used as the counter electrode and Ag/AgCl electrodes used as the reference electrode, and deposition was performed at 60 °C for 1800 seconds at a potential of -1.1V. Typical EC tests including CV (cyclic voltammogram), and galvanostatic charge/discharge (GC/D) were performed in 1 M KOH electrolyte. The increase in capacitance was attributed to (1) the new adhesive-free electrode which reduces resistance between the load material and the current collector, (2) the addition of GNS provides high conductivity and an open structure, allowing for fast and efficient ion charge transfer and electron transport.

3.2 Bimetal oxides/sulfides composites

3.2.1 Nickel molybdenum oxide based supercapacitors

Lou et al. reported that the asymmetric SCs electrodes of NiMoO$_4$ NS/NR (nanosheet/nanorod) arrays on various substrates (with conductive) have superior EC property and the high energy density of 60.9 Wh kg^{-1} at a power density of 850 W kg^{-1} [72]. Manickam et al. obtained a unique α-NiMoO$_4$ SCs with 1517 F g^{-1} (high specific capacitance) and 52.7 Wh kg^{-1} (energy density) at a current density of 1.2 A g^{-1} [73]. Wang et al. reported the electrodes of NiMoO$_4$ NWs/carbon cloth with an energy density of 70.7 Wh k/g, specific capacitance of 1587 F g^{-1} (the current density, 5 mA/cm^2), and power density of 16000W kg^{-1}. Furthermore, the wall-like layered NiMoO$_4$ grown on carbon cloth has been used for outstanding SC electrodes. The specific capacitance was 1483 F g^{-1} at current density 2 A g^{-1} and capacitance retention was 93.1% after 2000 cycles [74]. The hierarchical nanostructure composed of two-dimensional NSs and the directly grown active materials acted as a binder-free electrode on the conductive carbon cloth substrate resulting in NiMoO$_4$ possess excellent capacitive behavior.

3.2.2 Cobalt molybdenum oxide based supercapacitors

CoMoO$_4$ can replace Ru-based materials in SCs applications due to its lower cost and minimal toxicity. Composite electrodes based on CoMoO$_4$ have been investigated and reported, such as, MnMoO$_4$/CoMoO$_4$ [75], CoMoO$_4$-3D graphene [76], and Co$_3$O$_4$@CoMoO$_4$ [77]. Hao et al. [78] obtained CoMoO$_4$/graphene composites via hydrothermal method, simultaneously, it have high capacitance of 394.5 F g^{-1} at 1 mV s^{-1}. On account of CoMoO$_4$ possess strong redox activity and ultra-high theoretical capacitance it's considered to be a candidate PDC electrode material. Besides, the Faradaic process can operate according the following redox equation:

$$3[Co(OH)_3]^- \leftrightarrow Co_3O_4 + 4H_2O + OH^- + 2e^- \tag{5}$$

$$Co_3O_4 + H_2O + OH^- \leftrightarrow 3CoOOH^- + e^- \tag{6}$$

$$CoOOH + OH^- \leftrightarrow CoO_2 + H_2O + e^- \tag{7}$$

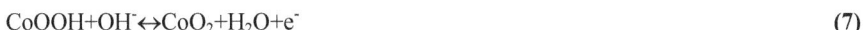

3.2.3 Iron cobalt oxygen based supercapacitors

It has been found that the conductivity and rate characteristics of Co_3O_4 were not very good. Therefore, through continuous and in-depth research, it was observed that the EC properties of Co_3O_4 can be improved by changing the microstructure, but the improvement has not been satisfactory. Therefore, it has been proposed to synthesize a bimetallic oxide MCo_2O_4 (M=Ni, Zn, Cu), also known as a binary transition metal oxide, by introducing a new metal ion (M^{2+}) instead of Co(II), such as $ZnCo_2O_4$, $NiCo_2O_4$, $MnCo_2O_4$, etc. [79-81]. Such structural variation allows the two metal elements to combine with each other, and benefits from the synergistic effect of M ions and Co ions, increasing more active sites, making the redox reaction easier to occur. Therefore, compared to single metal oxides, MCo_2O_4 with multiple oxidation states not only has better conductivity but also possesses a very high theoretical specific capacitance and cycle stability. For-example, Yuan et al. used a two-step hydrothermal way to synthesize $NiCo_2O_4$ ultrathin NSs with mesopores on a foamed nickel (Ni) substrate. The test showed good PDC and reached to 1450 F g^{-1} at a high current density of 20 A g^{-1}, at the same time, the cycle stability is also excellent [82]. $FeCo_2O_4$ is a typical bimetal oxide, compared with other bimetal oxides, $FeCo_2O_4$ has a high synergistic effect between cobalt and iron during operation, which improves the conductivity and specific capacity [83]. Furthermore, different morphologies would be obtained by adjusting different reaction temperatures and times (Figure 5). Mohamed et al. [84] prepared $FeCo_2O_4$ NSs by hydrothermal synthesis and assembled them into symmetric SCs, the voltage window and specific capacity were 1.6 V and 433 F g^{-1} (at 0.1 A g^{-1}), respectively, meanwhile, its coulomb efficiency can be maintained at 100% in the stability cycle of 2,500 cycles. Therefore, $FeCo_2O_4$ has a wonderful future in energy storage devices.

3.2.4 Iron cobalt sulfur based supercapacitors

Recently, people have found that sulfides have better conductivity than oxides, and therefore have more stable rate characteristics. Therefore, in order to find energy storage materials with better EC properties, many studies have focused on transition MSs, compared to single MSs such as nickel sulfide (NiS), cobalt sulfide (Co_9S_8). Bimetal sulfides such as $NiCo_2S_4$, $CuCo_2S_4$, and $FeCo_2S_4$ have higher specific capacity because of richer oxidation states [21, 85-87]. Compared to bimetallic oxides, bimetallic sulfides have higher conductivity and rate characteristics due to their lower electronegativity and higher EC activity [88]. Tang et al. [89] have synthesized $CuCo_2S_4$ nanoparticles following solvothermal method. It was found in the polysulfide electrolyte that the active material can reach a specific capacity of 5030 F g^{-1} at a current density of 20 A g^{-1}, and it can still reach to the specific capacity of 1365 F g^{-1} at high current density of 70 A g^{-1}. Zhu et al. [4] successfully developed $FeCo_2S_4$-$NiCo_2S_4$ hybrid NTs on the fabric after silver ion sputtering by hydrothermal reaction, which can possess a capacity of 1519 F/g^1 at a current density of 5 mA/cm^2, there was still 92% capacity retention after the constant current charge and discharge of 3000 cycles.

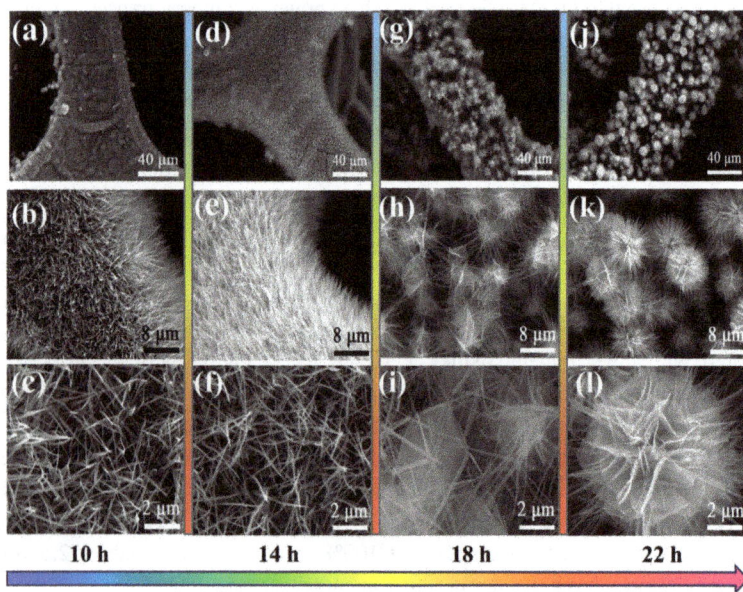

Figure 5. SEM images of as-prepared $FeCo_2O_4$ on Ni foams by controlling the reaction times of hydrothermal processes: (a-c) 10 h; (d-f) 14 h; (g-i) 18 h; (j-l) 22 h[83].

4.   Electrolytes

The electrolyte plays an important role in complementing the capacitance or performance of the electrode, charge conduction and electrode particle adhesion. In general, the ideal electrolyte requirements in SCs are: wide potential windows, high ionic conductivity, adequate EC stability, low resistivity, and wide operating temperature range. The aqueous electrolyte and organic electrolyte are the main electrolytes used in SCs.

4.1   Aqueous electrolytes

Aqueous electrolytes include KOH, H_2SO_4, Na_2SO_4, KCl, etc., and have the advantages of low cost, incombustibility, safety, and convenient installation. However, the potential window of aqueous electrolyte (~1.23V) is much lower than organic electrolyte [90]. The low-potential windows are the cause of low energy density and limiting their market applications. Therefore, the focus is to expand the voltage window based on aqueous-electrolyte SCs to further increase the energy density. In some symmetrical SCs, the introduction of a redox reaction in the water electrolyte can effectively enhance the reaction between the electrode and the electrolyte. Recently, AC//AC symmetric SCs using 0.5 M Na_2SO_4 aqueous electrolytes have demonstrated voltage windows up to 1.6 V and excellent cycle life during constant current charging and discharging [91].

4.2   Organic electrolytes

A high operating voltages of organic electrolytes can reach to 4V [92]. A major advantage of organic electrolytes is the high potential window. Acetonitrile (AN), propylene carbonates (PC) and organic salts are the most common in organic electrolyte solvents. However, the organic electrolytes have the following problems: (1) the power density of the capacitor is limited by high resistance; (2) the operating voltage of the capacitor is limited by high water content; (3) If the organic electrolyte is used improperly or for a long time, it is very dangerous to leak and corrode, for example, it may cause a fire in the capacitor devices.

5.   Synthesis approach for electrode materials

The synthesis methods used in SCs mainly include hydrothermal/solvothermal method, chemical bath deposition, in situ polymerization, direct coating, chemical vapor deposition (CVD), vacuum filtration technique, and so on.

5.1   In-Situ polymerization

In-situ polymerization, as the name suggests, involves filling of the reaction monomer into the layers, and allowing it to polymerize between the layers. Hu et al.[93] have reported synthesis MoO_3/PANI via in-situ polymerization of method for use as SCs and its specific capacitance was 714 F/g at 1 mV/s in 1 M H_2SO_4 electrolyte. The β-MnO_2/polypyrrole NR (nanorod) composites with highly specific capacitance were obtained by Li et al. using the in situ polymerization method [94]. In-situ polymerization has been the most widely used method for producing nanocomposites. Since the materials produced by this technology have better barrier properties, mechanical properties and lighter weight, they have become the development focus of flexible packaging in the future [95, 96].

5.2   Direct coating

Feng et al. [97] reported that ultra-flexible micro-SCs with high capacitance of 5.4 mF cm^{-2} via direct coating method. Zhang et al. [98] have been adopted directly coating the conductive polymer (PANi) on the RGO sheets (reduced graphene oxide) for SCs, and its specific capacitance is 361 F g^{-1} (at 0.3 A g^{-1}). The direct coating method is simple and easy to use, so it was choosed to make electrodes.

5.3   Chemical vapor deposition (CVD)

Compared with other deposition methods, CVD technology has the advantages of simple equipment, convenient operation and maintenance, and high flexibility. Zhang et al. [99] obtained 3D graphene networks via CVD method for SCs with the specific capacitance of 816 F/g at a scan rate of 5 mV/s. Lee et al. [100] have reported the synthesis of $NiCo_2O_4$/3D graphene foam for high performance SCs electrodes by CVD method. Which has the following advantages: (1) deposition can be performed at atmospheric pressure (atmospheric pressure) or below atmospheric pressure; (2) plasma and laser-assisted techniques can significantly promote chemical reactions, allowing deposition to be carried out at lower temperatures; (3) the chemical composition of the coating can be varied to obtain a gradient deposit or a mixed coating; (4) Control on the density and purity of the coating. But this technique also has certain disadvantages, such as higher reaction temperature and lower deposition rate. Furthermore, the gas involved in the sedimentation reaction can cause certain toxicity.

5.4   Hydrothermal synthesis

Simple and convenient hydrothermal technology has attracted the interest of scientists and technicians in different disciplines, especially during the past 15 years. Preparation of a variety of transition MOs hydrothermal synthesis is one of the most commonly used one-pot synthesis technology [83]. Figure 6 is a schematic diagram showing the growth process of $FeCo_2O_4$/Ni at different reaction times. Hydrothermal method has the following advantages: (a) provides highly mono-dispersed particles with controllable size and morphology, (b) possibility of large-scale production, and (c) production of high purity materials. But this method needs high temperature and high pressure.

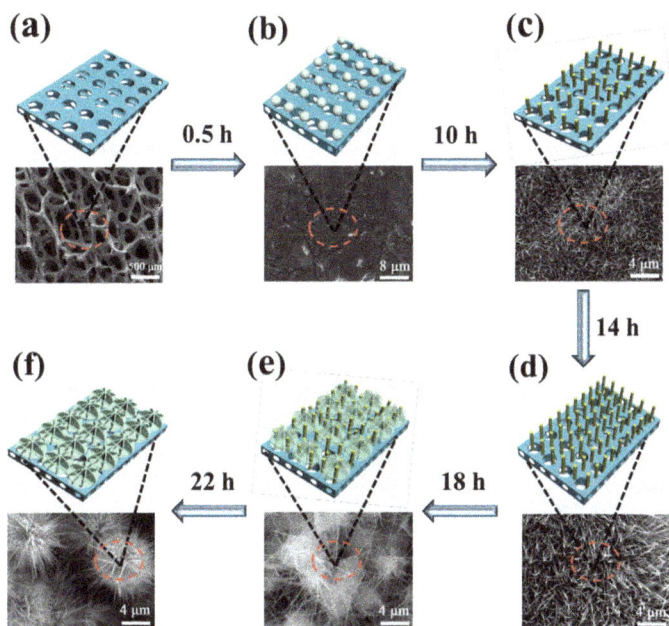

Figure 6. Schematic illustration of the growth process of FeCo₂O₄ on Ni foam and their corresponding morphologies. (a) Bare Ni foam substrate; (b) Growth of nanoparticles arrays on Ni foam; (c, d) growth of FeCo₂O₄ nanowires arrays on Ni foam; (e) Growth of nanoflake@nanowire hetero-structure array on Ni foam; and (f) Growth of hierarchical nanospheres on Ni foam[83].

5.5   Vacuum filtration technique

Lian et al. [101] used vacuum filtration technique to obtain pillared graphene-paper for SCs with highly specific capacitance. Similarly, Zhou et al. reported nanocarbon-based architectures for flexible SCs with excellent EC performance via vacuum filtration method [102]. Because hydrothermal and CVD methods are simple, low-cost, and have short synthesis times, and hence there are more widely used compared to the vacuum filtration techniques.

6.   Performances of supercapacitors

The performances of SCs include specific capacitance ($F\ g^{-1}$), power density ($W\ kg^{-1}$), and energy density ($W\ h\ kg^{-1}$), GC/D and (EIS) EC impedance spectroscopy etc. Testing is performed at a specific current density (A g^{-1}). Therefore, these EC measurements that assess the EC properties of electrode materials or SCs devices are summarized below. The capacitance of the electrode material can be calculated according to the formula:

$$C = \frac{i}{v} \tag{8}$$

C represents differential capacitance (F g^{-1}), v: scan rate (V s^{-1}), i: current density (A g^{-1}). The average capacitance of EDLC behavior and the typical PDC behavior of a rectangular CV curve can be evaluated via CV (cyclic voltammetry) measurements. Since the Faraday behavior of the battery-type electrode or the inserted pseudo-capacitor behavior shows a significant redox peak (on the CV curve), it's impossible to get the corresponding average capacitance directly from the CV measurement. Charge/discharge test is the most effective measurement method in capacitance evaluation. The capacitance of electrode materials can be based on equation:

$$C = \frac{\int I \times \Delta t}{m \times \Delta V} \tag{9}$$

In which m is the mass of active electrode materials, ΔV is the voltage change from discharging, I is the constant current and Δt is the discharge time.

EIS (EC impedance spectroscopy) tests are typically performed by collecting SCs impedance data for a specific potential with small voltage amplitude of 5/10 mV in a wide frequency range of 0.01 Hz-100 KHz.

Indicators often used to evaluate the overall performance of SCs are energy density and power density. The energy density values for SCs devices can be calculated according to the following equation:

$$E = \int_{t1}^{t2} IV dt = \frac{1}{2}C(V_1 + V_2)(V_2 - V_1) \tag{10}$$

C: capacitance (F g^{-1}), V_1: end-of-charge voltage, V_2: end-of-discharge voltage and ($V_2 - V_1$) represents the specific voltage window about the capacitive characteristics of SCs device. When $V_1 = 0$, this formula becomes:

$$E = \frac{CV_2^2}{2}. \tag{11}$$

According to the formula, the power density value of the SCs device can be acquired:

$$P = \frac{E}{t} \tag{12}$$

In which P: power density (W kg^{-1}), t: discharge time (h), and E: energy density (W h kg^{-1}).

Furthermore, to a large extent, the mass load energy of the electrode active material determines the power/energy density. In this point of view, Gogotsi et al. have made a very good discussion [103].

7. Trends, challenges, and future tasks in supercapacitors

SCs electrodes, which have been poorly understood, now can be detected at the molecular level using a variety of experimental and simulation techniques. Presently, SCs are used in power electronics integrated circuits, automobiles (stop/start systems) and public transportation systems such as trams and buses. The next challenge will be to push the limits of SCs further, to make them more widely used in energy storage or other fields. The update of energy storage technology has been slow, with an average annual increase of about 7% of specific energy and about 10% of energy density in the last 20 years, and the materials and structures used in these devices have not changed much. In essence, materials and manufacturing processes are mostly the same as they were 20 years ago. Therefore, if these energy storage devices are to be further developed, it is necessary to carry out major reforms of the assembled materials, and innovate the manufacturing and assembly processes of the materials. At present, new electronic devices such as flexible displays, curved cell phones, etc. are springing up instead of a simple energy storage device, but a power supply can be seamlessly integrated with these neoteric devices.

Summary

From previous studies, it has been clearly confirmed that BMOs/BMSs composites are suitable for SCs electrodes, benefit from the synergistic effect of M ions and Co ions, increasing more active sites, making redox reactions easier to occur. BMOs/BMSs composites-based SCs technologies are still in start-up phase, and the research and development of future should eventually produce high-performance and low-cost capacitors. We eagerly anticipate the development, in the near future, a breakthrough in BMOs/BMSs composites.

Supercapacitor Technology: Materials, Processes and Architectures   Materials Research Forum LLC
Materials Research Foundations **61** (2019) 233-262   https://doi.org/10.21741/9781644900499-10

Acknowledgements

This work was supported by the Provincial Natural Science Foundation of Hunan (No. 2016JJ2132), Science and Technology Program of Xiangtan (No. CXY-ZD20172002), and the Program for Changjiang Scholars and Innovative Research Team in University (IRT_17R91).

References

[1] Q. Wang, J. Yan, Z. Fan, Carbon materials for high volumetric performance supercapacitors: design, progress, challenges and opportunities, Energy Environ. Sci. 9 (2016) 729-762. https://doi.org/10.1039/C5EE03109E

[2] Q. Meng, K. Cai, Y. Chen, L. Chen, Research progress on conducting polymer based supercapacitor electrode materials, Nano Energy 36 (2017) 268-285. https://doi.org/10.1016/j.nanoen.2017.04.040

[3] A. González, E. Goikolea, J.A. Barrena, R. Mysyk, Review on supercapacitors: Technologies and materials, Renew. Sustain. Energy Rev. 58 (2016) 1189-1206. https://doi.org/10.1016/j.rser.2015.12.249

[4] J. Zhu, S. Tang, J. Wu, X. Shi, B. Zhu, X. Meng, Wearable high-performance supercapacitors based on silver-sputtered textiles with $FeCo_2S_4$-$NiCo_2S_4$ composite nanotube-built multitripod architectures as advanced flexible electrodes, Adv. Energy Mater. 7 (2017) 1601234-1601245. https://doi.org/10.1002/aenm.201601234

[5] J. Jiang, B. Liu, G. Liu, D. Qian, C. Yang, J. Li, A systematically comparative study on $LiNO_3$ and Li_2SO_4 aqueous electrolytes for electrochemical double-layer capacitors, Electrochim. Acta 274 (2018) 121-130. https://doi.org/10.1016/j.electacta.2018.04.097

[6] M. Winter, R.J. Brodd, What are batteries, fuel cells, and supercapacitors?, Chem. Rev. 35 (2004) 4245-4270. https://doi.org/10.1021/cr020730k

[7] E. Lim, C. Jo, J. Lee, A mini review of designed mesoporous materials for energy-storage applications: from electric double-layer capacitors to hybrid supercapacitors, Nanoscale 8 (2016) 7827-7833. https://doi.org/10.1039/C6NR00796A

[8] S. Tang, B. Zhu, X. Shi, J. Wu, X. Meng, General controlled sulfidation toward achieving novel nanosheet-built porous square-$FeCo_2S_4$-tube arrays for high-performance asymmetric all-solid-state pseudocapacitors, Adv. Energy Mater. 7 (2017) 1601985-1601996. https://doi.org/10.1002/aenm.201601985

[9] Z. Bo, C. Li, H. Yang, K. Ostrikov, J. Yan, K. Cen, Design of supercapacitor electrodes using molecular dynamics simulations, Nano-Micro Lett. 10 (2018) 33-56. https://doi.org/10.1007/s40820-018-0188-2

[10] G. Yury, Materials science: Energy storage wrapped up, Nature 509 (2014) 568-570. https://doi.org/10.1038/509568a

[11] P. Simon, Y. Gogotsi, B. Dunn, Where do batteries end and supercapacitors begin?, Sci. 343 (2014) 1210-1211. https://doi.org/10.1126/science.1249625

[12] Y. Zhang, H. Feng, X. Wu, L. Wang, A. Zhang, T. Xia, H. Dong, X. Li, L. Zhang, Progress of electrochemical capacitor electrode materials: A review, Int. J. Hydrogen Energy 34 (2009) 4889-48992. https://doi.org/10.1016/j.ijhydene.2009.04.005

[13] A. Borenstein, O. Hanna, A. Ran, S. Luski, T. Brousse, D. Aurbach, Carbon-based composite materials for supercapacitor electrodes: A review, J. Mater. Chem. A 5 (2017) 12653-12672. https://doi.org/10.1039/C7TA00863E

[14] M. Salanne, B. Rotenberg, K. Naoi, K. Kaneko, P.L. Taberna, C.P. Grey, B. Dunn, P. Simon, Efficient storage mechanisms for building better supercapacitors, Nat. Energy 1 (2016) 16070-16080. https://doi.org/10.1038/nenergy.2016.70

[15] K. Jost, G. Dion, Y. Gogotsi, Textile energy storage in perspective, J. Mater. Chem. A 2 (2014) 10776-10787. https://doi.org/10.1039/c4ta00203b

[16] R. Yan, M. Antonietti, M. Oschatz, Toward the experimental understanding of the energy storage mechanism and ion dynamics in ionic liquid based supercapacitors, Adv. Energy Mater. 8 (2018) 1800026-1800038. https://doi.org/10.1002/aenm.201800026

[17] N. Hao, Z. Dan, Y. Xue, L. Xin, W. Qian, F. Qu, Towards three-dimensional hierarchical ZnO nanofiber@Ni(OH)$_2$ nanoflake core-shell heterostructures for high-performance asymmetric supercapacitors, J. Mater. Chem. A 3 (2015) 18413-18421. https://doi.org/10.1039/C5TA04311E

[18] M.Q. Zhao, Q. Zhang, W.Z. Qian, F. Wei, J.Q. Huang, G.L. Tian, T.C. Chen, Towards high purity graphene/single-walled carbon nanotube hybrids with improved electrochemical capacitive performance, Carbon 54 (2013) 403-411. https://doi.org/10.1016/j.carbon.2012.11.055

[19] Z. Liu, H. Huang, B. Liang, X. Wang, Z. Wang, D. Chen, G. Shen, Zn_2GeO_4 and $In_2Ge_2O_7$ nanowire mats based ultraviolet photodetectors on rigid and flexible substrates, Opt. Express 20 (2012) 2982-2991. https://doi.org/10.1364/OE.20.002982

[20] F. Zhu, L. Yu, Y. Ming, W. Shi, Construction of hierarchical $FeCo_2O_4@MnO_2$ core-shell nanostructures on carbon fibers for high-performance asymmetric supercapacitor, J. Colloid Inter. Sci. 512 (2017) 419-427. https://doi.org/10.1016/j.jcis.2017.09.093

[21] S. Liu, G. Xu, J. Li, B. Wang, Z. Huang, Q. Chen, X. Qi, Iron-cobalt bi-metallic sulfide nanowires on ni foam for applications in high-performance supercapacitors, ChemElectroChem 5 (2018) 2250-2255. https://doi.org/10.1002/celc.201800486

[22] S. Karmakar, S. Varma, D. Behera, Investigation of structural and electrical transport properties of nano-flower shaped $NiCo_2O_4$ supercapacitor electrode materials, J. Alloys Com. 757 (2018) 49-59. https://doi.org/10.1016/j.jallcom.2018.05.056

[23] Z. Gao, N. Song, Y. Zhang, X. Li, Cotton Textile Enabled, All-solid-state flexible supercapacitors, Rsc Adv. 5 (2015) 15438-15447. https://doi.org/10.1039/C5RA00028A

[24] J. Xiao, L. Wan, S. Yang, F. Xiao, S. Wang, Design hierarchical electrodes with highly conductive $NiCo_2S_4$ nanotube arrays grown on carbon fiber paper for high-performance pseudocapacitors, Nano Lett. 14 (2014) 831-838. https://doi.org/10.1021/nl404199v

[25] L. Bao, J. Zang, X. Li, Flexible Zn_2SnO_4/MnO_2 Core/shell nanocable–carbon microfiber hybrid composites for high-performance supercapacitor electrodes, Nano Lett. 11 (2011) 1215-1220. https://doi.org/10.1021/nl104205s

[26] R. Kotz, M. Carlen, Principles and applications of electrochemical capacitors, Electrochim. Acta 45 (2000) 2483-2498. https://doi.org/10.1016/S0013-4686(00)00354-6

[27] F. Bonaccorso, L. Colombo, G. Yu, M. Stoller, V. Tozzini, A.C. Ferrari, R.S. Ruoff, V. Pellegrini, Graphene, related two-dimensional crystals, and hybrid systems for energy conversion and storage, Science 347 (2015) 1246501-1246512. https://doi.org/10.1126/science.1246501

[28] C. Yuan, X. Zhang, L. Su, G. Bo, L. Shen, Facile synthesis and self-assembly of hierarchical porous NiO nano/micro spherical superstructures for high performance supercapacitors, J. Mater. Chem. 19 (2009) 5772-5777. https://doi.org/10.1039/b902221j

[29] N.B. Mendoza-Sã, Y. Gogotsi, Synthesis of two-dimensional materials for capacitive energy storage, Adv. Mater. 28 (2016) 6104-6135. https://doi.org/10.1002/adma.201506133

[30] W. Qiong, X. Yuxi, Y. Zhiyi, L. Anran, S. Gaoquan, Supercapacitors based on flexible graphene/polyaniline nanofiber composite films, ACS Nano 4 (2010) 1963-1970. https://doi.org/10.1021/nn1000035

[31] Y. Huang, T. Shi, S. Jiang, S. Cheng, X. Tao, Y. Zhong, G. Liao, Z. Tang, Enhanced cycling stability of $NiCo_2S_4$@NiO core-shell nanowire arrays for all-solid-state asymmetric supercapacitors, Sci. Rep. 6 (2016) 38620-38630. https://doi.org/10.1038/srep38620

[32] Z. Yanwu, M. Shanthi, M.D. Stoller, K.J. Ganesh, C. Weiwei, P.J. Ferreira, P. Adam, R.M. Wallace, K.A. Cychosz, T. Matthias, Carbon-based supercapacitors produced by activation of graphene, Science 332 (2011) 1537-1541. https://doi.org/10.1126/science.1200770

[33] K.A. Jost, D. Stenger, C.R. Perez, J.K. Mcdonough, K. Lian, Y. Gogotsi, G. Dion, Knitted and screen printed carbon-fiber supercapacitors for applications in wearable electronics, Energy Environ. Sci. 6 (2013) 2698-2705. https://doi.org/10.1039/c3ee40515j

[34] P.L. Taberna, P. Simon, J.F. Fauvarque, Electrochemical characteristics and impedance spectroscopy studies of carbon-carbon supercapacitors, J. Electrochem. Soc. 150 (2003) A292-A300. https://doi.org/10.1149/1.1543948

[35] M. Kaempgen, C.K. Chan, J. Ma, Y. Cui, G. Gruner, Printable thin film supercapacitors using single-walled carbon nanotubes, Nano Lett. 9 (2009) 1872-1876. https://doi.org/10.1021/nl8038579

[36] K.S. Novoselov, A.K. Geim, S.V. Morozov, D. Jiang, Y. Zhang, S.V. Dubonos, I.V. Grigorieva, A.A. Firsov, Electric field effect in atomically thin carbon films, Sci. 306 (2004) 666-669. https://doi.org/10.1126/science.1102896

[37] H. Pan, C.K. Poh, Y.P. Feng, J. Lin, Supercapacitor electrodes from tubes-in-tube carbon nanostructures, Chem. Mater. 19 (2007) 6120-6125. https://doi.org/10.1021/cm071527e

[38] X. Liu, X. Qi, Z. Zhang, L. Ren, G. Hao, Y. Liu, Y. Wang, K. Huang, X. Wei, J. Li, Electrochemically reduced graphene oxide with porous structure as a binder-free electrode for high-rate supercapacitors, Rsc Adv. 4 (2014) 13673-13679. https://doi.org/10.1039/c3ra46992a

[39] J.F. Shen, J. Ji, P. Dong, R. Baines, Z. Zhang, P.M. Ajayan, M. Ye, Novel $FeNi_2S_4$/TMDs-based ternary composites for supercapacitor applications, J. Mater. Chem. A 4 (2016) 8844-8850. https://doi.org/10.1039/C6TA03111K

[40] P. Xiao, F. Bu, G. Yang, Y. Zhang, Y. Xu, Integration of Graphene, Nano sulfur, and conducting polymer into compact, flexible lithium-sulfur battery cathodes with ultrahigh volumetric capacity and superior cycling stability for foldable devices, Adv. Mater. 29 (2017) 1703324-1703332. https://doi.org/10.1002/adma.201703324

[41] H.B. Li, M.H. Yu, F.X. Wang, P. Liu, Y. Liang, J. Xiao, C.X. Wang, Y.X. Tong, G.W. Yang, Amorphous nickel hydroxide nanospheres with ultrahigh capacitance and energy density as electrochemical pseudocapacitor materials, Nat. Comm. 4 (2013) 1894-1901. https://doi.org/10.1038/ncomms2932

[42] C.C. Hu, K.H. Chang, M.C. Lin, Y.T. Wu, Design and tailoring of the nanotubular arrayed architecture of hydrous RuO_2 for next generation supercapacitors, Nano Lett. 6 (2006) 2690-2695. https://doi.org/10.1021/nl061576a

[43] T. Liu, G.W. Pell, B.E. Conway, Self-discharge and potential recovery phenomena at thermally and electrochemically prepared RuO_2 supercapacitor electrodes, Electrochim. Acta 42 (1997) 3541-3552. https://doi.org/10.1016/S0013-4686(97)81190-5

[44] O. Barbieri, M. Hahn, A. Foelske, R. Kötz, Effect of Electronic Resistance and Water Content on the Performance of RuO_2 for Supercapacitors, J. Electrochem. Soc. 153 (2006) A2049-A2054. https://doi.org/10.1149/1.2338633

[45] J. Yan, W. Tong, C. Jie, Z. Fan, M. Zhang, Preparation and electrochemical properties of lamellar MnO_2 for supercapacitors, Mater. Res. Bull. 45 (2010) 210-215. https://doi.org/10.1016/j.materresbull.2009.09.016

[46] M. Toupin, T. Brousse, D. Bélanger, Charge storage mechanism of MnO_2 electrode used in aqueous electrochemical capacitor, Chem. Mater. 16 (2004) 3184-3190. https://doi.org/10.1021/cm049649j

[47] C. Sheng, Z. Junwu, W. Xiaodong, H. Qiaofeng, W. Xin, Graphene oxide--MnO_2 nanocomposites for supercapacitors, ACS Nano 4 (2010) 2822. https://doi.org/10.1021/nn901311t

[48] L. Jinping, J. Jian, C. Chuanwei, L. Hongxing, Z. Jixuan, G. Hao, F. Hong Jin, Co_3O_4 Nanowire@MnO_2 ultrathin nanosheet core/shell arrays: a new class of high-performance pseudocapacitive materials, Adv. Mater. 23 (2011) 2076-2081. https://doi.org/10.1002/adma.201100058

[49] D. Shin, J. Shin, T. Yeo, H. Hwang, S. Park, W. Choi, Scalable synthesis of triple-core-shell nanostructures of TiO_2@MnO_2@C for high performance supercapacitors

using structure-guided combustion waves, Small 14 (2018) 1703755-1703768. https://doi.org/10.1002/smll.201703755

[50] C. Meng, C. Liu, L. Chen, C. Hu, S. Fan, Highly flexible and all-solid-state paperlike polymer supercapacitors, Nano Lett. 10 (2010) 4025-4031. https://doi.org/10.1021/nl1019672

[51] M. Qorbani, T.C. Chou, Y.H. Lee, S. Samireddi, N. Naseri, A. Ganguly, A. Esfandiar, C.H. Wang, L.C. Chen, K.H. Chen, Correction: Multi-porous Co_3O_4 nanoflakes@sponge-like few-layer partially reduced graphene oxide hybrids: towards highly stable asymmetric supercapacitors, J. Mater. Chem. A 5 (2017) 12569-12577. https://doi.org/10.1039/C7TA00694B

[52] S. Liu, S.C. Lee, U.M. Patil, C. Ray, K.V. Sankar, K. Zhang, A. Kundu, S. Kang, J.H. Park, S. Chan Jun, Controllable sulfuration engineered NiO nanosheets with enhanced capacitance for high rate supercapacitors, J. Mater. Chem. 5 (2017) 4543-4549. https://doi.org/10.1039/C6TA11049E

[53] C.K. Brozek, D. Zhou, H. Liu, X. Li, K.R. Kittilstved, D.R. Gamelin, Soluble supercapacitors: Large and reversible charge storage in colloidal iron-doped ZnO nanocrystals, Nano Lett. 18 (2018) 3297-3302. https://doi.org/10.1021/acs.nanolett.8b01264

[54] W. He, C. Wang, H. Li, X. Deng, T. Zhai, Ultrathin and porous $Ni_3S_2/CoNi_2S_4$ 3D-network structure for superhigh energy density asymmetric supercapacitors, Adv. Energy Mater. 7 (2017) 1700983-1700994. https://doi.org/10.1002/aenm.201700983

[55] J. Chen, W. Xu, J. Wang, P.S. Lee, Sulfidation of NiMn-layered double hydroxides/graphene oxide composites toward supercapacitor electrodes with enhanced performance, Adv. Energy Mater. 6 (2016) 1501745-1501453. https://doi.org/10.1002/aenm.201501745

[56] M. Guo, J. Balamurugan, X. Li, N.H. Kim, J.H. Lee, Hierarchical 3D cobalt-doped Fe_3O_4 nanospheres@NG hybrid as an advanced anode material for high-performance asymmetric supercapacitors, Small 13 (2017) 1701275-1701287. https://doi.org/10.1002/smll.201701275

[57] X.H. Xia, J.P. Tu, Y.J. Mai, X.L. Wang, C.D. Gu, X.B. Zhao, Self-supported hydrothermal synthesized hollow Co_3O_4 nanowire arrays with high supercapacitor capacitance, J. Mater. Chem. 21 (2011) 9319-9325. https://doi.org/10.1039/c1jm10946d

[58] S.K. Meher, G.R. Rao, Ultralayered Co_3O_4 for high-performance supercapacitor applications, J. Phy. Chem. C 115 (2011) 15646-15654. https://doi.org/10.1021/jp201200e

[59] K.C. Liu, M.A. Anderson, Porous Nickel Oxide Films for Electrochemical Capacitors, Mrs Proceedings 393 (1995) 124-130. https://doi.org/10.1557/PROC-393-427

[60] E.E. Kalu, T.T. Nwoga, V. Srinivasan, J.W. Weidner, Cyclic voltammetric studies of the effects of time and temperature on the capacitance of electrochemically deposited nickel hydroxide, J. Power Sources 92 (2001) 163-167. https://doi.org/10.1016/S0378-7753(00)00520-6

[61] K.O. Ukoba, A.C. Elokaeboka, F.L. Inambao, Review of nanostructured NiO thin film deposition using the spray pyrolysis technique, Renew. Sustain. Energy Rev. 82 (2017) 2900-2915. https://doi.org/10.1016/j.rser.2017.10.041

[62] L. Fang, M. Qiu, Q. Xiang, L. Yang, J. Yin, G. Hao, F. Xiang, J. Li, J. Zhong, Electrochemical properties of high-power supercapacitors using ordered NiO coated Si nanowire array electrodes, Appl. Phys. A 104 (2011) 545-550. https://doi.org/10.1007/s00339-011-6412-2

[63] W.U. Mengqiang, J. Gao, S. Zhang, A.I. Chen, Comparative studies of nickel oxide films on different substrates for electrochemical supercapacitors, J. Power Sources 159 (2006) 365-369. https://doi.org/10.1016/j.jpowsour.2006.04.013

[64] K. Krishnamoorthy, G.K. Veerasubramani, S. Radhakrishnan, J.K. Sang, One pot hydrothermal growth of hierarchical nanostructured Ni_3S_2 on Ni foam for supercapacitor application, Chem. Eng. J. 251 (2014) 116-122. https://doi.org/10.1016/j.cej.2014.04.006

[65] M. Wang, Y. Wang, H. Dou, G. Wei, X. Wang, Enhanced rate capability of nanostructured three-dimensional graphene/Ni_3S_2 composite for supercapacitor electrode, Ceram. Int. 42 (2016) 9858-9865. https://doi.org/10.1016/j.ceramint.2016.03.085

[66] G. Li, Y. Cong, C. Zhang, H. Tao, Y. Sun, Y. Wang, Hierarchical nanosheet-based Ni_3S_2 microspheres grown on Ni foam for high-performance all-solid-state asymmetric supercapacitors, Nanotechnology 28 (2017) 425401-425415. https://doi.org/10.1088/1361-6528/aa829d

[67] Z. Zhen, Z. Huang, R. Long, Y. Shen, Q. Xiang, J. Zhong, One-pot synthesis of hierarchically nanostructured Ni_3S_2 dendrites as active materials for supercapacitors, Electrochim. Acta 149 (2014) 316-323. https://doi.org/10.1016/j.electacta.2014.10.097

[68] Z. Zhang, X. Liu, X. Qi, Z. Huang, J. Zhong, Hydrothermal synthesis of Ni_3S_2/graphene electrode and its application in a supercapacitor, RSC Adv. 4 (2014) 37278-37283. https://doi.org/10.1039/C4RA05078A

[69] Y. Peihua, X. Xu, L. Yuzhi, D. Yong, Q. Pengfei, T. Xinghua, M. Wenjie, L. Ziyin, W. Wenzhuo, L. Tianqi, Hydrogenated ZnO core-shell nanocables for flexible supercapacitors and self-powered systems, ACS Nano 7 (2013) 2617-2626. https://doi.org/10.1021/nn306044d

[70] J. Hao, L. Ji, K. Wu, N. Yang, Electrochemistry of ZnO@reduced graphene oxides, Carbon 130 (2018) 480-486. https://doi.org/10.1016/j.carbon.2018.01.018

[71] Z. Zhen, R. Long, W. Han, L. Meng, X. Wei, Q. Xiang, J. Zhong, One-pot electrodeposition synthesis of ZnO/graphene composite and its use as binder-free electrode for supercapacitor, Ceram. Int. 41 (2015) 4374-4380. https://doi.org/10.1016/j.ceramint.2014.11.127

[72] S. Peng, L. Li, B.W. Hao, S. Madhavi, W.D.L. Xiong, Controlled growth of $NiMoO_4$ nanosheet and nanorod arrays on various conductive substrates as advanced electrodes for asymmetric supercapacitors, Adv. Energy Mater. 5 (2015) 1401172-1401179. https://doi.org/10.1002/aenm.201401172

[73] B. Senthilkumarab, K.V. Sankara, R.K. Selvan A, M. Danielleb, M. Manickamb, Nano α-$NiMoO_4$ as a new electrode for electrochemical supercapacitors, RSC Adv. 3 (2012) 352-357. https://doi.org/10.1039/C2RA22743F

[74] Z. Huang, Z. Zhang, X. Qi, X. Ren, G. Xu, P. Wan, X. Sun, H. Zhang, Wall-like hierarchical metal oxide nanosheet arrays grown on carbon cloth for excellent supercapacitor electrodes, Nanoscale 8 (2016) 13273-13279. https://doi.org/10.1039/C6NR04020A

[75] L.Q. Mai, F. Yang, Y.L. Zhao, X. Xu, L. Xu, Y.Z. Luo, Hierarchical $MnMoO_{(4)}$/$CoMoO_{(4)}$ heterostructured nanowires with enhanced supercapacitor performance, Nat. Commun. 2 (2011) 381-386. https://doi.org/10.1038/ncomms1387

[76] Y. Xinzhi, L. Bingan, X. Zhi, Super long-life supercapacitors based on the construction of nanohoneycomb-like strongly coupled $CoMoO_{(4)}$-3D graphene hybrid electrodes, Adv. Mater. 26 (2014) 1044-1051. https://doi.org/10.1002/adma.201304148

[77] W. Jing, Z. Xiang, Q. Wei, H. Lv, Y. Tian, Z. Tong, X. Liu, H. Jian, H. Qu, J. Zhao, 3D self-supported nanopine forest-like $Co_3O_4@CoMoO_4$ core–shell architectures for high-energy solid state supercapacitors, Nano Energy 19 (2016) 222-233. https://doi.org/10.1016/j.nanoen.2015.10.036

[78] X. Xia, L. Wu, Q. Hao, W. Wang, W. Xin, One-step synthesis of $CoMoO_4$ /graphene composites with enhanced electrochemical properties for supercapacitors, Electrochim. Acta 99 (2013) 253-261. https://doi.org/10.1016/j.electacta.2013.03.131

[79] S.G. Mohamed, Y.Q. Tsai, C.J. Chen, Y.T. Tsai, T.F. Hung, W.S. Chang, R.S. Liu, Ternary spinel MCo_2O_4 (M＝Mn, Fe, Ni, and Zn) porous nanorods as bifunctional cathode materials for lithium-O_2 batteries, ACS Appl. Mater. Inter. 7 (2015) 12038-12046. https://doi.org/10.1021/acsami.5b02180

[80] Z. Yan, L. Hu, S. Zhao, L. Wu, Preparation of $MnCo_2O_4@Ni(OH)_2$ core-shell flowers for asymmetric supercapacitor materials with ultrahigh specific capacitance, Adv. Funct. Mater. 26 (2016) 4085-4093. https://doi.org/10.1002/adfm.201600494

[81] L. Bin, L. Boyang, W. Qiufan, W. Xianfu, X. Qingyi, C. Di, S. Guozhen, New energy storage option: toward $ZnCo_2O_4$ nanorods/nickel foam architectures for high-performance supercapacitors, ACS Appl. Mater. Inter. 5 (2013) 10011-10017. https://doi.org/10.1021/am402339d

[82] C. Yuan, J. Li, L. Hou, X. Zhang, L. Shen, W.D.L. Xiong, Ultrathin mesoporous $NiCo_2O_4$ nanosheets supported on Ni foam as advanced electrodes for supercapacitors, Adv. Funct. Mater. 22 (2012) 4592–4597. https://doi.org/10.1002/adfm.201200994

[83] G. Xu, Z. Zhang, X. Qi, X. Ren, S. Liu, Q. Chen, Z. Huang, J. Zhong, Hydrothermally synthesized $FeCo_2O_4$ nanostructures: Structural manipulation for high-performance all solid-state supercapacitors, Ceram. Int. 44 (2018) 120-127. https://doi.org/10.1016/j.ceramint.2017.09.146

[84] S.G. Mohamed, C.-J. Chen, C.K. Chen, S.-F. Hu, R.-S. Liu, High-performance lithium-ion battery and symmetric supercapacitors based on $FeCo_2O_4$ nanoflakes electrodes, ACS Appl. Mater. Inter. 6 (2014) 22701-22708. https://doi.org/10.1021/am5068244

[85] Q. Liu, J. Jin, J. Zhang, $NiCo_2S_4@$graphene as a bifunctional electrocatalyst for oxygen reduction and evolution reactions, ACS Appl. Mater. Inter. 5 (2013) 5002-5008. https://doi.org/10.1021/am4007897

[86] L. Shen, J. Wang, G. Xu, H. Li, H. Dou, X. Zhang, $NiCo_2S_4$ Nanosheets grown on nitrogen-doped carbon foams as an advanced electrode for supercapacitors, Adv. Energy Mater. 5 (2015) 1400977-1400984. https://doi.org/10.1002/aenm.201400977

[87] L.L. Liu, K.P. Annamalai, Y.S. Tao, A hierarchically porous $CuCo_2S_4$ /graphene composite as an electrode material for supercapacitors, New Carbon Mater. 31 (2016) 336-342. https://doi.org/10.1016/S1872-5805(16)60017-3

[88] Y. Zhu, Z. Wu, M. Jing, X. Yang, W. Song, X. Ji, Mesoporous $NiCo_2S_4$ nanoparticles as high-performance electrode materials for supercapacitors, J. Power Sources 273 (2015) 584-590. https://doi.org/10.1016/j.jpowsour.2014.09.144

[89] J. Tang, Y. Ge, J. Shen, M. Ye, Facile synthesis of $CuCo_2S_4$ as a novel electrode material for ultrahigh supercapacitor performance, Chem. Commun. 52 (2015) 1509-1512. https://doi.org/10.1039/C5CC09402J

[90] L. Demarconnay, E. Raymundo-Piñero, F. Béguin, A symmetric carbon/carbon supercapacitor operating at 1.6 V by using a neutral aqueous solution, Electrochem. Commun. 12 (2010) 1275-1278. https://doi.org/10.1016/j.elecom.2010.06.036

[91] G. Qiang, L. Demarconnay, E. Raymundo-Piñero, F. Béguin, Exploring the large voltage range of carbon/carbon supercapacitors in aqueous lithium sulfate electrolyte, Energy Environ. Sci. 5 (2012) 9611-9617. https://doi.org/10.1039/c2ee22284a

[92] C. Zhao, W. Zheng, A Review for aqueous electrochemical supercapacitors, Frontiers in Energy Research 3 (2015) 23-34. https://doi.org/10.3389/fenrg.2015.00023

[93] F. Jiang, W. Li, R. Zou, Q. Liu, K. Xu, L. An, J. Hu, MoO_3/PANI coaxial heterostructure nanobelts by in situ polymerization for high performance supercapacitors, Nano Energy 7 (2014) 72-79. https://doi.org/10.1016/j.nanoen.2014.04.007

[94] J. Zang, X. Li, In situ synthesis of ultrafine β-MnO_2/polypyrrole nanorod composites for high-performance supercapacitors, J. Mater. Chem. 21 (2011) 10965-10969. https://doi.org/10.1039/c1jm11491c

[95] H. Peng, G. Ma, K. Sun, J. Mu, Z. Lei, High-performance supercapacitor based on multi-structural CuS@polypyrrole composites prepared by in situ oxidative polymerization, J. Mater. Chem. A 2 (2014) 3303-3307. https://doi.org/10.1039/c3ta13859c

[96] S. Ghosh, O. Inganäs, Conducting polymer hydrogels as 3D electrodes: Applications for supercapacitors, Adv. Mater. 11 (1999) 1214-1218.

https://doi.org/10.1002/(SICI)1521-4095(199910)11:14<1214::AID-ADMA1214>3.0.CO;2-3

[97] Z. Liu, Z.S. Wu, S. Yang, R. Dong, X. Feng, K.M. Llen, Ultraflexible in-plane micro-supercapacitors by direct printing of solution-processable electrochemically exfoliated graphene, Adv. Mater. 28 (2016) 2217-2222. https://doi.org/10.1002/adma.201505304

[98] J. Zhang, X.S. Zhao, Conducting polymers directly coated on reduced graphene oxide sheets as high-performance supercapacitor electrodes, J. Phy. Chem. C 116 (2012) 5420–5426. https://doi.org/10.1021/jp211474e

[99] C. Xiehong, S. Yumeng, S. Wenhui, L. Gang, H. Xiao, Y. Qingyu, Z. Qichun, Z. Hua, Preparation of novel 3D graphene networks for supercapacitor applications, Small 7 (2011) 3163-3168. https://doi.org/10.1002/smll.201100990

[100] C. Zhang, T. Kuila, N.H. Kim, S.H. Lee, J.H. Lee, Facile preparation of flower-like $NiCo_2O_4$/three dimensional graphene foam hybrid for high performance supercapacitor electrodes, Carbon 89 (2015) 328-339. https://doi.org/10.1016/j.carbon.2015.03.051

[101] W. Gongkai, S. Xiang, L. Fengyuan, S. Hongtao, Y. Mingpeng, J. Weilin, L. Changsheng, L. Jie, Flexible pillared graphene-paper electrodes for high-performance electrochemical supercapacitors, Small 8 (2012) 452-459. https://doi.org/10.1002/smll.201101719

[102] Z. Niu, L. Liu, Z. Li, W. Zhou, X. Chen, S. Xie, Programmable nanocarbon-based architectures for flexible supercapacitors, Adv. Energy Mater. 5 (2016) 1500677-1500697. https://doi.org/10.1002/aenm.201500677

[103] Y. Gogotsi, P. Simon, . Materials science. True performance metrics in electrochemical energy storage, Science 334 (2011) 917-918. https://doi.org/10.1126/science.1213003

[104] K. Naoi, P. Simon, New materials and new configurations for advanced electrochemical capacitors, J. Electrochem. Soci. (JES) 17 (2008) 34-37.

Keyword Index

About the Editors

Dr. Inamuddin is currently working as Assistant Professor in the Chemistry Department, Faculty of Science, King Abdulaziz University, Jeddah, Saudi Arabia. He is a permanent faculty member (Assistant Professor) at the Department of Applied Chemistry, Aligarh Muslim University, Aligarh, India. He obtained Master of Science degree in Organic Chemistry from Chaudhary Charan Singh (CCS) University, Meerut, India, in 2002. He received his Master of Philosophy and Doctor of Philosophy degrees in Applied Chemistry from Aligarh Muslim University (AMU), India, in 2004 and 2007, respectively. He has extensive research experience in multidisciplinary fields of Analytical Chemistry, Materials Chemistry, and Electrochemistry and, more specifically, Renewable Energy and Environment. He has worked on different research projects as project fellow and senior research fellow funded by University Grants Commission (UGC), Government of India, and Council of Scientific and Industrial Research (CSIR), Government of India. He has received Fast Track Young Scientist Award from the Department of Science and Technology, India, to work in the area of bending actuators and artificial muscles. He has completed four major research projects sanctioned by University Grant Commission, Department of Science and Technology, Council of Scientific and Industrial Research, and Council of Science and Technology, India. He has published 147 research articles in international journals of repute and eighteen book chapters in knowledge-based book editions published by renowned international publishers. He has published 60 edited books with Springer (U.K.), Elsevier, Nova Science Publishers, Inc. (U.S.A.), CRC Press Taylor & Francis Asia Pacific, Trans Tech Publications Ltd. (Switzerland), IntechOpen Limited (U.K.), and Materials Research Forum LLC (U.S.A). He is a member of various journals' editorial boards. He is also serving as Associate Editor for journals (Environmental Chemistry Letter, Applied Water Science and Euro-Mediterranean Journal for Environmental Integration, Springer-Nature), Frontiers Section Editor (Current Analytical Chemistry, Bentham Science Publishers), Editorial Board Member (Scientific Reports-Nature), Editor (Eurasian Journal of Analytical Chemistry), and Review Editor (Frontiers in Chemistry, Frontiers, U.K.) He is also guest-editing various special thematic special issues to the journals of Elsevier, Bentham Science Publishers, and John Wiley & Sons, Inc. He has attended as well as chaired sessions in various international and national conferences. He has worked as a Postdoctoral Fellow, leading a research team at the Creative Research Initiative Center for Bio-Artificial Muscle, Hanyang University, South Korea, in the field of renewable energy, especially biofuel cells. He has also worked as a Postdoctoral Fellow at the Center of Research Excellence in Renewable Energy, King Fahd University of Petroleum and Minerals, Saudi Arabia, in the field of polymer electrolyte membrane fuel cells and computational fluid dynamics of polymer electrolyte membrane fuel cells. He is a life member of the Journal of the Indian Chemical Society. His research interest includes ion exchange materials, a sensor for heavy metal ions, biofuel cells, supercapacitors and bending actuators.

Dr. Rajender Boddula is currently working for Chinese Academy of Sciences President's International Fellowship Initiative (CAS-PIFI) at National Center for Nanoscience and Technology (NCNST, Beijing). His academic honors includes University Grants Commission National Fellowship and many merit scholarships, and CAS-PIFI. He has published many scientific articles in international peer-reviewed journals and has authored twenty book chapters. He is also serving as an editorial board member and a referee for reputed international peer-reviewed journals. He has published edited books with Springer (UK), Elsevier, Materials Research Forum LLC (USA) and CRC Press Taylor & Francis Asia Pacific, Trans Tech Publications Ltd. (Switzerland). His specialized areas of research are energy conversion and storage, which include sustainable nanomaterials, graphene, polymer composites, heterogeneous catalysis for organic transformations, environmental remediation technologies, photoelectrochemical water-splitting devices, biofuel cells, batteries and supercapacitors.

Dr. Mohd Imran Ahamed received his Ph.D degree on the topic "Synthesis and characterization of inorganic-organic composite heavy metals selective cation-exchangers and their analytical applications", from Aligarh Muslim University, Aligarh, India in 2019. He has published several research and review articles in the journals of international recognition. He has also edited various books which are published by Springer, CRC Press Taylor & Francis Asia Pacific and Materials Research Forum LLC, U.S.A. He has completed his B.Sc. (Hons) Chemistry from Aligarh Muslim University, Aligarh, India, and M.Sc. (Organic Chemistry) from Dr. Bhimrao Ambedkar University, Agra, India. His research work includes ion-exchange chromatography, wastewater treatment, and analysis, bending actuator and electrospinning.

Prof. Abdullah M. Asiri is the Head of the Chemistry Department at King Abdulaziz University since October 2009 and he is the founder and the Director of the Center of Excellence for Advanced Materials Research (CEAMR) since 2010 till date. He is the Professor of Organic Photochemistry. He graduated from King Abdulaziz University (KAU) with B.Sc. in Chemistry in 1990 and a Ph.D. from University of Wales, College of Cardiff, U.K. in 1995. His research interest covers color chemistry, synthesis of novel photochromic and thermochromic systems, synthesis of novel coloring matters and dyeing of textiles, materials chemistry, nanochemistry and nanotechnology, polymers and plastics. Prof. Asiri is the principal supervisors of more than 20 M.Sc. and six Ph.D. theses. He is the main author of ten books of different chemistry disciplines. Prof. Asiri is the Editor-in-Chief of King Abdulaziz University Journal of Science. A major achievement of Prof. Asiri is the research of tribochromic compounds, a new class of compounds which change from slightly or colorless to deep colored when subjected to small pressure or when grind. This discovery was introduced to the scientific community as a new terminology published by International Union of Pure and Applied Chemistry (IUPAC) in 2000. This discovery was awarded a patent from European Patent office and from UK patent. Prof. Asiri involved in many committees at the KAU level and on the national level. He took a major role in the advanced materials committee working for King Abdulaziz City for Science and Technology (KACST) to identify the national plan for science and technology in 2007. Prof. Asiri played a major role in advancing the chemistry education and research in KAU. He has been awarded the best researchers from KAU for the past five years. He also awarded the Young Scientist Award from the Saudi Chemical Society in 2009 and also the first prize for the distinction in science from the Saudi Chemical Society in 2012. He also received a recognition certificate from the American Chemical Society (Gulf region Chapter) for the advancement of chemical science in the Kingdome. He received a Scopus certificate for the most publishing scientist in Saudi Arabia in chemistry in 2008. He is also a member of the editorial board of various journals of international repute. He is the Vice- President of Saudi Chemical Society (Western Province Branch). He holds four USA patents, more than one thousand publications in international journals, several book chapters and edited books.